Feature Detectors and Motion Detection in Video Processing

Nilanjan Dey
Techno India College of Technology, Kolkata, India

Amira Ashour
Tanta University, Egypt

Prasenjit Kr. Patra
Bengal College of Engineering and Technology, India

A volume in the Advances in Multimedia and Interactive Technologies (AMIT) Book Series

www.igi-global.com

Published in the United States of America by
IGI Global
Information Science Reference (an imprint of IGI Global)
701 E. Chocolate Avenue
Hershey PA, USA 17033
Tel: 717-533-8845
Fax: 717-533-8661
E-mail: cust@igi-global.com
Web site: http://www.igi-global.com

Library of Congress Cataloging-in-Publication Data

Names: Dey, Nilanjan, 1984- editor. | Ashour, Amira, 1975- editor. | Patra, Prasenjit Kr., 1987- editor.
Title: Feature detectors and motion detection in video processing / Nilanjan Dey, Amira Ashour, and Prasenjit Kr. Patra, editors.
Description: Hershey PA : Information Science Reference, [2017] | Series: Advances in multimedia and interactive technologies | Includes bibliographical references and index.
Identifiers: LCCN 2016033138| ISBN 9781522510253 (hardcover) | ISBN 9781522510260 (ebook)
Subjects: LCSH: Motion detectors. | Image processing--Digital techniques. | Image analysis.
Classification: LCC TK7882.M68 F43 2017 | DDC 681/.2--dc23 LC record available at https://lccn.loc.gov/2016033138

This book is published in the IGI Global book series Advances in Multimedia and Interactive Technologies (AMIT) (ISSN: 2327-929X; eISSN: 2327-9303)

British Cataloguing in Publication Data
A Cataloguing in Publication record for this book is available from the British Library.

Advances in Multimedia and Interactive Technologies (AMIT) Book Series

Joel J.P.C. Rodrigues
National Institute of Telecommunications (Inatel), Brazil & Instituto de Telecomunicações, University of Beira Interior, Portugal

ISSN:2327-929X
EISSN:2327-9303

Mission

Traditional forms of media communications are continuously being challenged. The emergence of user-friendly web-based applications such as social media and Web 2.0 has expanded into everyday society, providing an interactive structure to media content such as images, audio, video, and text.

The Advances in Multimedia and Interactive Technologies (AMIT) Book Series investigates the relationship between multimedia technology and the usability of web applications. This series aims to highlight evolving research on interactive communication systems, tools, applications, and techniques to provide researchers, practitioners, and students of information technology, communication science, media studies, and many more with a comprehensive examination of these multimedia technology trends.

Coverage

- Internet Technologies
- Multimedia technology
- Mobile Learning
- Digital Watermarking
- Social Networking
- Digital Communications
- Multimedia Services
- Multimedia Streaming
- Gaming Media
- Web Technologies

IGI Global is currently accepting manuscripts for publication within this series. To submit a proposal for a volume in this series, please contact our Acquisition Editors at Acquisitions@igi-global.com or visit: http://www.igi-global.com/publish/.

The Advances in Multimedia and Interactive Technologies (AMIT) Book Series (ISSN 2327-929X) is published by IGI Global, 701 E. Chocolate Avenue, Hershey, PA 17033-1240, USA, www.igi-global.com. This series is composed of titles available for purchase individually; each title is edited to be contextually exclusive from any other title within the series. For pricing and ordering information please visit http://www.igi-global.com/book-series/advances-multimedia-interactive-technologies/73683. Postmaster: Send all address changes to above address.

Titles in this Series

For a list of additional titles in this series, please visit: www.igi-global.com

Applied Video Processing in Surveillance and Monitoring Systems
Nilanjan Dey (Techno India College of Technology, Kolkata, India) Amira Ashour (Tanta University, Egypt) and Suvojit Acharjee (National Institute of Technology Agartala, India)
Information Science Reference • copyright 2017 • 321pp • H/C (ISBN: 9781522510222) • US $215.00 (our price)

Intelligent Analysis of Multimedia Information
Siddhartha Bhattacharyya (RCC Institute of Information Technology, India) Hrishikesh Bhaumik (RCC Institute of Information Technology, India) Sourav De (The University of Burdwan, India) and Goran Klepac (University College for Applied Computer Engineering Algebra, Croatia & Raiffeisenbank Austria, Croatia)
Information Science Reference • copyright 2017 • 520pp • H/C (ISBN: 9781522504986) • US $220.00 (our price)

Emerging Technologies and Applications for Cloud-Based Gaming
P. Venkata Krishna (VIT University, India)
Information Science Reference • copyright 2017 • 314pp • H/C (ISBN: 9781522505464) • US $195.00 (our price)

Digital Tools for Computer Music Production and Distribution
Dionysios Politis (Aristotle University of Thessaloniki, Greece) Miltiadis Tsalighopoulos (Aristotle University of Thessaloniki, Greece) and Ioannis Iglezakis (Aristotle University of Thessaloniki, Greece)
Information Science Reference • copyright 2016 • 291pp • H/C (ISBN: 9781522502647) • US $180.00 (our price)

Contemporary Research on Intertextuality in Video Games
Christophe Duret (Université de Sherbrooke, Canada) and Christian-Marie Pons (Université de Sherbrooke, Canada)
Information Science Reference • copyright 2016 • 363pp • H/C (ISBN: 9781522504771) • US $185.00 (our price)

Trends in Music Information Seeking, Behavior, and Retrieval for Creativity
Petros Kostagiolas (Ionian University, Greece) Konstantina Martzoukou (Robert Gordon University, UK) and Charilaos Lavranos (Ionian University, Greece)
Information Science Reference • copyright 2016 • 388pp • H/C (ISBN: 9781522502708) • US $195.00 (our price)

Emerging Perspectives on the Mobile Content Evolution
Juan Miguel Aguado (University of Murcia, Spain) Claudio Feijóo (Technical University of Madrid, Spain & Tongji University, China) and Inmaculada J. Martínez (University of Murcia, Spain)
Information Science Reference • copyright 2016 • 438pp • H/C (ISBN: 9781466688384) • US $210.00 (our price)

Table of Contents

Detailed Table of Contents

Section 1
Introduction to Video Processing and Mining

Video processing has various applications in several domains. This section highlights the fundamental concepts and algorithms of video processing in the medical domain. The real time frames for the acquired images from different image modalities is used. Another context is introduced for data mining that has great benefit in data extraction various applications including educational and medical domains.

In today's medical environments, imaging technology is extremely significant to provide information for accurate diagnosis. An increasing amount of graphical information from high resolution 3D scanners is being used for diagnoses. Improved medical data quality become one of the major aims of researchers. This leads to the development of various medical modalities supported by cameras that can provide videos for the human body internal for surgical purposes and more information for accurate diagnosis. The current chapter studied concept of the video processing, and its application in the medical domain. Based on the highlighted literatures, it is convinced that video processing and real time frame will have outstanding value in the clinical environments.

Section 3
Motion Detection in Video Applications and Miscellaneous Related Topics

Real time dynamic scene has been employed in a number of applications, which can be useful for three-dimensional (3D) animation videos in electronic games, 3D television, motion analysis, and gesture recognition. This process is proposed and discussed in this section. In addition, the developed artificial vision system allows allocating the position of objects located on the robot's Cartesian work plane. The frame by frame acquisition of images allows positioning in real time the links of a manipulator robot. However, cameras speed of frame acquisition restricts such operation, which is discussed in this section. Finally, the section includes other related topics.

This chapter introduces a system for acquiring synchronized multi-view color and depth (RGB-D) video data using multiple off-the-shelf Microsoft Kinect and methods for reconstructing temporally coherent 3D animation from the multi-view RGB-D video data. The acquisition system is very cost-effective and provides a complete software-based synchronization of the camera system. It is shown that the data acquired by this framework can be registered in a global coordinate system and then can be used to reconstruct the 360-degree 3D animation of a dynamic scene. In addition, a number of algorithms to reconstruct a temporally-coherent representation of a 3D animation without using any template model or a-prior assumption about the underlying surface are also presented. It is shown that despite some limitations imposed by the hardware for the synchronous acquisition of the data, a reasonably accurate reconstruction of the animated 3D geometry can be obtained that can be used in a number of applications.

The fundamental principles of the coding/decoding H.264/AVC standard are introduced emphasizing the role of motion estimation and motion compensation (MC) in error concealment using intra- and inter-frame motion estimates, along with other features such as the integer transform, quantization options, entropy coding possibilities, deblocking filter, among other provisions. Efficient MC is one of the certain reasons for H.264/AVC superior performance compared to its antecedents. The H.264/AVC has selective intra-prediction and optimized inter-prediction methods to reduce temporal and spatial redundancy more efficiently. Motion compensation/prediction using variable block sizes and directional intra-prediction to choose the adequate modes help decide the best coding. Unfortunately, motion treatment is a computationally-demanding component of a video codec. The H.264/AVC standard has solved problems its predecessors faced when it comes to image quality and coding efficiency, but many of its advantages require an increase in computing complexity.

Machine Safety is a growing technical discipline with a strong basis in the development of electrical and electronic devices, commonly known as Safety Protective Devices (SPDs). SPDs are designed to avoid or at least mitigate those risks associated with a particular human-machinery interaction. Ranging from conceptually simple electromechanical Emergency Stop Devices (ESDs) to the more complex Active Optoelectronic Protective Devices (AOPDs), a place for Real-Time Digital Video Processing has recently been open for research in Machine Safety. This chapter is intended to explore the standardized features of the so-called Vision-Based Protective Devices (VBPDs), their current technical development and principal applications in Machine Safety, with a stress on prominent vision-related implementation issues.

This article reviews the various image processing techniques in MATLAB and also hardware implementation in FPGA using Xilinx system generator. Image processing can be termed as processing of images using mathematical operations by using various forms of signal processing techniques. The main aim of image processing is to extract important features from an image data and process it in a desired manner and to visually enhance or to statistically evaluate the desired aspect of the image. This article provides an insight into the various approaches of Digital Image processing techniques in Matlab. This article also provides an introduction to FPGA and also a step by step tutorial in handling Xilinx System Generator. The Xilinx System Generator tool is a new application in image processing and offers a friendly environment design for the processing. This tool support software simulation, but the most important is that can synthesize in FPGAs hardware, with the parallelism, robust and speed, this features are essentials in image processing. Implementation of these algorithms on a FPGA is having advantage of using large memory and embedded multipliers. Advances in FPGA technology with the development of sophisticated and efficient tools for modelling, simulation and synthesis have made FPGA a highly useful platform.

The design and implementation stages of a redundant robotized manipulator with six Degrees Of Freedom (DOF), controlled with visual feedback by means of computational software, is presented. The various disciplines involved in the design and implementation of the manipulator robot are highlighted in their electric as well as mechanical aspects. Then, the kinematics equations that govern the position and orientation of each link of the manipulator robot are determined. The programming of an artificial vision system and of an interface that control the manipulator robot is designed and implemented. Likewise, the type of position control applied to each joint is explained, making a distinction according to the assigned task. Finally, functional mechanical and electric tests to validate the correct operation of each of the systems of the manipulator robot and the whole robotized system are carried out.

Vaccines build a defense mechanism against the disease causing agents through the immune system stimulation and disease agents' imitation. Some of the vaccines contain a part of the disease causing agents that are either weakened or dead. Along with using vaccines with viral infections, it can be used against the various types of cancers for both therapy and prevention. The use of cancer's vaccines in cancer therapies is called immunotherapy. It can be done either by specific cancer vaccine or universal cancer vaccine that contains tumor antigens, which stimulate the immune system. This in turn initiates various mechanisms that terminate tumor cells and prevents recurrence of these tumors. The present work proposed an Insilico approach in epitope prediction and analysis of antigenecity and Immunogenecity of Haemophilus influenzae strains. It was interested with the design of a novel vaccine delivery system with better adjuvancity, where vaccine adjuvant is significant for the improvement of the antigens' immunogenicity that present in the vaccines. The conducted insilico approaches selected the best strain target proteins, m strain selection, epitope prediction, antigenicity and immunogenicity prediction of target proteins to find out the best targets.

Preface

Multimedia is a media and content uses an arrangement of diverse content forms including sound, audio, graphics, animation, and video. The most significant form of multimedia is video as it has various relevance in different aspects. Motion detection is the main module for several video applications including security, visual localization and mapping (SLAM), visual tracking of vehicles ahead for safety/ auto-driving, and Augmented Reality (AR). However, motion detection systems/algorithms have specific requirements based on the application to achieve accurate, robust, and fast detection in real time problems.

In a digital video, the information/features extracted form a picture is digitized both spatially and temporally. Digital image and video processing are applied for improving the captured videos/images quality. The most crucial task in video processing is to divide the long video sequences into a number of shots. Afterward, discover a key frame of each shot for supplementary video information retrieval tasks, where a frame is known as an electronically coded still image. Background subtraction principle is mainly applied in motion detection systems to extract the objects from the background. The extracted features from the video are constantly tracked. Features are known as the points of interest for image description, such as lines, edges, and corners. The temporal features' information and their motion behaviors are used for identification, image alignment, 3D reconstruction, object recognition, motion tracking and robot navigation. Local invariant features provide good representation that allows efficient matching for local structures between images. The first phase for the local feature extraction pipeline is finding a set of reliable localized key points under varying viewpoint, changes imaging conditions, and in the presence of noise. Moreover, the extraction approach should realize the same feature locations even if translation or rotation has occurred. Hessian detector, Harris detector, Laplacian-of-Gaussian (LoG) detector, Harris-Affine and Hessian-Affine detectors are all examples for efficient detectors. Once a set of the interest regions are extracted from an image, their corresponding content are encoded in a descriptor, which is suitable for discriminative matching. Such descriptors include the Scale Invariant Feature Transform (SIFT), Speeded-Up Robust Features (SURF).

Segmentation of the video track into smaller items facilitates the succeeding processing procedures on video shots, for instance semantic representation/ tracking of the selected video information, video indexing and recognizing the frames where a transition occurred from one shot to another. The low level features can be extracted from the segmented video. The video data must be manipulated appropriately for efficient information retrieval. The main challenging task is the retrieval of information from the video data. The majority task is to transform the unstructured data into structured one for video data processing. Prior to processing the video frames noise elimination and illumination changes should be removed. Video processing has several applications including:

- Motion capturing of an athlete,
- Motion pictures' analysis,
- Rehabilitation to assess the locomotion abilities,
- Robot control, and
- Educational programs biometrics.

Consequently, this book focuses on exploring different applications for the feature detectors and descriptors as well as motion detection in video processing applications.

OBJECTIVE

This book thoughtful the foremost feature detectors and descriptors algorithms for video/image processing. It deals chiefly with methods and approaches that involve video analysis and retrieval, automatic shot detection, etc. This book grants significant frameworks and the most contemporary empirical research outcomes in the feature detectors and descriptors algorithms. As well as it introduces variety of motion detection in video processing applications in a wide range. It is written for professionals and researchers working in the field of video and imaging in various disciplines, e.g. Software/Hardware video security monitoring, medical devices engineering, researchers, academicians, advanced-level students, and technology developers.

ORGANIZATION

The book consists of an introduction followed by twelve chapters that organized in three sections as shown in Table 1. The first two chapters focus on the concept and applications for the image/video processing as well as the mining concept. The second section consists of Chapters 3 through 6 that extensively deploy feature detectors and descriptors. The last six chapters are included in section 3, which focuses on different motion detection in video applications and miscellaneous related topics.

INTRODUCTION

Object tracking is an estimating problem for the positions and other relevant information of moving objects within an image sequences (video). This chapter deployed concept of motion detection. Additionally, it includes the motion detection and tracking in different video processing applications that can be improved in new research aspects in the future.

Table 1.

<table>
<tr><td colspan="6">Section 1: Introduction to Video Processing and Mining</td></tr>
<tr><td colspan="3">Chapter 1
Medical Video Processing: Concept and Applications</td><td colspan="3">Chapter 2
Educational Data Mining and Indian Technical Education System: A Review</td></tr>
<tr><td colspan="6">Section 2: Feature Detectors and Descriptors</td></tr>
<tr><td>Chapter 3
Feature Detectors and Descriptors Generations with Numerous Images and Video Applications: A Recap</td><td colspan="2">Chapter 4
Analysis of Different Feature Description Algorithm in Object Recognition</td><td>Chapter 5
A Study on Different Edge Detection Techniques in Digital Image Processing</td><td colspan="2">Chapter 6
A Nearest Opposite Contour Pixel Based Thinning Strategy for Character Images</td></tr>
<tr><td colspan="6">Section 3: Motion Detection in Video Applications and Miscellaneous Related Topics</td></tr>
<tr><td>Chapter 7
Multi-view RGB-D Synchronized Video Acquisition and Temporally Coherent 3D Animation Reconstruction Using Multiple Kinects</td><td>Chapter 8
On the Use of Motion Vectors for 2D and 3D Error Concealment in H.264/AVC Video</td><td>Chapter 9
Vision-Based Protective Devices</td><td>Chapter 10
A Study on Various Image Processing Techniques and Hardware Implementation using Xilinx System Generator</td><td>Chapter 11
New Redundant Manipulator Robot with Six Degrees of Freedom Controlled with Visual Feedback</td><td>Chapter 12
Insilico Approach for Epitope Prediction Toward Novel Vaccine Delivery System Design</td></tr>
</table>

SECTION 1: INTRODUCTION TO VIDEO PROCESSING AND MINING (CHAPTERS 1-2)

This section elaborated the fundamental concepts and algorithms of video processing in the medical domain. Another context is introduced related to data mining that has great benefit in data extraction various applications including educational and medical domains.

Chapter 1

This chapter included an overview on the medical video processing. Recently, various medical modalities are supported by cameras to provide videos for the human body internal for surgical purposes. In addition, more information is acquired from such medical videos for accurate diagnosis. In this current chapter, it was convinced that video processing and real time frame will have outstanding value in the clinical environments.

Chapter 2

Educational Data Mining (EDM) is emerged as a powerful tool to explore the unique types of data in educational settings for better understand students. Clustering, classification, association rule mining and decision trees are employed to improve teaching and learning systems. This chapter is carried out to explore the most relevant studies in the data mining domain in higher and technical education sectors to probably portray a pathway towards the improvement of the quality education in technical institutions.

SECTION 2: FEATURE DETECTORS AND DESCRIPTORS (CHAPTERS 3-6)

Feature detectors and descriptors are the milestone in numerous applications including video camera calibrations, object recognition, biometrics, medical applications and image/video retrieval. Extract the points of interest between two similar scenes, objects, images/video shots are the main task of the feature detectors and descriptors. This section highlighted the different feature detectors and descriptors types that involved in various applications.

Chapter 3

This current chapter introduced a synopsis about the feature detectors, such as Moravec, Hessian, Harris and Features from Accelerated Segment Test. It addressed their generation over time, their concept, and their applications in image/video processes. Additionally, some recent feature detectors are addressed with comparison to illustrate their respective strengths and weaknesses. The advanced feature detectors indicated that combining a novel heuristic for feature detection with machine learning methods provided a robust, high-speed corner detection algorithm that can be use in real-time image processing applications.

Chapter 4

Local feature description and global feature description algorithms have a vital role in the object recognition process. This chapter analyzed the foremost feature detector/descriptor algorithm and evaluated their performance in different circumstances including rotational, scaling, illumination, and blurring effects. Moreover, the speed of each algorithm in different situations was measured.

Chapter 5

Edge detection is one of the fundamental techniques for image segmentation. The image edges include a good number of rich information that is very significant for obtaining the image characteristic image analysis. Thus, edge detection techniques transform the original images into edge images using the changes of grey tones in the image. This chapter studied the theory of edge detection for image segmentation using various computing approaches. It was stated that edge detection techniques are a combination of image smoothing and differentiation plus a post-processing for edge labeling.

Chapter 6

Removal of strokes or deformities in thinning of character images is a big challenge. This chapter proposed a nearest opposite contour pixel based thinning strategy used for performing skeletonization of printed and handwritten character images. Shape characteristics of text were used to get a skeleton of nearly same as the true character shape. This method assists to preserve the local features and true shape of the character images. The proposed algorithm produces one pixel-width thin skeleton. The results were conducted on printed English and Bengali characters and compared to other thinning methods without any post-processing.

SECTION 3: MOTION DETECTION IN VIDEO APPLICATIONS AND MISCELLANEOUS RELATED TOPICS (CHAPTERS 7-12)

Real time dynamic scene has been employed in a number of applications, such as the three- dimensional (3D) animation videos in electronic games, 3D television, motion analysis, and gesture recognition. This process was proposed and discussed in this section. In addition, the developed artificial vision system allows allocating the position of objects located on the robot's Cartesian work plane. The frame by frame acquisition of images allows positioning in real time the links of a manipulator robot. However, cameras speed of frame acquisition restricts such operation, which was discussed in this section. Finally, the section included other related topics.

Chapter 7

This chapter proposed a system for acquiring synchronized multi-view color and depth (RGB-D) video data using multiple off-the-shelf Microsoft Kinect and reconstructing methods for coherent 3D animation from the multi-view RGB-D video data. The data acquired by this framework can be registered in a global coordinate system and then can be used to reconstruct the 360-degree 3D animation of a dynamic scene.

Chapter 8

Motion estimation and motion compensation (MC) have a significant role in the coding/decoding H.264/AVC standard for error concealment. The H.264/AVC has selective intra-prediction and optimized inter-prediction methods to reduce temporal and spatial redundancy more efficiently. This chapter depicted that motion compensation/prediction using variable block sizes and directional intra-prediction assisted the decision for the best coding. Unfortunately, motion treatment is a computationally-demanding component of a video codec. The H.264/AVC standard has solved problems its predecessors faced when it comes to image quality and coding efficiency, but many of its advantages require an increase in computing complexity.

Chapter 9

Machine Safety is a technical discipline with a strong basis in the development of electrical and electronic devices. Recently, real-time digital video processing is involved in machine safety. This chapter intended to explore the standardized features of the vision-based protective devices (VBPDs), their technical development and principal applications in the machine safety domain with a stress on prominent vision-related implementation issues.

Chapter 10

This chapter was extensively discussed the various image processing techniques in MATLAB and the hardware implementation in FPGA using Xilinx system generator. The Xilinx system generator tool is a new application in image processing and offers a friendly environment design for the processing. This tool supports software simulation as well as synthesizes in FPGAs hardware with the parallelism, robust and speed.

Chapter 11

This chapter depicted the design and implementation stages of a redundant robotized manipulator with six degrees of freedom (DOF) controlled with visual feedback by means of computational software. The artificial vision system/interface programming are designed and implemented. Finally, functional mechanical and electric tests to validate the correct operation of each of the systems of the manipulator robot and the whole robotized system were carried out.

Chapter 12

Vaccines through mimicking disease agents and stimulating the immune system builds a defense mechanism against the disease causing agents. The present chapter proposed an Insilico technique in epitope prediction and analysis of antigenecity and Immunogenecity of Haemophilus influenzae strains. The conducted insilico approaches selected the best strain target proteins, m strain selection, epitope prediction, antigenicity and immunogenicity prediction of target proteins to find out the best targets.

Nilanjan Dey
Techno India College of Technology – Kolkata, India

Amira S. Ashour
Tanta University, Egypt & Taif University, Saudi Arabia

Prasenjit K. Patra
Bengal College of Engineering and Technology – Durgapur, India

Acknowledgment

There is no substitute for hard work. – Thomas Edison

We are greatly thankful to our parents and families for endless support, and love through all our life. We dedicate this book to all of them. Our appreciation is directed to the all peoples who support, share, read, wrote, and offered comments through the book journey. Moreover, we are grateful to all the authors who provide in time perfect blend of knowledge and skills with patience, and perseverance.

Special thanks to the IGI-publisher team, who showed us the ropes to start and continue as well as our readers, who gave us their thrust and hope our work inspired and guide them.

Nilanjan Dey
Techno India College of Technology – Kolkata, India

Amira S. Ashour
Tanta University, Egypt & Taif University, Saudi Arabia

Prasenjit K. Patra
Bengal College of Engineering and Technology – Durgapur, India

Introduction

MOTION DETECTION AND TRACKING IN VIDEO PROCESSING APPLICATIONS

Introduction

Motion detection and tracking to locate moving objects over time using a camera have several applications (Cucchiara et al., 2001, 2002, 2003). Such applications include video communication and compression, traffic control, human-computer interaction, surveillance and security, and medical imaging (Dey et al., 2012; Acharjee et al., 2012; Bose et al., 2014; Ikeda et al., 2014; Pal et al., 2015; Dey et al., 2016). However, due to the huge data amount in the consecutive video frames, moving objects detection and video tracking are considered to be time consuming processes. Moreover, for fast moving targets relative to the frame rate, or if the tracked objects change their orientation over time; the association between the tracked objects will be complex. Thus, video analysis and processing are required, where video refers to visual information in sequence of still image or time-varying images, where one of the main differences between the processing of still images and the digital video processing is that the later contains a substantial amount of temporal correlation between the frames. Multi-frames processing empowers the improvement of effective algorithms, such as motion-compensated prediction and motion compensated filtering. Video processing is defined as the content analysis of the video to extract information of the observed scene. Video processing includes object recognition, video indexing, video compression, video segmentation and video tracking. Consequently, for objects detection and tracking, a motion model is employed in the video tracking systems to describe motions of the objects within the video frames. Different algorithms are developed to analyses the sequential video frames to determine the targets' movement between the frames (Li et al., 2003; Prati et al., 2003).

Traditional video monitoring by human operator is infeasible and impractical. Thus, researchers are interested with the development of automated video sequences analysis for detecting and tracking objects and their motion (Nascimento et al., 2006; Elhabian et al., 2008). Typically, the main step for video analysis is the moving object detection. Any tracking technique involves an object detection process either when the first appearance of an object in the video or in every frame. These systems handle higher level processing such as segmentation/classification of moving objects from still background objects (Paragios & Deriche; 2000). However, owing to the different environmental conditions such as shadow, and illumination changes; object segmentation/classification becomes complex. The main idea for object's detection is to extract information in a single frame through extracting the feature points from each frame and then match them across the consecutive frames. However, in order to reduce the false detections, some object detection techniques use the temporal information that highlights dynamically changed regions in consecutive frames (Elgammal et al., 2002).

The organization of the remaining chapter is as follows. A related work to the motion detection and tracking systems concepts is introduced. Afterward, several methods including the background subtraction, optical flow, feature point-based tracking, contour-based object tracking, region-based object tracking are presented. Several applications for the motion detection and tracking are introduced followed by the conclusion.

Related Work

Motion detection and tracking systems have gained many interests in the last few years toward integrated improved performance for motion detection and visual tracking. Motion detection is carried out using an algorithm for combining temporal variance with background modeling techniques. Moreover, the tracking algorithm gathers motion and extracted information to track the object in subsequent frames. Yoneyama et al. (1999) extended the capabilities to MPEG2 (Motion Picture Experts Group 2) streams for tracking objects selected by a user throughout the sequence. The tracking was performed by the exploitation of the already present motion information in the bit stream. Hariharakrishnan et al. (2003) suggested a tracking algorithm by predicting the object boundary using motion vectors. Afterward, contour was updated using occlusion/disocclusion detection. For estimating motion between consecutive frames, an adaptive block-based scheme was used. An effective modulation approach was used to control the gap between frames that used for object tracking. Hernandez et al. (2009) proposed an automatic movement detection system based on the motion vector estimation for a video sequence. The authors used filter to eliminate the noisy and distorted vectors due to illumination variations and background movement. In addition, the proposed algorithm discriminated between relevant and non-relevant movements to allow the consideration of the movements whose direction is from outside to inside the restricted zone. The results established that by using a motion vector estimation, the proposed algorithm was also able to tract the trajectory of any person whose movement is from inside to outside of the restricted zone.

Feature description and interest point detection are considered the basis of feature-based tracking systems. Gauglitz et al. (2011) designed a dataset of video sequences of planar textures with ground truth that included various lighting conditions, geometric changes, and different levels of motion blur to be used as a testbed for tracking-related problems. In addition, an inclusive evaluation of detector-descriptor-based visual camera tracking was carried out. The experimental results included the evaluation of individual algorithm parameters impact and compared the algorithms for both detection and description in isolation as well as all detector-descriptor combinations for tracking solution. For accurate and robust detection of shot boundaries in video analysis, Warhade et al. (2013) addressed an approach for shot boundary detection in the fast object motion conditions, illumination change, and fast camera motion. The proposed method used dual-tree complex wavelet transform to extract the structure features from each video frame. Then, spatial domain structure similarity was computed between adjacent frames. The shot boundaries were determined based on the selected thresholds. The results were performed on a number of videos with significant illumination change and fast motion of camera and objects.

A slow motion replays captured from standard cameras as well as high speed cameras was extracted by Farn et al. (2003) form basketball videos. The dominate color of soccer field was considered; though it was not applicable in basketball videos due to the basketball court. A support vector machine (SVM) was employed to classify slow motion replays and normal shots by Wang et al. (2004) after extracting motion-related features in basketball videos. Chen and Chen (2015) proposed a novel approach to detect slow motion in basketball videos replay. A scoreboard was employed to filter large amount of non-replay frames in order to improve the detection accuracy. Afterward, every consecutive non-scoreboard frame sequence that bounded by scoreboard frames were considered as a non-scoreboard segment. In order to create the features, the characteristics of replays and non-replays were observed to detect replays and prune non-replays from non-scoreboard segments. The results depicted that the proposed replay detection method was applicable for kinds of basketball videos with/without TV commercials.

Motion Detection and Tracking

In digital image processing, motion detection and analysis are the most popular and recent research topics. The objects' movement is the imperative part to detect the objects' motion from the background image in a video sequence. Motion analysis includes detection, recognition, and tracking of the humans' behavior along with some other objects in motion from the video frame. Afterward, the detected objects can be classified in several categories including humans, vehicles or other moving objects (Parekh et al., 2014). Reliable motion detection algorithms are significant in any tracking system. Such algorithms based on partition the pixels of the image sequence in every frame into two main classes, namely

1. Background that correspond to pixels belonging to the static scene, and
2. Foreground that correspond to pixels of the moving objects (Lacassagne et al., 2009).

Typically, the motion detection algorithms should accurately discriminate the moving objects from the background, without being sensitive to the changing conditions or to the velocities and/or sizes of the objects. Due to the great amount of the extracted data from the video frames, the motion detection is considered the most computational process in the tracking systems. In the tracking systems, the region of interest is segmented from the video frames and thus keeps tracking for the motion and position of the object of interest. The main steps for the tracking system are illustrated in Figure 1.

As illustrated in Figure 1, an object detection algorithm such as frame differncing, optical flow or the background subtration are applied on the input video sequence. Afterward, a classification method is used to classify the objects into categories based on the objects' shape, color, texture, motion, or a combination of them. Finally, the tracking process using point-based, kernel-based, or silhouette-based is used to track the objects under concern. Such tracking methods derive the motion trajectory of the object over time in the video sequences, while handling the presence of noise and blur in the video, luminance/intensity changes, the object's abrupt motion, partial/full object occlusion, and the variation that may occur in the shape/size of the object from frame to frame. The most significant motion detection and tracking algorithms can be discussed as follows.

Figure 1. Object tracking steps

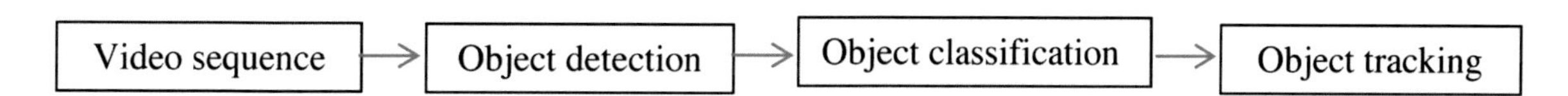

Background Subtraction Method

The background subtraction method simple to implement and widely used. It based on subtracting the existing frame from previous frame (reference frame) to obtain the threshold value. However, any background subtraction based motion detection system have to handle some critical situations including:

1. Noisy images that results from poor quality image source;
2. Small movements of non-static objects;
3. Variations of the lighting conditions and intensity in the scene;
4. Sudden changes in the light conditions;
5. Movements of the objects in the background;
6. Shadow regions; and
7. Multiple objects moving in the scene (Rakibe & Patil, 2013).

The background subtraction algorithm has the following steps (see Algorithm 1).

Optical Flow

Optical flow (optic flow) is the pattern of objects' superficial motion in a visual scene caused by a relative motion between the observer and the scene. The optical flow technique for motion detection describes the direction and time rate of the pixels in a time sequence of two consequent images/frames. For each pixel in a given place in the image, a two dimensional velocity vector are assigned. Generally, the optical flow determines the motion estimation by calculating the gradient, Laplacian, and subsequently the velocities of each pixel in a parallel to speed up the computations. It is an Intensity-based differential technique that has several methods to perform its calculations for the partial derivatives of the image signal. Such methods are the Lucas-Kanade and the Horn-Schunck (Aslani & Mahdavi-Nasab, 2013).

Feature Point-Based Tracking

Feature point tracking is a standard algorithm in numerous applications including motion analysis, navigation, and scene monitoring (Kouwenberg et al., 2016). In this model, feature points are used to describe the objects through three basic steps:

Algorithm 1. Steps of background subtraction algorithm

Background Subtraction Method Algorithm
Start
Input the frames of the video sequences
Separate the frame/ images in the current frame image and background frame image
Perform background subtraction
Detect the moving object
Perform background updating
Remove the noise
Analyze the shape
Stop

1. Recognize and track the object by extracting elements,
2. Cluster the objects into higher level features, and
3. Match the extracted features between images in successive frames.

Feature extraction and matching are the most important steps in the feature based object tracking. The challenging difficulty in feature point based tracking is the feature correspondence, where a feature point in one image may have other similar points in another image, which leads to ambiguity.

Contour-Based Object Tracking

The active contour model is applied to find the object's outline from an image/video (Serby et al., 2004). The objects are tracked using their outlines as boundary contours in the contour-based tracking algorithm. Subsequently, these contours are updated dynamically in the successive frames. The discrete version of the contour-based method is represented in active contour model. This algorithm is highly sensitive to the tracking initialization, making it complex to start tracking automatically.

Region-Based Object Tracking

In this method the regions are used as primitives in the tracking systems. The objects' tracking based on the region model is mainly depends on the color distribution of the tracked object (Kumar et al., 2005). Thus, it is computationally efficient. However, it suffers from degradation in the efficiency, when several objects move together in the image sequences. It is impossible to achieve accurate tracking in the presence of multiple objects that move due to occlusion. Furthermore, in the absence of object shape information, the object tracking is dependent on the background model.

Motion Detection Applications

The previous motion detection and tracking methods are applied in several applications as follows.

Traffic Monitoring System

Smart transportation system offers an effective alternative to the traditional traffic system. It provides more safe and fast tracking system for traffic through using video cameras. Moving objects detection is the first relevant step in this system. Yu and Chen (2009) proposed a novel method to detect moving objects in a complex, non-stationary background for an automatic traffic monitoring system. In order to compensate the disturbance caused by shaking of camera, the square neighborhood algorithm is adopted. Afterward, an enhanced temporal difference method was applied to determine the moving areas. Some post-processes were used for optimized detection by eliminating noise from the moving areas. The proposed method proved its superiority in real-time problems. The experiment results established that this technique can detect the moving objects professionally and accurately form the video recorded by a shaking camera under changing background and noises conditions.

Video Surveillance System

Video surveillance is a procedure of analyzing video sequences for monitoring and security. Joshi and Thakore (2012) presented a survey of several methods related to video surveillance system to improve the security using moving object detection and object tracking methods. Detection of suspicious human behavior is imperative in the automated video surveillance applications. However, due to the human movements' random nature, reliable classification of suspicious human movements can be complex. Boubekeur et al. (2014) deployed a non-parametric technique for background subtraction and moving object detection. The object detection was based on adaptive threshold using successive squared differences and frame differencing process. The adaptive threshold and dependent distance calculation using a weighted estimation procedure were proposed. The presented experimental results established that the proposed algorithm was successfully extracting the moving foreground efficiently with achieved robustness to noise. Rahangdale and Kokate (2016) proposed an approach to solve the problem of automatically tracking people and detecting unusual or suspicious movements in Closed Circuit TV (CCTV) videos for surveillance systems installed in indoor environments. The presented framework processed the video data obtained from a CCTV camera fixed at a specific location. The foreground objects were obtained by using background subtraction method, where the gray level intensity was used due to speed matters in real time applications. These foreground objects were then classified into people and suspicious objects. Afterward, these objects were tracked using a blob matching technique. Activities were classified by considering the spatial and temporal properties of these blobs using semantics-based approach. This proposed algorithm could increase the object detection efficiency as well as the accuracy.

Future Scope

The techniques used for objects tracking in video sequences are based on algorithms to analyze the sequential video frames and to produce the target movement between the frames as an output. Such techniques for object tracking in videos require improvement to overcome the real time conditions, such as:

- Noisy environments, the non-static objects movement, the light conditions variation, the movements of the objects in the background, the effect of the shadow regions; and the multiple objects moving.
- In addition, the segmentation and tracking of multiple humans in crowded locations is complex by inter-object occlusion. Thus, a model-based approach to interpret the image observation is required to handle such situations.
- Tracking objects underwater is a very inflexible task due to the water presence that leads to hostile environment. Hence, several parameters should be considered to reduce the effect of the hostility.
- Develop new techniques to reduce the computational time for the motion detection that results from the huge amount of the video data. These new methods can depend on extract the more significant features for the object under concern and using new feature selection methods.

CONCLUSION

Tracking is performed in the context of higher-level applications, which require the location and/or shape of the object in every frame to be defined. Recently, real time object tracking becomes significant in the video analysis and processing fields. Object tracking is concerned with monitoring and tracking an object or multiple objects over a video sequence. It can also be defined as a segmenting process of an object of interest from a video scene while keeping track of its motion, occlusion, orientation, etc. in order to extract useful information. Several challenges in objects tracking can be arise due to changing the appearance of the object in the different frames, abrupt object motion, non-rigid object structures, and the camera motion.

The traditional algorithms often cannot track moving objects accurately in real time, in order to overcome. Motion detection and object tracking can be applied to video surveillance systems, medical systems, traffic monitoring systems and more other applications. Every tracking method requires an object detection algorithm either in every frame or when the object first appears in the video. Using the information in a single frame is the common approach for object detection. However, some object detection techniques use the temporal information in the form of frame differencing to reduce the number of false detections. The tracking approaches were divided into three categories, namely contour based, region based and feature based approach. Tracking object motion can be performed by object detection and then using tracking approach.

Nilanjan Dey
Techno India College of Technology – Kolkata, India

Amira S. Ashour
Tanta University, Egypt & Taif University, Saudi Arabia

REFERENCES

Acharjee, S., Dey, N., Biswas, D., Das, P., & Chaudhuri, S. S. (2012, November). A novel Block Matching Algorithmic Approach with smaller block size for motion vector estimation in video compression. In *Intelligent Systems Design and Applications (ISDA), 2012. 12th International Conference on* (pp. 668-672). IEEE. doi:10.1109/ISDA.2012.6416617

Aslani, S., & Mahdavi-Nasab, H. (2013). Optical flow based moving object detection and tracking for traffic surveillance. *International Journal of Electrical, Electronics, Communication Energy Science and Engineering*, *7*(9), 789–793.

Bachir, B. M., Tarek, B., SenLin, L., & Hocine, L. (2014). Weighted Samples Based Background Modeling for the Task of Motion Detection in Video Sequences. *TELKOMNIKA Indonesian Journal of Electrical Engineering*, *12*(11), 7778–7784. doi:10.11591/telkomnika.v12i11.6545

Bose, S., Chowdhury, S. R., Sen, C., Chakraborty, S., Redha, T., & Dey, N. (2014, November). Multi-thread video watermarking: A biomedical application. In *Circuits, Communication, Control and Computing (I4C), 2014 International Conference on* (pp. 242-246). IEEE. doi:10.1109/CIMCA.2014.7057798

Chen, C. M., & Chen, L. H. (2015). A novel method for slow motion replay detection in broadcast basketball video. *Multimedia Tools and Applications*, *74*(21), 9573–9593. doi:10.1007/s11042-014-2137-5

Cucchiara, R., Grana, C., Neri, G., Piccardi, M., & Prati, A. (2002). The Sakbot system for moving object detection and tracking. In Video-Based Surveillance Systems (pp. 145-157). Springer US. doi:10.1007/978-1-4615-0913-4_12

Cucchiara, R., Grana, C., Piccardi, M., & Prati, A. (2003). Detecting moving objects, ghosts, and shadows in video streams. *Pattern Analysis and Machine Intelligence. IEEE Transactions on*, *25*(10), 1337–1342.

Cucchiara, R., Grana, C., Piccardi, M., Prati, A., & Sirotti, S. (2001). Improving shadow suppression in moving object detection with HSV color information. In Intelligent Transportation Systems, 2001. Proceedings. 2001 IEEE (pp. 334-339). IEEE. doi:10.1109/ITSC.2001.948679

Dey, N., Bose, S., Das, A., Chaudhuri, S. S., Saba, L., Shafique, S., & Suri, J. S. et al. (2016). Effect of watermarking on diagnostic preservation of atherosclerotic ultrasound video in stroke telemedicine. *Journal of Medical Systems*, *40*(4), 1–14. doi:10.1007/s10916-016-0451-3 PMID:26860914

Dey, N., Das, P., Roy, A. B., Das, A., & Chaudhuri, S. S. (2012, October). DWT-DCT-SVD based intravascular ultrasound video watermarking. In *Information and Communication Technologies (WICT), 2012 World Congress on* (pp. 224-229). IEEE. doi:10.1109/WICT.2012.6409079

Elgammal, A., Duraiswami, R., Harwood, D., & Davis, L. S. (2002). Background and foreground modeling using nonparametric kernel density estimation for visual surveillance. *Proceedings of the IEEE*, *90*(7), 1151–1163. doi:10.1109/JPROC.2002.801448

Elhabian, S. Y., El-Sayed, K. M., & Ahmed, S. H. (2008). Moving object detection in spatial domain using background removal techniques-state-of-art. *Recent Patents on Computer Science*, *1*(1), 32–54. doi:10.2174/1874479610801010032

Farn, E. J., Chen, L. H., & Liou, J. H. (2003). A new slow-motion replay extractor for soccer game videos. *International Journal of Pattern Recognition and Artificial Intelligence*, *17*(08), 1467–1481. doi:10.1142/S0218001403002964

Gauglitz, S., Höllerer, T., & Turk, M. (2011). Evaluation of interest point detectors and feature descriptors for visual tracking. *International Journal of Computer Vision*, *94*(3), 335–360. doi:10.1007/s11263-011-0431-5

Hariharakrishnan, K., Schonfeld, D., Raffy, P., & Yassa, F. (2003, September). Video tracking using block matching. In *Image Processing, 2003. ICIP 2003. Proceedings. 2003 International Conference on* (Vol. 3). IEEE. doi:10.1109/ICIP.2003.1247402

Hernandez, J., Morita, H., Nakano-Miytake, M., & Perez-Meana, H. (2009). *Movement detection and tracking using video frames*. Progress in Pattern Recognition, Image Analysis, Computer Vision, and Applications. doi:10.1007/978-3-642-10268-4_123

Ikeda, N., Araki, T., Dey, N., Bose, S., Shafique, S., El-Baz, A., & Suri, J. S. (2014). Automated and accurate carotid bulb detection, its verification and validation in low quality frozen frames and motion video. *International Angiology: A Journal of the International Union of Angiology*, *33*(6), 573-589.

Joshi, K. A., & Thakore, D. G. (2012). A survey on moving object detection and tracking in video surveillance system. *International Journal of Soft Computing and Engineering*, *2*(3).

Kouwenberg, J. J., Ulrich, L., Jäkel, O., & Greilich, S. (2016). A 3D feature point tracking method for ion radiation. *Physics in Medicine and Biology*, *61*(11), 4088–4104. doi:10.1088/0031-9155/61/11/4088 PMID:27163162

Kumar, P., Ranganath, S., Weimin, H., & Sengupta, K. (2005). Framework for real-time behavior interpretation from traffic video. *Intelligent Transportation Systems. IEEE Transactions on*, *6*(1), 43–53.

Lacassagne, L., Manzanera, A., Denoulet, J., & Mérigot, A. (2009). High performance motion detection: Some trends toward new embedded architectures for vision systems. *Journal of Real-Time Image Processing*, *4*(2), 127–146. doi:10.1007/s11554-008-0096-7

Li, L., Huang, W., Gu, I. Y., & Tian, Q. (2003, November). Foreground object detection from videos containing complex background. In *Proceedings of the eleventh ACM international conference on Multimedia* (pp. 2-10). ACM. doi:10.1145/957013.957017

Nascimento, J. C., & Marques, J. S. (2006). Performance evaluation of object detection algorithms for video surveillance. *Multimedia. IEEE Transactions on*, *8*(4), 761–774.

Pal, G., Acharjee, S., Rudrapaul, D., Ashour, A. S., & Dey, N. (2015). Video segmentation using minimum ratio similarity measurement. *International Journal of Image Mining*, *1*(1), 87–110. doi:10.1504/IJIM.2015.070027

Paragios, N., & Deriche, R. (2000). Geodesic active contours and level sets for the detection and tracking of moving objects. *Pattern Analysis and Machine Intelligence. IEEE Transactions on*, *22*(3), 266–280.

Parekh, H. S., Thakore, D. G., & Jaliya, U. K. (2014). A survey on object detection and tracking methods. *International Journal of Innovative Research in Computer and Communication Engineering*, *2*(2).

Prati, A., Mikic, I., Trivedi, M. M., & Cucchiara, R. (2003). Detecting moving shadows: Algorithms and evaluation. *Pattern Analysis and Machine Intelligence. IEEE Transactions on*, *25*(7), 918–923.

Rahangdale, K., & Kokate, M. (2016). *Event detection using background subtraction for surveillance systems*. Academic Press.

Rakibe, R. S., & Patil, B. D. (2013). Background subtraction algorithm based human motion detection. *International Journal of Scientific and Research Publications, 3*(5).

Serby, D., Meier, E. K., & Van Gool, L. (2004, August). Probabilistic object tracking using multiple features. In *Pattern Recognition, 2004. ICPR 2004.Proceedings of the 17th International Conference on* (*Vol. 2*, pp. 184-187). IEEE.

Wang, L., Liu, X., Lin, S., Xu, G., & Shum, H. Y. (2004, October). Generic slow-motion replay detection in sports video. In *Image Processing, 2004. ICIP'04. 2004 International Conference on* (Vol. 3, pp. 1585-1588). IEEE.

Warhade, K. K., Merchant, S. N., & Desai, U. B. (2013). Shot boundary detection in the presence of illumination and motion. *Signal. Image and Video Processing*, *7*(3), 581–592. doi:10.1007/s11760-011-0262-4

Yoneyama, A., Nakajima, Y., Yanagihara, H., & Sugano, M. (1999). Moving object detection from MPEG video stream. *Systems and Computers in Japan*, *30*(13), 1–12. doi:10.1002/(SICI)1520-684X(19991130)30:13<1::AID-SCJ1>3.0.CO;2-G

Yu, Z., & Chen, Y. (2009). A real-time motion detection algorithm for traffic monitoring systems based on consecutive temporal difference.*Proceedings of the 7th Asian Control Conference*.

KEY TERMS AND DEFINITIONS

Feature Extraction: The transforming the existing features into a lower dimensional space.

Features: Inherent properties of data, independent of coordinate frames.

Image Segmentation: The process of partitioning a digital image into multiple segments as a set of pixels or super-pixels. It assists to locate objects and boundaries in images.

Motion Detection: The procedure used to determine the presence of relevant motion in the observed scene.

Object Detection: Used to define the presence of an object or entity.

Video Processing: The manipulation of video content to resize, clarify or compress it for further information extraction.

Video Segmentation: The process that separates the foreground/background in a video.

Video Tracking: Used to determine the location of persons or objects in the video sequence time by time.

Section 1
Introduction to Video Processing and Mining

Video processing has various applications in several domains. This section highlights the fundamental concepts and algorithms of video processing in the medical domain. The real time frames for the acquired images from different image modalities is used. Another context is introduced for data mining that has great benefit in data extraction various applications including educational and medical domains.

Chapter 1

Medical Video Processing:
Concept and Applications

Srijan Goswami
Institute of Genetic Engineering, India

Payel Roy
St. Mary's Technical Campus – Kolkata, India

Urmimala Dey
JIS College of Engineering, India

Amira Ashour
Tanta University, Egypt

Nilanjan Dey
Techno India College of Technology – Kolkata, India

ABSTRACT

In today's medical environments, imaging technology is extremely significant to provide information for accurate diagnosis. An increasing amount of graphical information from high resolution 3D scanners is being used for diagnoses. Improved medical data quality become one of the major aims of researchers. This leads to the development of various medical modalities supported by cameras that can provide videos for the human body internal for surgical purposes and more information for accurate diagnosis. The current chapter studied concept of the video processing, and its application in the medical domain. Based on the highlighted literatures, it is convinced that video processing and real time frame will have outstanding value in the clinical environments.

INTRODUCTION

Video processing is an essential part for various technologies, including video surveillance, robotics, medical applications and multimedia. In the clinical environments, medical imaging is significant as multiple imaging modalities, such as X-ray angiography, magnetic resonance imaging (MRI), or endoscopy assemble information on the patient's medical status. Some of the medical modalities captured videos for the body internal, thus image/video processing become essential. For the medical consultant, visual information is vital to confirm the diagnosis and to control a therapy. Generally, the human visual system has bounded spatio-temporal sensitivity; however, below this capacity, several signals can be instructive. Moreover, Liu et al. (2005) revealed that motion with low spatial amplitude is impossible

DOI: 10.4018/978-1-5225-1025-3.ch001

for humans to see, thus it can be magnified to provide remarkable mechanical behaviour. This motivates researchers to develop tools that extract invisible signals in videos. The extracted signals can be employed to redisplay the videos in an indicative manner. Aachet al. (2002) studied the coronary angiography, in the coronary arteries where the inflow and outflow of a contrast agent to provide decisive information.

Thus, image/video processing algorithms are required for image enhancement, analysis, interpretation, and displaying as well as other processing phases for an accurate diagnosis. Video processing can be defined as the analysis of video content in order to obtain an understanding of the scene that it describes. Generally, image frames are acquired from a medical video source. From the basic science perspective, the methods in video analysis are motivated by the necessity to develop machine algorithms that can mimic the capabilities of human visual systems.

Various functions in video analysis were grouped into several categories/levels, namely Low Level, Mid-Level and High Level (Beyong et al., 2007). However, there are certain uncertainties about which function belongs to which level. From the single image point of view, two fundamental basic tasks are employed for video processing:

1. The computation of gradients, which highlights sections of images that have significant change in the image intensity. Identification of edges in the image is used to detect lines and corners of the image.
2. Image Intensity analysis such as gray scale or color value.

The fundamental structures like edges, lines and corners can be implemented as the building blocks for the identification of higher level structures like shapes of the object.

Detection is also a basic low-level image analysis task as it can assist the identification of interesting objects in the scene, which can then lead to an understanding of the scene or an analysis of the actions of the objects. One of the factors that affect the performance of image/video analysis algorithms is the quality of the image, where different sources of noise, including sensor noise, environmental conditions like lighting, and occluding objects can temporarily mask the objects of interest. This requires statistical modelling of image quality, machine learning based approaches to compensate for the variation in quality, and physics-based approaches that model the environmental factors to account for their effects.

Alain et al. (2008) provided a sequence of images (i.e., a video), which are usually highly correlated, an additional task is to compute the motion of the objects over the video. Optical flow is a scheme for estimating the motion of each individual pixel. Combined with segmentation, it can provide a sense of how each part of the scene is changing over time. Additionally, tracking involves computing the location of each object over time, given the detections of the objects in each frame. Bayesian tracking approaches like the Kalman filter or the particle filter, combined with suitable data association strategies, have been adapted to the video analysis tasks. Mayank et al. (2005) examined the effect of distributions of certain object characteristics, like image intensity value; change over time to compute a track of these objects. Although motion analysis, including both flow computation and tracking, has been the mainstay of video understanding research for some time, robustness to environmental variations, as well as scene occlusions and clutter, remains dominant challenges for existing methods.

Consequently, the overall goal of video analysis is to obtain an understanding of the scene and to extract/ track objects or organs under concern. Scene understanding requires recognition of objects and events. Object recognition can be achieved at the level of a single image, while recognition of activities and events usually requires multiple images. In fact, various time scales can be used for recognizing activities over different temporal horizons. Moreover, the complexity of the scene can define the kinds of activities/objects that to be recognized. Furthermore, video processing is vital in monitoring video systems in the medical applications.

The current chapter is concerned with the video processing in the medical domain. It repeated the modalities used for tracking and capturing videos in the human body. As follows a background based on the related work for several video processing applications.

LITERATURE REVIEW

Medical video processing magnetized the focus of several researchers. Vatsa et al. (2005) highlighted the bio-metric technology based on video sequences versus face, eye (eye/retina), and gait. Kang (2007) presented the concept of various elements of image and video processing along with the current technologies related to them for face detection. Tremeau et al. (2008) motivated an overview on the recent trends and future research directions in color image and video processing over the pints, such as:

- The choice of appropriate color space,
- The intensity content dropping,
- The reduction of the gap between low level features, and
- High level interpretation.

The authors used color acquisition systems, color image appearance. Quah et al. (2009) presented a video based marker less motion capture method to operate in natural scenarios such as occlusive and cluttered scenes. The used method was video based marker less motion capture method, where mechanisms of digital technologies and intensification of economic and organizations. Ong et al. (2009) suggested segmentation of objects from video sequences, temporal and spatial information and their appropriate. The automatic video object segmentation tool is a very useful tool that has very wide practical applications in everyday life contributing to efficiency, time, manpower, and cost savings.

Steen et al. (2010) proposed an algorithm, which used the background subtraction to remove or at least minimize the effects of the immobile parts of the cell. Mishra et al. (2011) provided the availability of medical expertise even to the remotest geographical location where a stranded and isolated person requires an emergency medical attention. The authors used agent based architecture to ensure Quality of service (QoS) for video medical services. Hewage et al. (2011) proposed a wireless health system which transmitted 3D surgical video over a 3Gpp LTE network which provided the provision of more natural viewing conditions and improved diagnosis. Zeljkovic et al. (2014) proposed an effective method for moving object detection to an outdoor environment, invariant to extreme illumination changes, which was presented as an improvement to the shading model method. Konstantopulous et al. (2014) presented a software system based on smart card technology for recording; monitoring and studying patients of any surgery specialty. It provided fast access to accurate information acted as portable data repository to speed the manual processes. Bharath et al. (2015) proposed architecture for PUS system whose abilities included automated kidney detection in real time. The proposed Viola-Jones algorithm was trained with a good set of kidney data consisting of diversified shapes and sizes for Automatic Kidney Detection and 3D Reconstruction. Table 1 includes a summarization for various techniques used in the video processing domain with different applications.

Table 1. Various related work for video processing

Author Name	Proposed Work	Used Technique	Drawbacks
Vatsa et al. (2005)	Highlighted the biometric technology which based on video sequences viz. face, eye (eye/retina), and gait.	Used a three step process- Capture, Process, Enroll.	Without the former registration biometric sample, the biometric technique is of no use.
Kang (2007)	Presented the concept of various elements of image and video processing along with the current technologies related to them.	Used the concept of digital image processing and face detection.	The new techniques have been discussed without discussing the practical use and working of the techniques of Image and Video Processing.
Tremeau et al. (2008)	Motivated an overview on the recent trends and future research directions in color image and video processing over the pints including choicc of appropriate color space, dropping the intensity content, reducing the gap between low level features and high level interpretation.	Used the color Acquisition Systems, Color Image Appearance.	The importance on the quality of images and displays was not discussed.
Quah et al. (2009)	Presented a video based marker less motion capture method that has the potential to operate in natural scenarios such as occlusive and cluttered scenes. The authors presented the digital technologies and the intensification of Economic and Organizational Mechanisms in Commercial Sport.	The video based marker less motion capture method was depicted. Mechanisms of digital technologies and intensification of economic and organizations were also employed.	Discussed a whole loathsome of techniques and methods without properly describing the techniques.
Ong et al. (2009)	Authors included extensively exhausted overview for segmentation of objects from video sequences, temporal and spatial information. It was concluded that the automatic video object segmentation tool is a very useful tool that has very wide practical applications in everyday life contributing to efficiency, time, manpower, and cost savings.	Used the Computer supported collaborative learning, virtual/online community, and Virtual Learning Environment techniques.	Fully automatic extraction of semantically meaningful objects in many practical applications but faces problems like limited domain of application and approaches, need of excessive parameter/ threshold setting and fine-tuning and overly complicated algorithms.
Steen et al. (2010)	Proposed an algorithm named MEDIC algorithm which used the background subtraction to remove or minimize the effects of the immobile parts of the cell.	Used the MEDIC algorithm for images with larger resolution.	Images with small resolution could not use this algorithm.
Mishra et al. (2011)	Provided the ease of availability of medical expertise even to the remotest geographical location where a stranded and isolated person requires an emergency medical attention.	Used agent based architecture to ensure QoS for video medic services.	This did not assure QoS for data provisioning and at the same time affected other services running at terminal equipment.
Hewage et al. (2011)	Envisaged a wireless health system which transmitted 3D surgical video over a 3Gpp LTE network providing the provision of more natural viewing conditions and improved diagnosis.	Used concepts of 4G, cellular network, 3D Computer graphics, and Rcmotc surgery.	Proper acknowledgement of several packets over the network was not maintained.

continued on next page

Table 1. Continued

Author Name	Proposed Work	Used Technique	Drawbacks
Zeljkovic et al. (2014)	Proposed an effective method for moving object detection to an outdoor environment, invariant to extreme illumination changes, which was presented as an improvement to the shading model method.	Shading model method was proposed.	The ethical implications of surveillance technologies on the privacy of citizens and the pragmatic approach to the central ethical concern about privacy and confidentiality protections and relevant issues of the informed consent process as well as the appropriate balance between law enforcement, national security, and civil liberties.
Konstantopulous et al. (2014)	Suggested a software system based on smart card technology for recording, monitoring and studying patients of any surgery specialty. It provided fast access to accurate information, acted as portable data repository, speeding manual processes.	Used the .Net platform and Java cards used for the development of the system and architectural model of the system.	The breakdown of the system may lead to the non-functioning of the whole Hospital, using this system.
Bharath et al. (2015)	Proposed architecture for PUS system whose abilities included automated kidney detection in real time. The authors trained the Viola-Jones algorithm with a good set of kidney data consisting of diversified shapes and sizes.	Used the concept of Automatic Kidney Detection and the 3D Reconstruction.	They used the bounding box which cannot capture the entire organ inside it. They did not utilize the signal processing algorithms to detect more organs in the image.

VIDEO PROCESSING: OVERVIEW

Video signal is basically any sequence of time varying images. A still image is a spatial distribution of intensities that remain constant with time, whereas a time varying image has a spatial intensity distribution that varies with time. Video signal is treated as a series of images called frames. Quing et al. (2011) stated that an illusion of continuous video is obtained by changing the frames in a faster manner which is generally termed as frame rate.

Video processing can be defined as video's content analysis, which performed in order to obtain an understanding of the scene under concern. It is an essential component of a number of technologies, including video surveillance, robotics, and multimedia. From a basic science perspective, methods in video analysis are motivated by the need to develop machine algorithms that can mimic the capabilities of human visual systems. Impressive growth of video processing over the past decade is primarily attributing to associated advances in video compression and wireless network technologies. The former allow real-time, robust and efficient encoding, while recent video coding standards also encompass a network abstraction layer for higher flexibility. The latter, facilitate constantly increasing data transfer rates, extended coverage, reduced latencies and transmission reliability.

Video processing can help medical practitioners in their diagnoses, as well as researchers working in various biological fields with automated analysis of larger volumes of data that is being collected, e.g., time-lapse microscopy images. Typically, video is a sequence of images which are displayed in order. Each of these images is called a frame. So, any video can be considered as a collection of frames. Since, video deals with multiple frames one after the other in a sequential manner, thus all the techniques that applies to image processing can be used for video processing. Consequently, image processing can be defined as subset of video processing.

Wiegand et al. (2003) stated that video processing technology has revolutionized the world of multimedia with products such as Digital Versatile Disk (DVD), the Digital Satellite System (DSS), high definition television (HDTV), digital still and video cameras. The different areas of video processing include:

1. Video de-noising,
2. Video compression,
3. Video indexing,
4. Video segmentation, and
5. Video tracking.

The most common video analysis/processing phase is the video de-noising. It is the process of removing noise from a video signal. Video denoising methods can be divided into:

1. Spatial video denoising methods, where image noise reduction is applied to each frame individually,
2. Temporal video denoising methods, where noise between frames is reduced, where motion compensation can be employed to avoid ghosting artifacts when blending together pixels from several frames, and
3. Spatial-temporal video denoising methods use a combination of spatial and temporal denoising.

Typically, video denoising methods are designed and tuned for specific types of noise. The most common video noise types include the analog noise (due to Radio channel artifacts, VHS artifacts, Film artifacts) and digital noise (due to blocking, ringing, blocks damage in case of losses in digital transmission channel).

A highlight on video coding standards and wireless transmission technologies evolution timeframe is as follows. In terms of video coding standards, the efficient video compression systems were introduced over the last decades. In early 1990's, the international telecommunication union-telecommunication sector (ITU-T) developed the first H.261 video coding standard, which was originally designed for vide otelephony and videoconferencing applications. The H.261 supported the quarter common intermediate format (QCIF-176x144) and the common intermediate format (CIF-352x288) video resolutions. Subsequently, H.262 released in 1995 facilitated interlaced video provision and increased video resolutions support to 4CIF (720x576) and 16CIF (1408x1152). Its successor, termed H.263 introduced in 1996 provided for improved quality at lower bit rates and also allowed lower, sub-QCIF (128x96) video resolution encoding. The highly successful H.264/AVC standard was released by ITU-T in 2003 and accounted for bit rate demands reductions of up to 50% for equivalent perceptual quality compared to its predecessor. The current state-of-the-art video coding standard is the High Efficiency Video Coding (HEVC) standard, standardized in 2013. HEVC supports video resolutions ranging from 128x96 to 8192x4320 and provides 50% bit rate gains for comparable visual quality compared to H.264/AVC.

Furthermore, video processing is involved in the wireless transmission technologies. The global system for mobile communications (GSM) signified the transition from analog 1st generation (1G) to digital 2nd generation (2G) technology of mobile cellular networks. In the past two decades, mobile telecommunication networks are continuously evolving. Milestone advances range from 2.5G (general packet radio service (GPRS) and enhanced data rates for GSM evolution (EDGE)) and 3G (universal mobile telecommunications system (UMTS)) wireless networks to the 3.5G (high speed downlink packet access (HSDPA), high speed uplink packet access (HSUPA), high speed packet access (HSPA), and HSPA+), mobile WiMAX networks, and long term evolution (LTE) systems. The afore-described wireless networks facilitate incremental data transfer rates while minimizing end- to- end delay. In other words, features in wireless communications follow those of wired infrastructure with a reasonable time gap of few years. The latter enables the development of responsive mHealth systems suitable for emergency telemedicine (Panayides et al., 2011). Evolving wireless communications networks' theoretical upload data rates range from 50 kbps - 86 Mbps. In practice, typical upload data rates are significantly lower. More specifically typical upload data rates range from

1. **GPRS:** 30-50 kbps,
2. **EDGE:** 80-160 kbps,
3. **Evolved EDGE:** 150-300 kbps,
4. **UMTS:** 200-300 kbps,
5. **HSPA:** 500 kbps - 2 Mbps,
6. **HSPA+:** 1-4 Mbps, and
7. **LTE:** 6-13 Mbps.

MEDICAL IMAGING/VIDEO PROCESSING MODALITIES

Medical imaging is the technique and process of creating visual representations of the body interior for clinical analysis and medical intervention, as well as visual representation of the function of some organs or tissues (physiology). Measurement and recording techniques such as electroencephalography (EEG), magneto encephalography (MEG), and electrocardiography (ECG) are not primarily designed to produce images. These technologies produce data susceptible to representation as a parameter graph versus time or maps which contain data about the measurement locations as suggested by Jurgan et al. (2014). In a limited comparison these technologies can be considered as forms of medical imaging in another discipline. In the clinical context, "invisible light" medical imaging is generally equated to radiology or "clinical imaging" and the medical practitioner responsible for interpreting (and sometimes acquiring) the images are a radiologist. "Visible light" medical imaging involves digital video or still pictures that can be seen without special equipment. Dermatology and wound care are two modalities that use visible light imagery. Diagnostic radiography designates the technical aspects of medical imaging and in particular the acquisition of medical images. The radiographer or radiologic technologist is usually responsible for acquiring medical images of diagnostic quality, although some radiological interventions are performed by radiologists.

Hangzai et al. (2006) suggested medical imaging to reveal internal structures hidden by the skin and bones, as well as to diagnose and treat disease. Medical imaging also establishes a database of normal anatomy and physiology to make it possible to identify abnormalities. Although imaging of removed organs and tissues can be performed for medical reasons, such procedures are usually considered part of pathology instead of medical imaging. As a discipline and in its widest sense, it is part of biological imaging and incorporates radiology which uses the imaging technologies of:

- X-ray radiography,
- Magnetic resonance imaging,
- Medical ultrasonography or ultrasound,
- Endoscopy,
- Elastography,
- Tactile imaging,
- Thermography,
- Medical photography and nuclear medicine functional imaging techniques as positron emission Tomography (PET), and
- Single-photon emission computed tomography (SPECT) suggested by Zeno et al. (2015).

Consequently, medical imaging and medical video processing are often perceived to designate techniques, which noninvasively produce images of the internal aspect of the body. In this restricted sense, medical imaging can be seen as the solution of mathematical inverse problems. In the case of medical ultrasonography, the probe consists of ultrasonic pressure waves and echoes that go inside the tissue to show the internal structure. In the case of projectional radiography, the probe uses X-ray radiation, which is absorbed at different rates by different tissue types such as bone, muscle and fat. The term non-invasive is used to denote a procedure where no instrument is introduced into a patient's body which is the case for most imaging techniques used. Generally, there are different modalities that can be used for image/video acquisition such as the following:

Radiography

Two forms of radiographic images can be used in medical imaging; projection radiography and fluoroscopy. These 2D techniques are still in wide use despite the advance of 3D tomography due to the low cost, high resolution, and depending on application, lower radiation dosages. Daniela et al. (2008) suggested an imaging modality utilizes a wide beam of x-rays for image acquisition and is the first imaging technique available in modern medicine.

Projectional radiographs is more commonly known as x-rays, which used to determine the type and extent of a fracture as well as for detecting pathological changes in the lungs. With the use of radio-opaque contrast media, such as barium, visualize the structure of the stomach and intestines or certain types of colon cancer become easier. However, fluoroscopy produces real-time images of the body internal structures in a similar fashion to radiography. Conversely, it employs a constant input of x-rays, at a lower dose rate. Contrast media, such as barium, iodine, and air are used to visualize internal organs' work. Moreover, fluoroscopy is used in image-guided procedures when constant feedback during a procedure is required. An image receptor is required to convert the radiation into an image after it has passed through the area of interest. Fluorescing screen provided a way to an image amplifier (IA) which was a large vacuum tube that had the receiving end coated with cesium iodide, and a mirror at the opposite endsuggested by Richard et al. (2014).

Magnetic Resonance Imaging (MRI)

A magnetic resonance imaging instrument (MRI scanner) traditionally creates a two dimensional image of a thin "slice" of the body, thus it is considered a tomographic imaging technique. Modern MRI instruments are capable of producing images in the form of 3D blocks, which may be considered a generalization of the single-slice, tomographic, concept. Unlike the computed tomography (CT), MRI does not use the ionizing radiation. Because CT and MRI are sensitive to different tissue properties, the appearance of the images obtained with the two techniques differ markedly. In CT, X-rays must be blocked by some form of dense tissue to create an image, so the image quality when looking at soft tissues will be poor. Rakesh et al. (2011) stated that in MRI, while any nucleus with a net nuclear spin is used, the proton of the hydrogen atom remains the most widely used, especially in the clinical setting, because it is so ubiquitous and returns a large signal. This nucleus, present in water molecules, allows the excellent soft-tissue contrast achievable with MRI.

Nuclear Medicine

Nuclear medicine encompasses both diagnostic imaging and treatment of disease. It can be referred to as molecular medicine or molecular imaging and therapeutics. Different from the typical concept of anatomic radiology, nuclear medicine enables assessment of physiology. Sarvazyan et al. (2011) approached to medical evaluation that was useful applications in most subspecialties, notably oncology, neurology, and cardiology. Gamma cameras and PET scanners are used in e.g. scintigraphy, SPECT and Positron emission tomography (PET) to detect regions of biologic activity that may be associated with disease. Isotopes are often preferentially absorbed by biologically active tissue in the body, and can be used to identify tumors or fracture points in bone. Images are acquired after collimated photons are detected by a crystal that gives off a light signal, which is in turn amplified and converted into count data.

The PET uses coincidence detection to image functional processes. Images of activity distribution throughout the body can show rapidly growing tissue, such as tumor, metastasis, or infection. PET images can be viewed in comparison to computed tomography scans to determine an anatomic correlate. Modern scanners may integrate PET, allowing PET-CT, or PET-MRI to optimize the image reconstruction involved with positron imaging. This is performed on the same equipment without physically moving the patient off of the gantry. The resultant hybrid of functional and anatomic imaging information is a useful tool in non-invasive diagnosis and patient management.

Ultrasound

Medical ultrasonography uses high frequency broadband sound waves in the megahertz range, which are reflected by tissue to varying degrees to produce (up to 3D) images. This is commonly associated with imaging the fetus in pregnant women. Uses of ultrasound are much broader, however. Other important uses include imaging the:

- Abdominal organs,
- Heart,
- Breast,
- Muscles,
- Tendons,
- Arteries, and
- Veins.

While it may provide less anatomical detail than techniques such as CT or MRI, it has several advantages which make it ideal in numerous situations, in particular that it studies the function of moving structures in real-time, emits no ionizing radiation, and contains speckle that can be used in elastography. Vesna et al. (2014) used ultrasound as a popular research tool for capturing raw data, for the purpose of tissue characterization and implementation of new image processing techniques. The concepts of ultrasound differ from other medical imaging modalities in the fact that it is operated by the transmission and receipt of sound waves. Egorov et al. (2008) stated that the high frequency sound waves sent into the tissue and depending on the composition of the different tissues; the signal will be attenuated and returned at separate intervals. It is very safe to use and does not appear to cause any adverse effects. It is also relatively inexpensive and quick to perform. The real time moving image obtained can be used to guide drainage and biopsy procedures. Doppler capabilities on modern scanners allow the blood flow in arteries and veins to be assessed.

Elastography

Elastography is a relatively new imaging modality that maps the elastic properties of soft tissue. It is useful in medical diagnoses, as elasticity can discern healthy from unhealthy tissue for specific organs/growths. Several elastographic techniques where suggested based on the use of ultrasound, magnetic resonance imaging and tactile imaging (Ophir et al., 1991; Weiss et al., 2008; Egorov et al., 2010; Parker et al., 2011). The wide clinical use of ultrasound elastography is a result of the implementation of technology in clinical ultrasound machines. Main branches of ultrasound elastography include:

- Quasistatic Elastography/Strain Imaging,
- Shear Wave Elasticity Imaging (SWEI),
- Acoustic Radiation Force Impulse imaging (ARFI),
- Supersonic Shear Imaging (SSI), and
- Transient Elastography.

Parker et al. (2011) proposed a steady increase of activities in the field of elastography, which demonstrated successful application of the technology in various areas of medical diagnostics and treatment monitoring.

Tactile Imaging

Tactile imaging is a medical imaging modality that translates the sense of touch into a digital image. The tactile image is a function of $P(x,y,z)$, where P is the pressure on soft tissue surface under applied deformation, where x,y,z are the coordinates and P represented the pressure. It is used for imaging of the prostate; furthermore myofascial trigger points in muscle was suggested by Montruccoli et al. (2004), Egorov et al. (2010) employed the Tactile imaging approach for breast imaging. In 2012, Turo et al. suggested vagina and pelvic floor support structures imaging using the same technique.

Echocardiography

When ultrasound is used to image the heart it is referred to as an echocardiogram. Weiss et al. (2008) stated that echocardiography allows detailed structures of the heart, including chamber size, heart function, the valves of the heart, as well as the pericardium (the sac around the heart) to be seen. Echocardiography uses 2D, 3D, and Doppler imaging to create pictures of the heart and visualize the blood flowing through each of the four heart valves. Andrei et al. (1996) suggested the echocardiography for the cases of shortness of breath or chest pain. In emergency situations, echocardiography is fast, easily accessible, and able to be performed at the bedside, making it the modality of choice for many physicians.

ENDOSCOPY

One of the most common medical video modality is the endoscopy. It is a nonsurgical procedure used to examine a person's digestive tract. The endoscope consists of a flexible tube with attached light camera. The physician can view pictures/videos of the patient's digestive tract on a color TV monitor. During an upper endoscopy, an endoscope is easily passed through the mouth and throat and into the esophagus, allowing the physician to view the esophagus, stomach, and upper part of the small intestine. Sarvazyan et al. (2011) stated that the endoscopes can be passed into the large intestine (colon) through the rectum to examine this area of the intestine. This procedure is called sigmoidoscopy or colonoscopy depending on the colon length to be examined. A special form of endoscopy called endoscopic retrograde cholangiopancreaticography (ERCP), which allows pictures of the pancreas, gallbladder, and related structures to be taken. Endoscopies fall into categories based on the area of the body that they investigate as illustrated in Figure 1.

Figure 1. Various endoscopy types

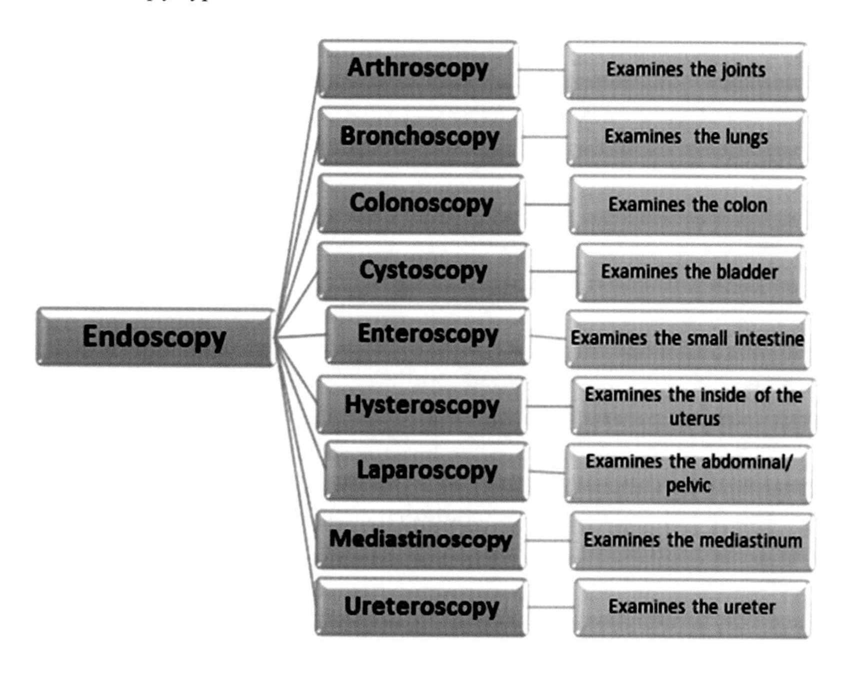

MEDICAL VIDEO PROCESSING: APPLICATIONS

Researchers are interested with developing various approaches in video processing that applied in the medical domain. Pless (2003) suggested a video representation using a low dimensional image space and a trajectory to analyse natural video sequences. Rodriguez et al. (2006) introduced the use of diagnostically-guided video compression. The video objects of interest were extracted using a novel AM-FM segmentation methods. It was noted that the near-field video and the region beyond the epicardium would require high bandwidth for quality transmission.

Suzuki & Hattori (2008) handled 3D human structures in an interactive way, where real-time imaging of medical 3D or 4D images can be used for diagnosis as well as novel medical treatments. The authors developed a surgery simulation and navigation systems to produce the haptic sensation of surgical maneuvers in the patient's fingers. Furthermore, the surgical navigation systems were explained to encourage the ability of robotic surgery and its trial for clinical case. The system was consisted of two large parts:

1. Assemblage of computer-controlled video cameras to gather time sequential video images, and
2. A computer system to control the camera array and to collect and process video images from multiple cameras analytically.

The proposed system was equipped with 65 computer-controlled video cameras to film the human locomotion from all angles concurrently. The video images were installed into a graphic workstation and translated into a time-spatial matrix. Using images recorded from all directions at a particular time, the system was able to acquire a function to reconstruct 4D models of the subject's moving human body surface. Thus, the system was able to visualize the inner structures including the skeletal/muscular systems. The authors recommended applying this imaging system to clinical observation in the area of orthopedics, rehabilitation and sports science.

Typically, endoscopic video analysis is directed towards the detection of abnormalities and uninformative frames. Gastrointestinal (GI) endoscopic videos suffer from a great number of poor-quality frames due to the out-of-focus blur, motion, or specular highlights, in addition to the artifacts caused by turbid fluid inside the GI tract. Consequently, post procedural analysis and processing of the GI endoscopic videos is considered a complex task. Atasoy et al. (2010) presented a low dimensional depiction of endoscopic videos based on a manifold learning (EVMs) scheme. The introduced EVMs enabled the clustering of poor-quality frames and alignment of different segments of the GI endoscopic video in an unsupervised way to assist subsequent visual assessment. The authors presented two new inter-frame similarity measures for manifold learning in order to generate structured manifolds from complex endoscopic videos. The results depicted that the proposed method achieved high precision and recall of 90.91% and 82.90% values; respectively.

Bharath et al. (2015), proposed an automatic kidney detection algorithm to detect the kidney in a real-time ultrasound video. Automated kidney detection was executed by training the Viola–Jones algorithm with a good set of kidney data, which consisted of diversified shapes and sizes. The proposed PUS with kidney detection algorithm is implemented on a single Xilinx Kintex-7 FPGA, integrated with a Raspberry Pi ARM processor running at 900 MHz. The system was evaluated on an FPGA-based PUS (portable ultrasound scanning) machine in real-time. The reconstructed images of the PUS were visually compared to the images of available platforms. It was established that the proposed algorithm achieved high accuracy.

Ikeda et al. (2014) proposed an automated system for carotid bulb detection in low quality frozen frames and motion video. The used database consisted of 155 ultrasound bulb images obtained from various ultrasound machines with varying resolutions and imaging conditions. Further a mixture database of 336 images consisted of bulbs and no-bulbs were used. The method estimated the lumen-intima borders accurately using classification paradigm with locating the transition points automatically based on curvature characteristics. The proposed system was able to detect the bulbs in the bulb database with 100% accuracy having 92% as close as to a neurologist's bulb location.

Ikeda et al. (2015) proposed a novel pilot study on predicting the SYNTAX (Synergy between percutaneous coronary intervention with TAXUS and cardiac surgery) score. The predication with the cIMT (carotid intima–media thickness) measures using computerized automated IMT all along the common carotid artery is compared to the manual IMT determined by Sonographers. The carotid artery includes bulb region and region proximal to the bulb. The correlation between automated cIMT that includes bulb plaque and SYNTAX score are compared to the correlation between the sonographer's IMT reading and SYNTAX score. Furthermore, the talk highlights on the innovative automated cIMT measurement system with a classification paradigm that used a combined global and local strategy involving texture based entropy and morphology.

FUTURE RESEARCH DIRECTIONS

From the preceding sections, it was established that image/video processing has the same procedures. Furthermore, they were engaged in various domains including medical, surveillance, and multimedia. In the medical domain, there are different modalities with attached cameras such as in the endoscopy that used to record videos for the internal parts of the body and used for surgical tasks. However, in hospitals and medical wards, it is required to develop more medical virtual reality devices that can be applied in everyday use. Thus, patients can undergo diagnosis and treatment with medical virtual reality devices as routinely as with CT and MRI devices.

Another trend in the medical video processing domain includes the real time frame as a solution for real time video processing in medical environments. It offers services for image acquisition from different image modalities. The data is processed through several phases include filtering, analysis, and saving frame sequences or entire videos in an arbitrary combination. Supplementary research is necessary for establishing diagnostic measures and video objects of diagnostic interest for 2D-mode and the Doppler modes. Moreover, it is important to implement automated systems that can concurrently, optimally compress video from combinations of different modes.

CONCLUSION

Digital video content is in continuous increase, thus automated interpretation become indispensable to guarantee optimal use of data in the shortest possible time. Massive extension of video processing applications becoming a reality and supplementary content-analysis-based applications are about to follow. Furthermore, the development of various medical modalities attracts researchers to develop systems based on video processing. As a result, the quality and reliability of the medical data analysis improves the diagnosis.

For biological imaging, different modalities are used such as:

- The X-ray radiography,
- MRI,
- Medical ultrasonography or ultrasound,
- Endoscopy,
- Tactile imaging,
- PET, and the SPECT.

The endoscope is an important medical modality that has a flexible tube with attached light camera. It supplies the physicians with videos of the patient's digestive tract on a color TV monitor. However, motion from different sources includes breathing or patient motion affects the acquired video/images. Thus, video processing steps such as de-nosing, enhancement, restoration, segmentation and classification; can be applied to enhance the video's frames and thus can be used for further processing phases. Various challenges encounter the researchers in the medical video processing, which attract them to develop more efficient and accurate automated systems.

REFERENCES

Aach, T., Mayntz, C., Rongen, P. M., Schmitz, G., & Stegehuis, H. (2002, May). Spatiotemporal multiscale vessel enhancement for coronary angiograms. In *Medical Imaging 2002* (pp. 1010–1021). International Society for Optics and Photonics. doi:10.1117/12.467056

Alain, T., & Shoji, T. (2008). Color in image and video processing: Most recent trends and future research directions. *EURASIP Journal on Image and Video Processing*.

Andrei, S., Mark, A. L., William, F. G., Gentaro, H., Mary, C. W., Etta, D. P., & Henry, F. (1996). Technologies for augmented reality systems: Realizing ultrasound guided needle biopsies. In *Proceeding from, 23rd Annual Conference On Computer Graphics and Interactive Techniques*.

Atasoy, S., Mateus, D., Lallemand, J., Meining, A., Yang, G. Z., & Navab, N. (2010). Endoscopic video manifolds. In *Medical Image Computing and Computer-Assisted Intervention–MICCAI 2010* (pp. 437–445). Springer Berlin Heidelberg. doi:10.1007/978-3-642-15745-5_54

Bharath, R., Kumar, P., Dusa, C., Akkala, V., Puli, S., Ponduri, H., & Desai, U. B. et al. (2015). FPGA-Based Portable Ultrasound Scanning System with Automatic Kidney Detection. *Journal of Imaging*, *1*(1), 193–219. doi:10.3390/jimaging1010193

Byeong, H. K. (2007). A review on image and video processing. *International Journal of Multimedia and Ubiquitous Engineering*, *2*(2).

Chaminda, T. E., Maria, G. M., & Nabeel, K. (2011). 3D medical video transmission over 4g networks. In *Proceedings of the 4th International Symposium on Applied Sciences in Biomedical and Communication Technologies*.

Chowdhury, R. A. K. (2014). Video processing—An overview. In Image, Video Processing and Analysis, Hardware, Audio, Acoustic and Speech Processing. Chennai: Academic Press.

Daniela, G. T., Luciana, P., & Benoit, M. (2008). *2008 ACM Symposium on Applied Computing*. ACM.

Egorov, V., & Sarvazyan, A. P. (2008). Mechanical imaging of the breast. *IEEE Transactions on Medical Imaging*, *27*(9), 1275–1287. doi:10.1109/TMI.2008.922192 PMID:18753043

Egorov, V., Van, R. H., & Sarvazyan, A. P. (2010). Vaginal tactile imaging. *IEEE Transactions on Bio-Medical Engineering*, *57*(7), 1736–1744. doi:10.1109/TBME.2010.2045757 PMID:20483695

Erickson, B. J., & Jack, C. R. Jr. (n.d.). Correlation of single photon emission CT with MR image data using fiduciary markers. *AJNR. American Journal of Neuroradiology*, *14*(3), 713–720. PMID:8517364

Hangzai, L., & Jianping, F. (2006). Building concept ontology for medical video annotation. In *Proceedings from14th ACM International Conference on Multimedia.*

Ikeda, N., Araki, T., Dey, N., Bose, S., Shafique, S., El-Baz, A., Cuadrado, G.E., Anzidei, M., Saba, L., & Suri, J. S. (2014). Automated and accurate carotid bulb detection, its verification and validation in low quality frozen frames and motion video. *International Angiology: A Journal of the International Union of Angiology, 33*(6), 573-589.

Ikeda, N., Gupta, A., Dey, N., Bose, S., Shafique, S., Arak, T., & Suri, J. S. et al. (2015). Improved correlation between carotid and coronary atherosclerosis SYNTAX score using automated ultrasound carotid bulb plaque IMT measurement. *Ultrasound in Medicine & Biology*, *41*(5), 1247–1262. doi:10.1016/j.ultrasmedbio.2014.12.024 PMID:25638311

James, A. P., & Dasarathy, B. V. (2014, September). Medical image fusion: A survey of state of the art. *Information Fusion*, *19*, 4–19. doi:10.1016/j.inffus.2013.12.002

Liu, C., Torralba, A., Freeman, W. T., Durand, F., & Adelson, E. H. (2005). Motion magnification. *ACM Transactions on Graphics*, *24*(3), 519–526. doi:10.1145/1073204.1073223

Manfred, J. P. (2014). Segmentation and indexing of endoscopic videos. In *Proceedings from22nd ACM International Conference on Multimedia.*

Mayank, V., & Richa, S. (2005). *Video biometrics. In Video Data Management and Information Retrieval* (pp. 149–176). Hershey, PA: IGI Global.

Michela, A., Beatrice, L., & Francesco, M. (2005). *2005 ACM Symposium on Applied Computing*. ACM.

Montruccoli, G. C., Montruccoli, S. D., & Casali, F. (2004). A new type of breast contact thermography plate: A preliminary and qualitative investigation of its potentiality on phantoms. *Physica Medica*, *20*(1), 27–31.

Nektarios, K., Vasileios, S., Vassilis, M., Ioannis, P., Nikolaos, A., & Elias, P. (2014). A smart card based software system for surgery specialties. *IJUDH International Journal of User-Driven Health Care*, *4*(1), 48–63. doi:10.4018/ijudh.2014010104

Ohm, J.-R., Sullivan, G. J., Schwarz, H., Tan, T. K., & Wiegand, T. (2012). Comparison of the coding efficiency of video coding standards – Including High Efficiency Video Coding (HEVC). *IEEE Transactions on Circuits and Systems for Video Technology*, *22*(12), 1669–1684. doi:10.1109/TCSVT.2012.2221192

Ong, E. P., & Weise, L. (2009). *Video object segmentation. In Encyclopedia of information communication technology* (pp. 809–816). Hershey, PA: IGI Global. doi:10.4018/978-1-59904-845-1.ch106

Ophir, J., Céspides, I., Ponnekanti, H., & Li, X. (1991). Elastography: A quantitative method for imaging the elasticity of biological tissues. *Ultrasonic Imaging*, *13*(2), 111–134. doi:10.1177/016173469101300201 PMID:1858217

Panayides, A., Pattichis, M. S., Pattichis, C. S., & Pitsillides, A. (2011). A tutorial for emerging wireless medical video transmission systems. *IEEE Antennas & Propagation Magazine*, *53*(2), 202–213. doi:10.1109/MAP.2011.5949369

Parker, K. J., Doyley, M. M., & Rubens, D. J. (2011). Imaging the elastic properties of tissue: The 20 year20-year perspective. *Physics in Medicine and Biology*, *56*(2), 513. doi:10.1088/0031-9155/56/2/513 PMID:21119234

Pless, R. (2003, October). *Image Spaces and video trajectories: Using Isomap to explore video sequences* (Vol. 3). ICCV.

Punit, R. B. K., Chandrashekar, D., Vivek, A., & Suresh, P. (2015). FPGA-based portable ultrasound scanning system with automatic kidney detection. *Journal of Imaging*, *1*(1), 193–219. doi:10.3390/jimaging1010193

Qiang, Z., & Baoxin, L. (2011). Video –based motion expertise analysis in simulation based surgical training using hierarchical Dirichlet Process Hidden Markov Model. In *Proceedings from2011 International ACM Workshop on Medical Multimedia Analysis and Retrieval.*

Quah, C. K., Michael, K., Alex, O., Hook, S. S., & Andre, G. (2009). Video based motion capture for measuring human movement. In Digital sport for performance enhancement and competitive evolution: Intelligent gaming technologies. Hershey, PA: IGI Global.

Rakesh, K., Sankhayan, C., & Nabendu, C. (2011). Optimizing mobile terminal equipment for video medic services. In *Proceedings of the 1st International Conference on Wireless Technologies for Humanitarian Relief*, (pp. 373-378).

Richard, D., Sara, G., Kholood, S., Gary, U., Graham, M., & Janet, E. (2014). Early response markers from video games for rehabilitation strategies. ACM SIGAPP Applied Computing Review, 36-43.

Rodriguez, P., Pattichis, M. S., Pattichis, C. S., Abdallah, R., & Goens, M. B. (2006). Object-based ultrasound video processing for wireless transmission in cardiology. In M-Health (pp. 491-507). Springer US. doi:10.1007/0-387-26559-7_37

Roobottom, C. A., Mitchell, G., Morgan-Hughes, G., & Mitchell, M.-H. (2010). Radiation-reduction strategies in cardiac computed tomographic angiography. *Clinical Radiology*, *65*(11), 859–867. doi:10.1016/j.crad.2010.04.021 PMID:20933639

Sarvazyan, A., Hall, T. J., Urban, M. W., Fatemi, M., Aglyamov, S. R., & Garra, B. S. (2011). Overview of elastography–An emerging branch of medical imaging. *Current Medical Imaging Reviews*, *7*(4), 255–282. doi:10.2174/157340511798038684 PMID:22308105

Suzuki, N., & Hattori, A. (2008). The road to surgical simulation and surgical navigation. *Virtual Reality (Waltham Cross)*, *12*(4), 281–291. doi:10.1007/s10055-008-0103-0

Turo, D., Otto, P., Egorov, V., Sarvazyan, A., Gerber, L. H., & Sikdar, S. (2012). Elastography and tactile imaging for mechanical characterization of superficial muscles. *Journal of the Acoustical Society of America, 132*(3).

Vesna, Z. (2014). *Illumination independent moving object detection algorithm. In Video surveillance techniques and technologies.* IGI Global.

Villringer, A., & Chance, B. (1997). Non-invasive optical spectroscopy and imaging of human brain function. *Trends in Neurosciences*, *20*(10), 435–442. doi:10.1016/S0166-2236(97)01132-6 PMID:9347608

Weiss, R. E., Egorov, V., Ayrapetyan, S., Sarvazyan, N., & Sarvazyan, A. (2008). Prostate mechanical imaging: A new method for prostate assessment. *Urology*, *71*(3), 425–429. doi:10.1016/j.urology.2007.11.021 PMID:18342178

Wells, P. N. T., & Liang, H.-D. (2011). Medical ultrasound: Imaging of soft tissue strain and elasticity. *Journal of the Royal Society, Interface*, *8*(64), 1521–1549. doi:10.1098/rsif.2011.0054 PMID:21680780

Wiegand, T., Sullivan, G. J., Bjontegaard, G., & Luthra, A. (2003). A.: Overview of theH.264/AVC video coding standard. *IEEE Transactions on Circuits and Systems for Video Technology*, *13*(7), 560–576. doi:10.1109/TCSVT.2003.815165

Zeno, A., Michael, R., Pal, H., Jiang, Z., Carsten, G., Ilangko, B., & Cathal, G. (2015). Expert Driven Semi-Supervised Elucidation Tool for Medical Endoscopic Videos. In *Proceedings from the 6th ACM Multimedia Systems Conference.*

KEY TERMS AND DEFINITIONS

Endoscopy: An instrument used to examine the interior of a hollow organ or cavity of the body.

Medical Image Processing: The process of creating visual representation of the body interior for clinical analysis and medical intervention.

Real Time Image Processing: The rapid acquisition and manipulation of the information from a scanning probe by electronic circuits to enable images to be shown on TV screens immediately.

Video Processing: A particular image processing case, where the input and output signals are video files/streams.

Chapter 2
Educational Data Mining and Indian Technical Education System: A Review

Nancy Kansal
Mewar University, India

Vijender Kumar Solanki
Institute of Technology and Science, Ghaziabad, UP, India

Vineet Kansal
ITS Engineering College, Greater Noida, UP, India

ABSTRACT

Educational Data Mining (EDM) is emerged as a powerful tool in past decade and is concerned with developing methods to explore the unique types of data in educational settings. Using these methods, to better understand students and the settings in which they learn. Different unknown patterns using classification, Clustering, Association rule mining, decision trees can be discovered from this educational data which could further be beneficial to improve teaching and learning systems, to improve curriculum, to support students in the form of individual counseling, improving learning outcomes in terms of students' satisfaction and good placements as well. Therefore a literature survey has been carried out to explore the most recent and relevant studies in the field of data mining in Higher and Technical Education that can probably portray a pathway towards the improvement of the quality education in technical institutions.

DOI: 10.4018/978-1-5225-1025-3.ch002

INTRODUCTION

Technical Educational institutions as a regular operation collect lots of information and data produced from day to day educational activities:

- Details of the courses offered,
- Students' enrollment data:
    - Course chosen,
    - Gender,
    - Age,
    - Father's name, and
    - Family details.
- Students' continuous assessment data:
    - Marks statement and grades.
- Teachers data viz.
    - Personal details,
    - Qualifications,
    - Courses taught, and
    - Student feedback.
- Placement data and many other similar ones which can be mined using Data mining tools and techniques.

Educational Data Mining (EDM) which is emerged as a powerful tool in last decade, is concerned with developing methods to explore the unique types of data in educational settings and, using these methods, to better understand students and the settings in which they learn (Baker and Yacef, 2009). Different unknown patterns using classification, Clustering, Association rule mining, decision trees can be discovered from this educational data which could further be beneficial to improve teaching and learning systems, to improve curriculum, to support students in the form of individual counseling, improving learning outcomes in terms of students' satisfaction and good placements as well (Naeimeh et al., 2005). Since Indian Technical Education system with an increased participation of unaided Private sector is dealing with improvement in the employability of trained graduates and postgraduates, coming out of these technical institutions, therefore a literature survey has been carried out to explore the most recent and relevant studies in the field of data mining in Higher and Technical Education that can probably portray a pathway towards the improvement of the quality education in technical education.

This chapter is organized as follows, a description of the technical education growth in India followed by including challenges emerging with increased unaided private sector participation in Technical education. Afterward, several related work in the field of data mining in higher/technical education were included. Finally, a discussion of the future possibilities of applying data mining in technical education system for the improved learning outcomes was carried out.

BACKGROUND

Data Mining has proved to be beneficial in extracting data in the field of education and thus reaching conclusions, assisting improvements in curriculum, predicting student's performance so as to take steps in order to receive desired results, and to determine factors affecting enrolments, dropping-out ratios, or the low placement rates.

Some work done in the field of education using data mining are discussed below.

Professors Srimani and Patil used Logistic Model Trees, Random Forest, J48 decision tree, RANDOM TREES, REPtree on teachers' data like designation, student and HOD feedback etc. to detect those faculty who have unusual behavior (Srimani and Balaji, 2014) such as:

- Erroneous actions,
- Low motivation,
- Attitudes towards superiors, colleagues and students, and
- Faculty interest in learning new things to provide a better approach to teaching.

Yadav and Pal worked on Information like stream, marks in graduation, students performance etc. using decision tree to select students for enrollment in MCA course and also to counsel them to choose the right course and improve performance as performance depends on the interest of student in a field (Yadav et al., 2012).

Usharani and Chandrasekaran applied Decision Tree (CHAID, CART, C4.5, and C5.), Link Analysis, Decision Fores on student records of enrollees to better target curriculum on student needs, used Decision Tree, Link analysis, Decision Forest on student enrollment data to assist course planning (Usharani and Chandrasekaran, 2010).

Siraj and Abdoulah used algorithm Crisp DM methodology, Neural Networks, Logistic regression, Decision tree, and cluster analysis on student enrollment data to help re-plan registration process/ enrollment processes (Siraj and Abdoulha, 2007). Erdogan, Timor used K-means and cluster analysis on Students' University Entrance Examination Results for Grouping students with similar characteristics (TIMOR and ERDOGAN, 2005).

Oladipupo and Oyelade used Aporiori-gen algorithm and Association rule mining technique on students' result repository of failed students to assist curriculum restructuring and monitoring student's ability (Oyelade and Oladipupo, 2010). Ramaswami and Bhaskaran used CHAID, Classification Trees on data from Questionnaire for students, and Student data from office of Chief Educational Officer to Predict slow learners and to study dominant factors on their academic performance (Ramaswami and Bhaskaran, 2006).

Yathongchai et al. used decision tree, classifiers algorithms (J48 and Naïve Bayes) on sample data from faculty of science which has the highest students drop out rate to deploy the analysis results to decrease of the number of drop out students, and to facilitate decision making of teachers, education management team etc (Yathongchai et al, 2003)

Ogor used C 5.0 -Rule induction and Artificial Neural Network on Students' Demographic information and course assessment data to Predict Final Achievement status of students upon Graduation (Ogor, 2007)

Neelam Naik has worked for enhancement of students; institutes and to provide feedback for improvement of teaching by faculty members of the higher education institutes; using previous academic records of students and applying various classification algorithm on the data. This data helped produce knowledge of students before admitting them to the course and also helped assist in increasing placement ratio (Naik and Purohit, 2012).

GROWTH OF TECHNICAL EDUCATION IN INDIA

Higher Education system of any country is central to economic and political development and plays a vital role as the change agent to prepare informed, responsible citizens who are able to work effectively in a global multicultural context. Higher education in general terms means University level education provided by universities, vocational universities, degree colleges, arts colleges, technical and medical colleges, and other institutions that offers various qualifications varying from Higher National Diplomas and Foundation Degrees to Honors, Masters Degrees and Doctorates (Kaul, 2006). According to the Indian educational system, Higher education is the highest level of the educational system which further aims to provide education that will yield high-caliber students with global proficiencies, responsible citizens feeling their responsibilities towards National developments and competencies in term of research and innovativeness (Mishra, 2007).

According to Department of Higher Education, MHRD, GOI "India has one of the largest Higher Education system in the world" and also has the capacity to become global hub in education. Looking into the statistics of capacity expansion in higher education, a manifold growth in the number of Central, State public/ private Universities, colleges, teachers and student enrollment has been noticed (refer to Figure1 and 2) (UGC Report, 2008, January 2011, 2012, 2013; MHRD Report 2012). In last few years (2006- 2013), an increase in the number of State Private universities can easily be estimated to approximately 15 times (refer to Figure 1)

Also, from 2004- 2012, the number of Colleges (both Public and Private) offering higher education has increased to 2.1 times with an increase of approximately 2 times in number of student enrollment and approximately 2 times in number of faculties (refer to Figure 2).

Higher education in general terms means University level education provided by

- Universities,
- Vocational universities,
- Degree colleges,
- Arts colleges,
- Technical and medical colleges, and
- Other institutions that offers various qualifications varying from higher national diplomas and foundation degrees to honors, Masters degrees and doctorates.

Moreover, Technical education as a part of higher education includes management, engineering, architecture, technology, pharmacy related courses, producing skilled manpower and improving the productivity of industries with improved quality of lives. In India, Technical education is imparted by three types of institutions:

Figure 1. Growth in the number of different types of universities (2006-2013)

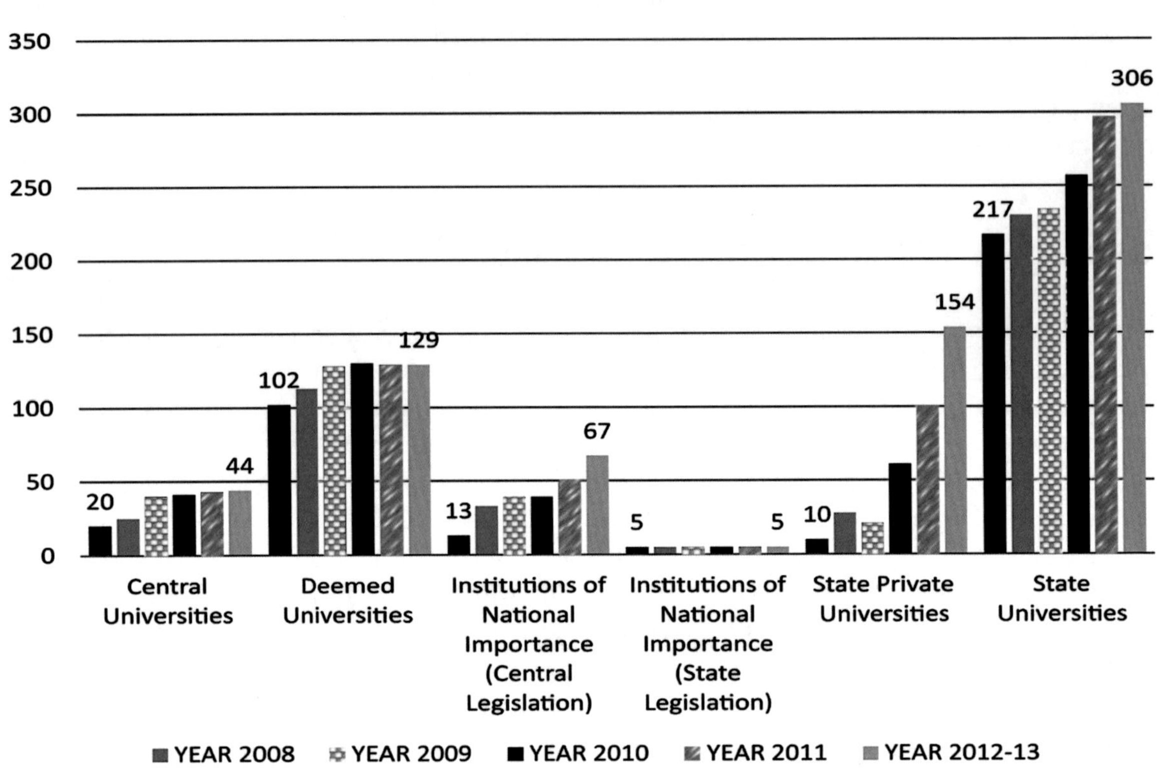

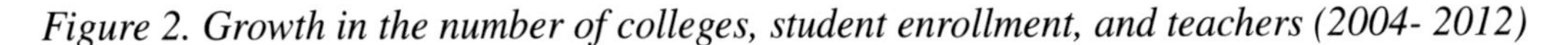

Figure 2. Growth in the number of colleges, student enrollment, and teachers (2004- 2012)

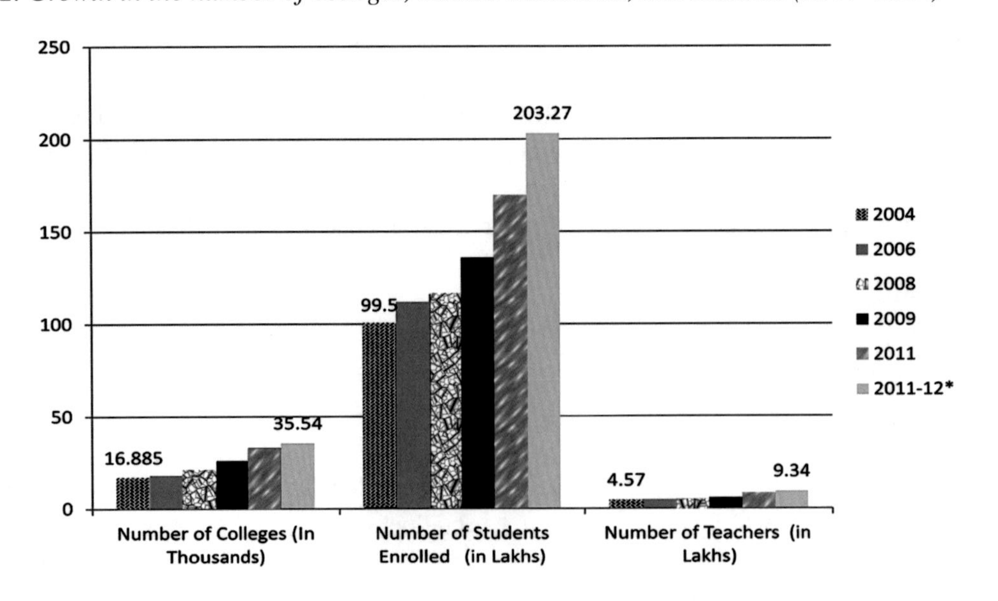

- Government funded institutions,
- State Government/State Private funded institutions, and
- Self-financed institutions.

In last few years, due to the participation of Private and Voluntary Organizations in the setting up of Technical Institutions on self-financing basis, an exceptional increase in the number of technical institutions approved by AICTE and student enrollment in the technical courses especially like engineering, Management and MCA has been seen (AICTE Handbook, 2014).

Figure 3. Growth of AICTE approved technical institutions in India

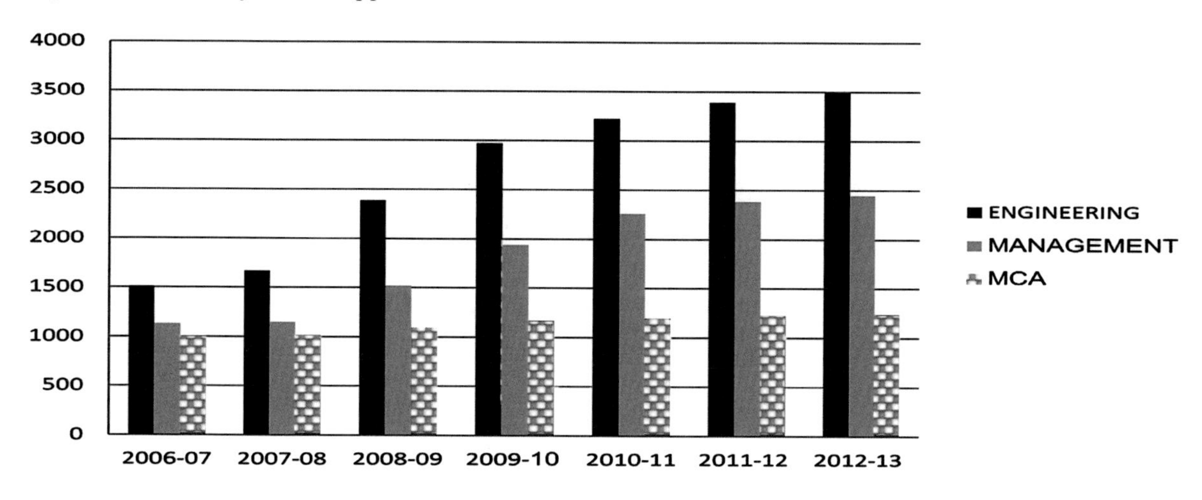

Figure 4. Student annual intake in popular technical courses

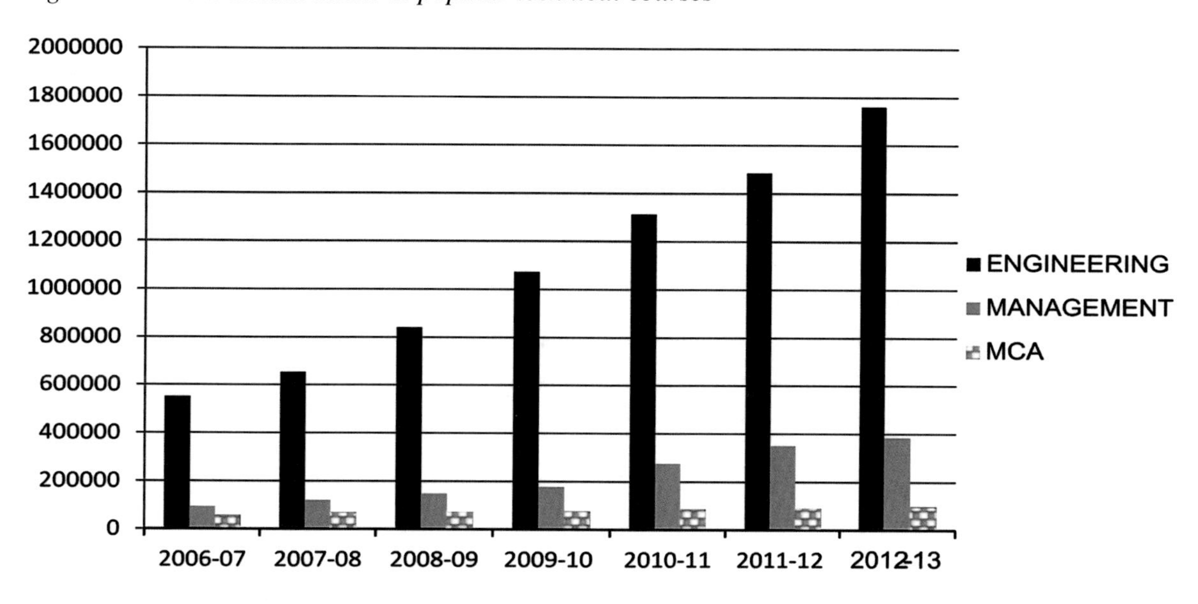

Challenges Emerged with Increased Unaided Private Sector Participation in Technical Education System

In year 2010-2011, 9701 colleges out of total 31,234 colleges were professional colleges those were offering courses in the field of Engineering, Management, Computer applications, Medical etc. and after considering the statistics produced by AICTE on the growth of technical institutions and student enrollment in the above mentioned courses; it is evident that in year 2013-14, approximately 11000 colleges are offering these technical courses. As a matter of fact, in last few years, there is an increase in the number of unaided private institutions on higher education and student enrollment in the unaided private institutions which are offering programmes in professional disciplines such as Engineering, Pharmacy, computer applications etc and after including both regular and distance education mode, Engineering/ Technology has got maximum percentage of total higher education student enrollment ie 2862439 (16.86 percent) after Arts, Science, Commerce/Management in year 2010-11 (UGC Report, 2012).

However, despite of producing more than 20 Lakhs technical professionals every year, India is lacking behind to bridge the gap between their knowledge resource and its actual implementation in the real world [Blom and Hiroshi, 2011]. Department of higher education in its Annual report 2009-2010 clearly declared "improvement in employability of trained graduates and postgraduates coming out of the technical institutes" as one of the major challenge in front of the Indian technical education system (DHE Annual Report, 2010). Andreas blom et al. in their study showcased that over 75 percent of IT graduates are not ready for jobs and do not have industry ready talent (Blom and Hiroshi, 2011). According to a report by employability Assessment Company Aspiring Minds, only about 17.45 percent of engineering graduates of the year 2011 were employable and nearly 92 per cent of engineering graduates' lacked in computer programming and algorithms skill whereas 56 per cent show lack of soft skills and cognitive skills that are required for IT product companies. It is also depicted that students at government colleges are doing much better than private colleges and there is a difference in their skills as well (AMNE Report, 2011).

In a report to people on education 2010-11, it is revealed that there are only 79 Central Government funded institutions along with State government funded and self-financing Institutions are supported by the government to strengthen the technical education system of the country, in terms of funds (AICTE Handbook, 2014) and rest others are unaided private self-financed institutions which are obviously having the maximum student enrollment with approximately 74 percent of the total enrollment into the regular class room teaching system instead of Distance education. Ministry of Human Resource Development and Planning commission along with University Grant Commission (UGC) are taking new initiatives and making new policies to sustain and improve the quality of higher education along with advisory bodies and agencies like All India Council of Technical Education (AICTE), National Board of Accreditation (NBA), and National Assessment and Accreditation Council (NAAC) who are continuously monitoring and controlling the development of technical education system by adopting essential measures and mechanisms including accreditation (AICTE, 2010). Under the new initiatives for enhancing Quality in Higher Education, UGC has made accreditation by National Assessment and Accreditation Council (NAAC) compulsory for all Universities/institutions of higher education and developed a scheme to set up an in-house Quality Assessment Cell in all universities/institutions to monitor their quality status (UGC Report, November 2011).

However, the initiativesand policies adopting by the government bodies/ agencies in the field of improving the quality of education are worth appreciating but the eye opening statistics stresses upon a need to review the quality of the technical education delivered by these unaided Private technical institutions. Educational Data Mining (EDM) which is emerged as a powerful tool in last decade, is concerned with developing methods to explore the unique types of data in educational settings and, using these methods, to better understand students and the settings in which they learn. Different unknown patterns using different data mining techniques viz. classification, Clustering, Association rule mining, decision trees can be discovered from this educational data which could further be beneficial to

- Improve teaching and learning systems,
- To improve curriculum,
- To support students in the form of individual counseling,
- Improving learning outcomes in terms of students' satisfaction, good placements and many other similar ones which can help the students, teachers to bridge the gap between the employability skills of required by the industry and the learning outcomes of the technical professionals.

SOLUTIONS AND RECOMMENDATIONS

Technical Education in Traditional Classroom System

In any conventional educational environment, there is a face to face interaction between the student and the teacher, therefore generates a feedback on both teaching mechanism and learning outcomes. Such environment also involves the processes related to assessing the student's performance, effectiveness of the teaching pedagogy, relevance of the course curriculum to the skills developed etc.

In any traditional classroom, Teaching and learning mechanisms are the ones that can bring a change in the attitude, perception, abilities of a learner apart from regular knowledge and skills. And it is evident that the overall quality outcomes of any institution are directly related to the quality of their learning environment, because the success of a student in the job market depends on his learning patterns. Therefore, it is extremely important to lift the teaching and learning methods in these educational institutions as these

Figure 5. Technical education system in traditional classroom environment

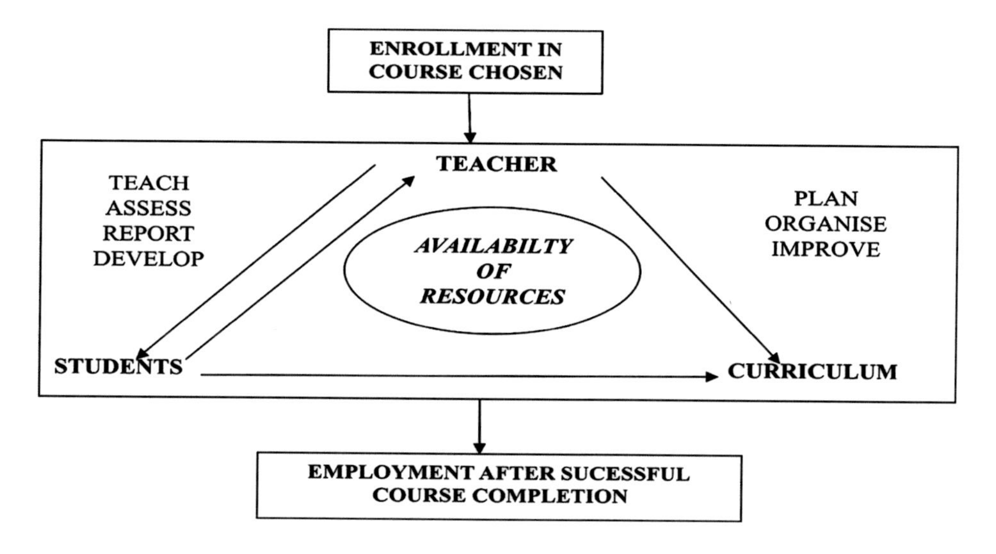

methods inculcate the basic skills along with technical knowledge required by a person to perform well in his taskforce. Technical education private institutions with a large no. of annual intake are yielding a huge amount of both structured and unstructured data on the daily basis which is but obvious generated from the conventional educational processes like class scheduling, student learning assessment, teacher teaching assessment, etc. This educational data may vary from institution to institution depending upon the methods adopted for the training and assessment, teaching pedagogy etc.

As the quality initiatives, such institutions need better assessment, analysis and prediction tools to analyze and predict student related issues like:

- Is the course curriculum meets the demands of the job market?
- Are the course structure and the content of units having correlation?
- Are teaching pedagogy adopted by the teachers based on the demanding scenario?
- Have the students learned in the same manner as the teacher planned them to learn?
- Have the leaning and teaching techniques inculcated the required employability skill sets?

In context of these issues, it is required to consider the Students' enrollment data viz.:

- Course chosen,
- Gender,
- Age,
- Father's name,
- Family details,
- Students' continuous assessment data,
- Marks statement and grades,

Teachers data viz.:

- Personal details,
- Qualifications,
- Courses taught and their student feedback,
- Class schedules,
- Course contents,
- Industry demands and other related data.

Efficiently extracting significant knowledge patterns for:

- Designing better course curriculum,
- Predicting students' performance,
- Improving attendance (or dropouts),
- Providing additional support where necessary,
- Allocating instructors in a better managed way,
- Solving problems related to student counseling, registration, evaluation,
- Allocating resources and staff, more effectively.

An explicit knowledge is required to be explore that can also help these institutions to make plans more strategically, taking better decisions with respect to:

- Improving curriculum,
- Counseling,
- Directing,
- Assessing students and their performances,
- Improving student's promotion rate, retention rate, transition rate,
- Increasing educational improvement ratio, student's success, student's learning outcome, as well as
- Predicting individual behaviors with higher accuracy.

Understanding Quality in Technical Education

Likewise any educational system, technical educational system also has different users or stakeholders. Since every education system must be student or learner centric, therefore the Student is identified as the first and important stakeholder of the Technical Education system. The teaching learning process is a dialog or interaction, between students and teacher that is strengthen by a Teacher effectively, the second hierarchical stakeholder of Technical Education system. The Management or administrators of Technical Education system, either in private or Government, are identified as the third stakeholder as they are the major resource generation agencies, which facilitate the whole teaching-learning process with required inputs. In Technical Education system, Industry is considered as the fourth and important stakeholder that hire the competent and skilled manpower or the lively product of the teaching-learning process i.e. student Finally, In India, unlike in western countries, because of the highly bonded family structure, parents feel themselves responsible for educating their wards up to the settlement in life. The parents are payers or financers to the Technical Education system in most of the cases, thus Parents are another important stakeholder of TE system.

Quality in Technical education may mean different things to different stakeholders. Teachers, Students, Employers, Parents, and Administrators may have different understanding and may have different expectations of Quality, from the technical education institutions. For students, quality in TEI may mean the educational facilities and the employment opportunities providing by the institution that will make them financially independent and would lead them towards respectable life. At the same time, teachers/ trainers may designate the effective teaching processes as the quality measure and may correlate learning outcomes with good results and student's feedback. Employers as another important stakeholder may associate the employability skills or competence of the technocrats with the quality education, imparting by a Technical institution. For parents, quality in higher education may be the one that adds prosperity to their family and improve their livelihood.

Quality, in general, is crucial for customer satisfaction, accountability, maintaining standards and attracting better stakeholders and finally surviving in cut-throat competition (Mishra, 2007), make it globally competitive. The expansion of any Technical Education Private Institution (TEPI) striving for quality in technical education can be defined in terms of the outcomes (satisfied number of pass outs, acceptable level of employment, acquisition of knowledge for the students) coming out from the educational processes (teaching, evaluation, counseling, directing etc) (Sharma and Kansal, 2011).

Figure 6. Stakeholders in technical education system

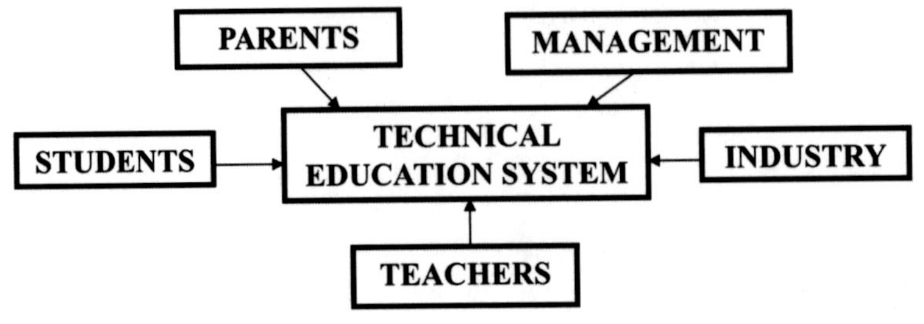

Educational Data Mining in Technical Educational System

In last decade, data mining in the field of higher education has emerged as a powerful tool to discover unknown hidden patterns from educational processes for the improvement of Quality education. Educational data mining (EDM) is an emerging interdisciplinary research area that deals with the development of methods to explore data originating in an educational context. It uses computational approaches to analyze educational data in order to study educational questions and exploits statistical, machine learning and data mining algorithms over the different types of educational data (Romero and Sebastian, 2010).

For the first time in mid-1990's, data mining was applied in traditional education domain to discover knowledge from a huge database of student records for finding the solution to the problem of declining university enrollment, retention of good students, and problem of drop out (Sanjeev et al., 1995). Data mining was introduced as a powerful decision support tool for seeking knowledge in higher education institutional research. In the project, the author selected a typical community college data warehouse in Silion valley on the west coast and prepared data using CRISP-DM (Cross Industry Standard Process for Data Mining) model. On the prepared data sets, he applied clustering and prediction algorithms (rule induction algorithms- Cand RT, C5.0, neural network and two step cluster technique) using SPSS Clementine to depict the best prediction model which finally concludes that such knowledge can be used to analyze available data with respect to the student problems like drop out, success rates, persistence etc. as well as can be used for alumni fund raising and Institutional marketing (Luan, 2002). Clustering, classification and association rule mining were examined to find out the characteristic patterns of previous students who took a particular major and the patterns of previous students which were likely to be good in a given major in order to improve the quality of graduates (Waiyamai, 2003). Decision Tree, Bayesian network algorithms and predication were being used to investigate the suitability of data mining tools for predicting academic performance (GPA) using two case studies and further achieved the patterns for predicting the academic performance of undergraduate and postgraduate students (Nghe et al., 2007). Romera and Ventura in their educational data mining survey from 1995 to 2005 briefed about the work applied in the area of data mining in various educational systems viz traditional classroom teaching, distance education, web based systems and the discovered knowledge from the traditional education environment can be used both by the education providers as well as by students. They clearly mentioned that as data source and objectives of both traditional and distance education are different, therefore both of them must be dealt, separately (Romero and Ventura, 2007). Later, an explicit knowledge as pattern of students' performance in accordance of their previous test scores was extracted using k-means clustering algorithm and it was by analyzing students' result data of a private Institution in Nigeria which was an effort towards the monitoring of academic performance progress of students in higher Institutions (Oyelade et al., 2010). Likewise the problem of drop outs in any educational institution was analyzed using decision tree analysis techniques (Quadri and Kalyankar, 2010) and the enrollment data to predict the most important factors for student success using:

- Classification and Regression Tree (CART),
- CHAID,
- Exhaustive CHAID, and
- QUEST classification tree methods (Koyacic, 2010).

Also, a CHAID predication model (Classification Tree Algorithm) was implemented to identify the slow learners and study the influence of the dominant factors on their academic performance (Ramaswami, 2010).

In these studies, the data mining tools, techniques like association, classification, prediction, clustering etc were used to extract the hidden patterns of student assessment, counseling of drop outs etc which can help in bridging the gap between the quality objectives and the quality achieved. It is a matter of fact that a in last decade, different authors carried out a number of studies to give an insight into the application of data mining algorithms and techniques in different educational systems including conventional educational system adopted by higher education institutions which involves entities like Student, lecturer, staff, alumni and management. A good enough work has been carried out in the field of educational data mining specifically in the traditional education systems, e learning systems, Web based educational systems, learning content management system which collects their own specific data for the education system (Romero and Ventura, 2007) and utilize existing online course environments which are used by large numbers of students worldwide, such as Moodle and WebCAT (Baker et al., 2009) but the need for a significant study in the field of technical education, keeping in mind its special characteristics needs to be explored yet and that has become the motivation of this study.

Table 1. Future research directions: educational tasks involved and DM algorithms

Educational Tasks Involved	Educational Data Chosen	Users/ Actors	Data Mining Algorithms Used/ Applied	Data Mining Technique	Authors	Knowledge/ Patterns Extracted Helps in:
Improving Student Performance	Students' result repository of failed students	Academic Planners	Apriori-gen	Association rule mining	Oladipupo, Oyelade (2010)	Curriculum restructuring and monitoring Student's ability
Student's Performance assessment and Monitoring	Students' Demographic information and course assessment data	Students, Academic planners	C 5.0 -Rule induction and Artificial Neural Network,	Prediction	E.N Ogor (2007)	Predicting Final Achievement status upon Graduation
Student Assessment	Students' University Entrance Examination Results	Academic Planners	K-Means	Cluster Analysis	Erdogan, Timor(2005)	Grouping students with similar characterstics

continued on next page

Table 1. Continued

Educational Tasks Involved	Educational Data Chosen	Users/ Actors	Data Mining Algorithms Used/ Applied	Data Mining Technique	Authors	Knowledge/ Patterns Extracted Helps in:
Student's Performance assessment-course wise	Students' Performance in C++ course assignments and other related	Teachers/ Instructors	Decision Tree	Classification	Qaseem et al. (2006)	Better predicting performance of a student in a specific course
Student Academic Performance Assessment and Monitoring	Students' complete records including demographics data, academic data	Teachers/ Instructors	Decision Tree, Bayesian Network Algo	Prediction	Nghe et al. (2007)	Predicting Student Academic Performance
Student Course Registration	Previous Students' Enrollment data, Demographics data, and their Academic records	Students	C 4.5 algo	Prediction	Cesar V et al. (2009)	Better Enrollment decisions by Students
Student Course enrolment	Student enrollment data	Registrar Office/ University administrator	Crisp DM methodology,Neural Networks, Logistic regression, Decision tree	Cluster Analysis	F. Siraj and M.A. Abdoulha	Re-planning registration process/ enrollment processes
Student Performance	Student enrollment data	Teachers/ Academic planners	Decision Tree, Link analysis, Decision Forest	Information Gain	C.Usharani and Rm.Chandrasekaran	Course planning
Improving Student performance	Questionnaire for students, Student data from office of Chief Educational Officer	Educators	CHAID, Classification Trees	Prediction, regression technique	M.Ramaswami and R. Bhaskaran	Predicting slow learners and studying dominant factors on their academic performance
Improving academic performance and reducing drop-out ratio	Educational data like class quizez, mid and final exam	Teachers	k-means clustering, Decision tree	Prediction	Shaeela Ayesha et al.	Predicting weak learners and improve performance
Improving Student's Academic Performance	Graded Point Average	Faculty, Academic planner	K-means Clustering	Prediction, Eucledian distance	Oyelade, O.J., Oladipupo, O.O.; Obagbuwa, I.C.	Monitoring the progression of academic performance of students so as to assist effective decision by academic planners

Many similar researches have been done to determine of academic performance of students so as to assist effective decision by academic planners and faculties to increase academic performance, assist weak students, reduce drop-out ratios etc. But not much work has been done to sort out employability issues of students. There is need to apply data mining in discovering various factors and techniques to inculcate traditional as well as other social skills needed to increase employability of graduates (especially students of professional courses). Therefore, work should be done on following aspects-

- Teaching pedagogy adopted by the teachers to conclude the most productive style (course wise) which would help in practical application of knowledge and bring the students closer to industry needs.
- Course curriculum adopted and their duration. Data has to be retrieved about the best average time period for students to grasp knowledge, the interval time and the entire day's teaching.
- Study should be done on bright students to study their learning style and thus apply it in the teaching pedagogy to help students learn faster.
- Work should be done on company employees, their characteristics and their unique abilities that are beneficial for the organization. Students should thus be prepared accordingly from the college level.
- Technical education demand extra skills and reasoning ability. So to improve academic performance and increase placement ratio, studies should be done to spot out the ways to inculcate these essentials in student's course curriculum.

Data mining tools should be used on these aspects as future research to bridge the gap between knowledge attained at technical institutes and their practical application in real world and thus increase the placement ratio of students of technical branch.

CONCLUSION

This paper surveys the most related work done in the field of Data Mining in Higher/Technical Education and discusses the future possibilities of applying data mining in technical education system for the improved learning outcomes. The growth of technical education in India is tremendous but the challenges emerging with increased unaided private sector participation in the field of technical education is to deal with improvement in the employability of trained graduates and postgraduates, coming out of these technical institutions. As the Technical Educational institutions as a regular operation collect lots of information and data produced from day to day educational activities viz details of the courses offered, Students' enrollment data like:

- Course chosen,
- Gender,
- Age,
- Father's name,
- Family details,
- Students' continuous assessment data,
- Marks statement and grades.

Teachers data (personal details, qualifications, courses taught and their student feedback), Placement data and many other similar ones, therefore the possibilities of can be mined using Data mining tools and techniques.

REFERENCES

All India Council for Technical Education. (2014). *Approval process handbook (2013 – 2014)*. Retrieved from http://www.informindia.co.in/education/Approval_Process_Handbook_091012.pdf

Aspiring Mind's National Employability Report- Engineering Graduate. (2011). Retrieved from http://www.aspiringminds.in/docs/national_employability_report_engineers_2011.pdf

Baker, R., & Yacef, K. (2009). The state of educational data mining in 2009: A review and future visions. *Journal of Educational Data Mining, 1*(1), 3-17.

Baker, R., & Yacef, K. (2009). The state of educational data mining in 2009: A review and future visions. J. Educ. *DataMining*, *1*(1), 3–17.

Blom, A. & Hiroshi, S. (2011). *Employability and skill set of newly graduated engineers in India*. Academic Press.

Department of Higher Education. (2010). *Annual Report 2009-2010*. Ministry of HRD, 151.

Kaul, S. (2006). *Higher education in India: seizing the opportunity*. The Indian Council for Research on International Economic Relations.

Kovacic, Z. (2010). Early prediction of student success: Mining students' enrolment data. *Proceedings of Informing Science & IT Education Conference* (InSITE).

Luan, J. (2002). Data mining and its applications in higher education. *New Directions for Institutional Research, 113*(113), 17–36. doi:10.1002/ir.35

Ministry of Human Resource Development. (2012). *Report to the people on Education- 2010-2011*. Retrieved from http://mhrd.gov.in/sites/upload_files/mhrd/files/RPE-2010-11.pdf

Mishra, S. (2007). *Quality assurance in higher education: An introduction. Commonwealth of Learning*. National Assessment and Accreditation Council.

Naeimeh, D., Beikzadeh, M. D., & Amnuaisuk, S. P. (2005). Application of enhanced analysis model for data mining processes in higher educational system. *Information Technology Based Higher Education and Training, 2005. ITHET 2005. 6th International Conference on*. IEEE.

Naik, N., & Purohit, S. (2012). Prediction of Final Result and Placement of Students using Classification Algorithm. *International Journal of Computers and Applications*, *56*(12).

Nghe, T. N., Janecek, P., & Haddawy, P. (2007). A comparative analysis of techniques for predicting academic performance. Frontiers In Education Conference-Global Engineering: Knowledge Without Borders, Opportunities Without Passports, 2007. FIE'07. 37th Annual. IEEE.

Ogor, E. N. (2007, September). Student academic performance monitoring and evaluation using data mining techniques. In *Electronics,Robotics and Automotive Mechanics Conference (CERMA 2007)* (pp. 354-359). IEEE. doi:10.1109/CERMA.2007.4367712

Oyelade, O. J., Oladipupo, O. O., & Obagbuwa, I. C. (2010). Application of k-Means Clustering algorithm for prediction of Students' Academic Performance. *International Journal of Computer Science and Information Security*, *7*(1).

Oyelade, O. J., Oladipupo, O. O., & Obagbuwa, I. C. (2010). Application of k-Means Clustering algorithm for prediction of Students' Academic Performance. *International Journal of Computer Science and Information Security*, *7*(1).

Quadri, M., & Kalyankar, N. V. (2010). Drop out feature of student data for academic performance using decision tree techniques. *Global Journal of Computer Science and Technology, 10*(2).

Ramaswami, M., & Bhaskaran, R. (2010). A CHAID Based Performance Prediction Model in Educational Data Mining. *International Journal of Computer Science Issues*, *7*(4).

Romero, C., & Sebastián, V. (2010). Educational data mining: a review of the state of the art. *Systems, Man, and Cybernetics, Part C: Applications and Reviews. IEEE Transactions on*, *40*(6), 601–618.

Romero, C., & Ventura, S. (2007). Educational data mining: A survey from 1995 to 2005. *Expert Systems with Applications*, *33*(1), 135–146. doi:10.1016/j.eswa.2006.04.005

Sanjeev, A. (1995). *Discovering Enrollment Knowledge in University Databases*. KDD.

Sharma, N., & Kansal, V. (2011). Identifying the capabilities of data mining in providing the Quality in Technical Education. *Proceedings of the 5th National Conference; INDIACom-2011,Computing For Nation Development*. BVICAM.

Siraj, F., & Abdoulha, M. A. (2007). Mining enrolment data using predictive and descriptive approaches. *Knowledge-Oriented Applications in Data Mining*, 53-72.

Srimani, P. K., & Balaji, K. (2014). A comparative study of different classifiers on search engine based educational data. *International Journal of Conceptions on computing and Information Technology, 2*, 6-11.

Timor, M., & Erdogan, S. Z. (2005). A data mining application in a student database. *Journal of Aeronautics and Space Technologies, 2*(2), 53-57.

UGC Report. (2008). *Higher education in India - Issues related to Expansion, Inclusiveness, Quality and Finance.* Retrieved from http://www.ugc.ac.in/oldpdf/pub/report/12.pdf

UGC Report. (2011a). *Higher Education in India - Strategies and Schemes during Eleventh Plan Period (2007-2012) for Universities and Colleges*. Retrieved from http://www.ugc.ac.in/oldpdf/pub/he/HEIstategies.pdf

UGC Report. (2011b). *Inclusive and qualitative expansion of higher education 12 Five-Year Plan, 2012-17.* Retrieved from http://www.ugc.ac.in/ugcpdf/740315_12FYP.pdf

UGC Report. (2012). *Higher education in India at a glance*. Retrieved from http://www.ugc.ac.in/ugcpdf/208844_HEglance2012.pdf

UGC Report. (2013). *Higher education in India at a glance*. Retrieved from http://www.ugc.ac.in/pdf-news/6805988_HEglance2013.pdf

Usharani, C., & Chandrasekaran, R. (2010). Course planning of higher education to meet market demand by using data mining techniques-a case of a Technical University in India. *International Journal of Computer Theory and Engineering*, *2*(5), 809–814. doi:10.7763/IJCTE.2010.V2.245

Waiyamai, K. (2003). *Improving Quality of Graduate Students by Data Mining*. Bangkok: Department of Computer Engineering, Faculty of Engineering, Kasetsart University.

Yadav, S. K., & Pal, S. (2012). Data mining application in enrollment management: A case study. *International Journal of Computers and Applications*, *41*(5).

Yathongchai, W., Yathongchai, C., Kerdprasop, K., & Kerdprasop, N. (2003). *Factor Analysis with Data Mining Technique in Higher Educational Student Drop Out*. Latest Advances in Educational Technologies.

Section 2

Feature Detectors and Descriptors

Feature detectors and descriptors have a vital role in numerous applications including video camera calibrations, object recognition, biometrics, medical applications and image/video retrieval. Extract point correspondences "Interest points" between two similar scenes, objects, images/video shots is their main task. This section outlines the different feature detectors and descriptors types that involved in various applications.

Chapter 3
Feature Detectors and Descriptors Generations with Numerous Images and Video Applications: A Recap

Nilanjan Dey
Techno India College of Technology – Kolkata, India

Amira S. Ashour
Tanta University, Egypt

Aboul Ella Hassanien
Cairo University, Egypt

ABSTRACT

Feature detectors have a critical role in numerous applications such as camera calibrations, object recognition, biometrics, medical applications and image/video retrieval. One of its main tasks is to extract point correspondences "Interest points" between two similar scenes, objects, images or video shots. Extensive research has been done concerning the progress of visual feature detectors and descriptors to be robust against image deformations and achieve reduced computational speed in real-time applications. The current chapter introduced an overview of feature detectors such as Moravec, Hessian, Harris and FAST (Features from Accelerated Segment Test). It addressed the feature detectors' generation over time, the principle concept of each type, and their use in image/video applications. Furthermore, some recent feature detectors are addressed. A comparison based on these points is performed to illustrate their respective strengths and weaknesses to be a base for selecting an appropriate detector according to the application under concern.

DOI: 10.4018/978-1-5225-1025-3.ch003

INTRODUCTION

Recently, a huge amount of data to be processed is increased due to the increasing popularity of camera-ready cellphones as well as the broadband wireless devices, which capature high-quality images and video content. Mainly, this information has the form of text, pictures, and graphics or integrated multimedia presentations. Digital images and digital video are pictures and movies, respectively that converted into a binary format. Typically, image means a still picture that does not change with time, while a video changes with time and contains moving and/or changing objects. The real world is rich in visual information, deducing the massive amount of data becomes a challenging process. Thus, presenting and processing enormous amount of image/video data become serious problem. Image/video content information are used in several applications and used in some extent as a search tools. The foremost motivation for extracting the information content is the accessibility problem. To solve this problem, the extraction of relevant information features for a given content domain is required. Vision systems deal with images captured to extract information to perform certain tasks. The process of tracking a moving object(s) continuously using a camera is known as visual object tracking. It is employed to determine the object's position in frames continuously and reliably in video. It is very imperative task in several computer vision applications. Computer vision is a discipline of artificial intelligence that provide computers with the ability to observe objects. Analysis of the extracted information different types depends on the application to be accomplished. The ultimate goal is to use the detected information to attain an understanding of different objects through their physical and geometrical attributes.

Video consists of a collection of video frames moving at certain speed (frame rate). Each frame is a picture image made of pixels. A sequence of frames which recorded in a single-camera operation is called shot, while the collection of consecutive shots which have semantic similarity in persons, objects, time and space is called scene. Meanwhile, video can be considered as a combination of images collected in frame.

Foremost, image features are used to distinguish image regions and to characterize the appearance/shape of any objects in the images. Features embrace point, line or compound features to be extensively used by numerous computer vision applications including recognition, object detection, camera calibration, stereo, tracking and 3D reconstruction.

In both computer vision and image processing, abundant applications depend upon the robust detection of image features and their parameters' estimation. Thus, feature detection is an essential concern in the intermediary levels vision applications, such as:

- Image registration stereo,
- Motion correspondence,
- Simultaneous localization and mapping (SLAM) for autonomous vehicles,
- Object recognition, and
- Stereoscopic vision (Lowe, 2004).

Moreover, it is considered a powerful tool that has been applied effectively in an extensive range of other systems and application domains include:

- Facial feature detection,
- Land mine detection,

- Medical applications,
- Video mining,
- Image retrieval, and
- Texture analysis (Tien et al., 2008; Shin, & Kim, 2014).

Primarily distinguishing between feature detectors and descriptors is essential. Detectors are operators that search two dimensional (2D) locations in the images (i.e. a point or a region) geometrically stable under different transformations and containing high information content that results 'interest points', 'corners', 'affine regions' or 'invariant regions'. While, descriptors analyze the image to provide a 2D vector of pixel information for certain positions (e.g. an interest point). This information can be used to classify the extracted points or in a matching process.

Consequently, feature detection is considered the first step in detecting and matching specific features in any application; as it identifies the interest points. The second step is the description known as the extracted vector feature descriptor surrounding each interest point. Finally, matching that determines the connection between descriptors in two scenes/ images.

Commonly, feature detectors (FD) are techniques used for computing abstractions of image information and for constructing local assessments at every image point regarding whether at that point there is an image feature of a given type or not (Tuytelaars & Mikolajczyk, 2008). FDs identify a set of image locations with visual information and well defined spatial location, thus they are named interest point detection/ keypoint detection. The detectors are used to extract the features from the image, e.g., points, edges, continuous curves or connected regions, corners, "blobs" (centers of roughly circular regions) or small image patches. Then, measurements are taken from a region centered on the detected features and converted into descriptors. For example, edges detected in aerial images often correspond to roads and blob detection can be used to identify impurities in some assessment task. Furthermore, to recognize objects/ scenes without using segmentation, a set of detected features can be used for image representation.

Obviously, a superior feature of an application may be ineffective in the context of the other problems, where each of the FD techniques obligates its own constraints. Therefore, searching for the suitable feature detectors for an application becomes a requirement. Noticeably, there are various properties that describe good feature detection and depend on the concrete application and settings. Generally, the properties of high-quality FDs are as follows:

1. Detect the same features of an object, even under different viewing conditions, occlusions and image clutter.
2. Provide an interesting description of the image content for object/scene recognition and image retrieval tasks.
3. Afford simple model approximations of the geometric between two images taken under different viewing conditions.
4. Have adequately large number of detected features even for small objects.
5. Localize accurately the detected features.
6. Have less frequency of false alarms and mismatches.
7. Be robust and less sensitive to deformations such as image noise, compression artifacts, blur, lighting changes, etc.

8. Have less computational cost for each step.

Diverse applications consider these obligations differently. For example, in object recognition and SLAM applications, viewpoint changes are more extensively than in image mosaicking. While, the false matches frequency is more critical in object recognition rather than in the case of wide-baseline stereo and mosaicing applications.

Organization of the remaining sections will be as follows: an overview on feature detectors followed by the concept/ development of the common feature detectors such as Hessian detector, Moravec's, Harris, and others; according to the invention year. The most recent feature detectors are then addressed followed by the discussion and the conclusion.

OVERVIEW ON FEATURE DETECTORS

An interest point has locality in space with no spatial extent. The existence of interest points can significantly reduce the computation time. A foremost interest point detector class is the contour curvature based technique. Typically, these were relevant to piecewise constant regions, line drawings, extract geometrically important corners (Kim, 2012). Numerous methods have been extended to discover corners by:

- Analyzing the sequence code,
- Changing in direction,
- Modifying in appearance,
- Varying in the direction at the large gradient locations, or
- Finding maxima of curvature.

Edges may be produced from diverse physical sources such as different surface properties (texture, color, and reflective properties), discontinuity in the distance as well as surface orientation and shadows. However, all the resultant edges display assorted degrees of discontinuities in image intensity. Therefore, edge detection is defined to be the procedure of recognizing and locating sharp discontinuities in an image. The discontinuities are rapid changes in pixel intensity which differentiate the boundaries of objects in an image or scene. Various methods for edge detection are founded on gray-level gradient and angular changes in the curves as proposed in (Rosenfeld & Thurston, 1971). Where, there are different types of edges as step, ridge, ramp and roof edge. Various variables are considered for edge detection operator selection, such as (Cagnoni, 2007):

- **Noisy Environment:** Both the noise/edges contain high-frequency components, thus edge detection become complex with noisy images. Operators used on noisy images try to average enough data to discount localized noisy pixels. This leads to less precise localization of the detected edges.
- **Edge Orientation:** The geometry of the operator establishes a characteristic direction which is most sensitive to edges. Operators can be developed to search for horizontal, vertical or diagonal edges.
- **Edge Structure:** Refraction or poor focus effects can cause objects with boundaries identified by a gradual change in intensity. The edge detection operators are responsive to such a gradual change in these cases.

This requires the presence of a variety of edge detection methods. Conventional methods of edge detection involve convolving the image with a 2-D filter sensitive to large gradients in the image although it returns zero values in the uniform regions. A tremendously large number of edge detection operators are designed to be perceptive to definite edge types. However, the two major categories of diverse methods are:

1. **Gradient Based Edge Detection (Hardie & Boncelet, 1994):** To discover the maximum and minimum in the first derivative of the image to detect the edges.
2. **Laplacian Based Edge Detection (Kumar & Saxena, 2013):** To search for zero crossings in the second derivative of the image to detect the edges.

Classical operators are such as Sobel Operator, Robert's cross operator, Prewitt's operator. While, Zero Crossing operators are as Laplacian of Gaussian (LoG) (Marr-Hildreth) and Second directional derivative. Gaussian and Canny's edge detection are other types of operators. The advantages and disadvantages of these techniques were presented in (Maini & Aggarwal, 2009). The authors concluded that Canny's edge detection algorithm is computationally more exclusive compared to Sobel, Prewitt and Robert's operator. Conversely, the Canny's edge detection algorithm achieves enhanced performance than all these operators under almost all scenarios. Assessment of the images indicated that under noisy conditions, Canny, LoG, Sobel, Prewitt, Roberts's demonstrate better performance; respectively.

Once the edges are detected, consequently the corners can be identified. Where, a corner can be considered as the intersection of two edges and an edge is a sharp change in the image brightness.

Feature detector is considered a significant early vision problem; earlier work on FD was done. The first local feature detector was produced in 1954, for corn and junction detection (Mikolajczyk & Schmid, 2004). Then, for line drawing FD algorithm is developed in 1969 (Bloomenthal, 1983). Through in 1978, Hessian detector was initiated, then Hart detector was introduced in 1973 . While, Moravec's operator has been emerged in the year1979. Accordingly, Sequent developed algorithms for feature:

- Harris detector (1988),
- Affine (1994),
- SUSAN (1995-1997),
- SIFT (Scale Invariant Feature Transform) (2004),
- SURF (Speeded Up Robust Feature) (2006), and recently
- The FAST (Features from Accelerated Segment Test).

The following section will discuss these different feature detectors concept, followed their applications.

VARIOUS FEATURE DETECTOR METHODS

The concept of the most principal feature detectors are mentioned as follows.

Hessian Detector

Hessian detector «Determinant of Hessian (DoH) method» (Beaudet, 1978) computed the corner strength as the determinant of the Hessian matrix $\left(I_{xx}I_{yy} - I_{xy}^2\right)$. The corners in the image are determined by the local maxima of the corner strength. The determinant is related to the Gaussian curvature of the image. This measure is invariant to rotation. A broadened version, called Hessian Laplace detects points that are invariant to rotation and scale (local maxima of the Laplacian-of-Gaussian).

The Hessian matrix is constituted from the second order partial derivatives developed from a Taylor series expansion. This matrix has been common for local image structure analysis. The 2x2 Hessian matrix $He\left(x,\sigma_w\right)$ is expressed as follows:

$$He\left(x,\sigma_w\right) = \begin{bmatrix} I_{xx}\left(x,\sigma_w\right) & I_{xy}\left(x,\sigma_w\right) \\ I_{yx}\left(x,\sigma_w\right) & I_{yy}\left(x,\sigma_w\right) \end{bmatrix} \quad (1)$$

where, σ_w is the standard deviation; I_{xx}, I_{yy} and I_{xy} are the second order derivatives that can be used to measure the curvature at a point when the image is treated as an intensity surface. These second order derivatives are computed using Gaussian kernels of σ_w. The eigenvalues stand for the extent of the curvature in those directions, while the eigenvectors of the matrix provide the directions for minimum and maximum curvature. Consequently, the Hessian matrix can be used to describe the local structure in a neighborhood around a point, while its determinant can be used to detect image structures that have strong variations in two directions. The determination of the Hessian is expressed as:

$$\det\left(He\right) = I_{xx}I_{yy} - I_{xy}^{\ 2} \quad (2)$$

Moravec's Detector

In fact, Hans P. Moravec was the first who initiated interest point detection using intensity variations in an image. He extended Moravec operator in 1977 for a research concerning the navigation of the Stanford Cart. However, Hans P. Moravec considered the interest points as points with a large intensity variation in every direction which is the case at corners. Consequently, the Moravec operator is considered a corner detector.

Moravec described the "points of interest" as points where the image intensity of a pixel at $\left(x,y\right)$ is $I\left(x,y\right)$ that varies in all directions. Moravec's corner detector considered a local window in the image, then it determined the average changes of image intensity that extended from shifting the window via a small amount in various directions. The minimum difference in intensity between a small window centered on the pixel and the same window shifted by a few pixels along each of the eight cardinal plus ordinal direction to measure the local intensity variation around each pixel are to be calculated. The coordinates of the local maxima are identified and extracted from the large values of which correspond to points with large intensity variations in all directions with interest points. Formally, Moravec corner detector is stated as shown as follows (Dey et al., 2012):

Moravec Corner Detector Algorithm

Input: Grayscale image, window size, threshold T
Step 1: For each pixel (x, y) in the image, calculate the intensity variation from a shift (ξ, ψ) by:

$$V_{\xi,\psi} = \sum_{\forall a,e \in window} \left(I\left(x+\xi+a, y+\psi+e\right) - I\left(x+a, y+e\right)\right)^2$$

Here a and e are points within the window; while ξ and ψ are the shifts given by: (1,0),(1,1),(0,1),(-1,1),(-1,0),(-1,-1),(0,-1),(1,-1).

Step 2: Construct the corner map by calculating the corner measure $C(x, y)$ for each pixel (x, y) using $C(x, y) = \min\left(V_{\xi,\psi}(x, y)\right)$.
Step 3: Set all $C(x, y)$ below a threshold T to zero. (To threshold the interest map).
Step 4: Perform non-maximal suppression to find local maxima. All non-zero points remaining in the corner map are corners.
Output: Map indicating position of each detected corner.

Moravec detector describes a corner as a point where there is a large intensity variation in every direction, but it only considers shifts in discrete 45 degree angle.

Harris Detector

Harris detector in 1988 is similar to Moravec operator with certain improvements as it calculates a matrix associated with the image autocorrelation function. Its objective is to determine the direction of fastest and lowest change for feature orientation using a covariance matrix of local directional derivatives. The squared first derivatives of the image are averaged over a window. Then, the eigenvalues of the resultant matrix are the principal curvatures of the auto-correlation function. If the two found curvatures are high, this indicates an interest point detection. Harris points are invariant to rotation. Developed versions of the Harris detector have been presented in (Mikolajczyk & Schmid, 2001) where the detected points are invariant to both scale and rotation.

Harris and Stephens removed the directional quantization inherent in Moravec's algorithm by expanding the intensity variation equation using a first order Taylor series. They showed that the intensity gradient structure local to a point is encapsulated by the structure tensor (or second moment matrix) averaged over the neighborhood of the point (Morgan, 2010):

$$H(x, y) = b(\sigma_h) * \begin{bmatrix} I_x(x, y)^2 & I_x(x, y) I_y(x, y) \\ I_x(x, y) I_y(x, y) & I_y(x, y)^2 \end{bmatrix} \quad (3)$$

This equation presents the basis for several interest point detectors. Where, '*' refers to the convolution, while H is fundamentally rotationally invariant, $I_{x(y)}$ is the partial derivative of the image with respect to $x(y)$, and b is the Gaussian window defined as follows:

$$b(\sigma) = e^{\left(\frac{-(x^2+y^2)}{2\sigma^2}\right)} \tag{4}$$

To provide averaging and reduce noise, the Gaussian function is convolved with the second moment matrix.

To determine the derivatives, Harris & Stephens employed Prewitt filter to develop the rotational invariance H and signal-to-noise. To obtain differentiation of the image, Gaussian derivative filter is used as:

$$I_x(x,y) = b(\sigma_d) * \frac{\partial I(x,y)}{\partial x} = \frac{\partial b(\sigma_d)}{\partial x} * I(x,y) \tag{5}$$

where, the standard deviations σ_d and σ_h are for differentiation and averaging; respectively.

Harris & Stephens (sometimes known as the Plessey operator) proposed the widely used feature response function defined as follows (Morgan, 2010):

$$G_{HS}(x,y) = \det(H(x,y)) - kTr^2(H(x,y)) \tag{6}$$

Here, $\det$ is the determinant, k is an adjusting parameter and Tr^2 is the trace. Thus, the maxima of G_{HS} symbolizes points to a corner-like, i.e. vertex-like, structure. Where the determinant stands for the product (sum) of the eigenvalues of H, so G_{HS} will be large if both eigenvalues are large.

SUSAN Detector

The basic principal of the Smallest Univalue Segment Assimilating Nucleus (SUSAN) detector is based on the concept that each image point has a local area of similar brightness associated with it. This local area contains much information about the image structure. Two dimensional features and edges can be detected by conducting the size, centroid and the second moments of the SUSAN. The characteristic advantage of this detector over the other ones is that no image derivatives are used and therefore no noise reduction is required.

Direct local measurements were used in Susan detector to analyze different regions disjointedly and to locate places where individual regional boundaries have high curvature (Smith & Brady, 1997). To define an area with similar brightness to the center, the brightness of each pixel in a circular mask is compared to the central pixel. The size, centroid and second moment of this area are to be calculated in order to detect the 2D interest features.

The SUSAN edge detector method takes an image and uses a predetermined window centered on each pixel in the image as the usual detectors. Then, it applies a set of rules to get the edge response. The SUSAN detector's mask is placed over each pixel of the image and its brightness is compared to the brightness of each pixel (nucleus) of the mask. The SUSAN corner detector is recognized by a circular mask. Then, an area of the mask "USAN" can be described which has similar brightness as the nucleus. Consistent with the area of the USAN, corners can be detected. As the nucleus is on the corner as the area of USAN is equal to the smallest value. To detect corners, the comparison function among every pixel within the mask and mask's nucleus follows equation (7) (Smith & Brady, 1997):

$$z\left(\nu,\nu_0\right)=\begin{cases}1 & \left|I\left(\nu\right)-I\left(\nu_o\right)\right|\leq Y\\ 0 & otherwise\end{cases} \quad (7)$$

Here, ν_0 and ν are the nucleus's coordinates and the other points within the mask coordinates; respectively. The result comparison function is $z\left(\nu,\nu_0\right)$, while $\mathrm{I}\left(\nu\right)$ is the gray value of the point and Y is the gray difference threshold that determines the smallest contrast detected by SUSAN detector. In fact, this equation is not stable in practice. The USAN region size $n\left(\nu_0\right)$ is defined as:

$$n\left(\nu_0\right)=\sum_{\nu\in z\left(\nu_0\right)} z\left(\nu,\nu_0\right) \quad (8)$$

The initial response to corners $R\left(\nu_0\right)$ is obtained as follows,

$$R\left(\nu_0\right)=\begin{cases}g-n\left(\nu_0\right), & n\left(\nu\right)<\mathrm{g}\\ 0, & n\left(\nu\right)\geq g\end{cases} \quad (9)$$

The smaller USAN region, the greater initial response to corners will be attained. Here, g is the geometric threshold that enhances the corner information of an image and establishes the sharp level of a corner. Finally, corners can be found via the non-maximum inhibition.

Affine Detectors

The mainly widespread affine region detectors (Tuytelaars & Van Gool, 2004) are:

1. **The Hessian-Affine Detector:** The Hessian matrix and the scale selection based on the Laplacian are used to estimate points with the elliptical regions using the eigenvalues of the second moment matrix of the intensity gradient.
2. **The Harris-Affine Detector:** (Named also the Harris-Laplace detector) is accustomed to decide the localization and scale. The second moment matrix of the intensity gradient concludes the affine neighborhood.

3. **The MSER (Maximally Stable Extremal Region) Detector:** Using the monotonic transformation of the image intensities as well as a continuous transformation of the image coordinates to detect closed regions.
4. **The Salient Regions Detector:** Detects regions through measuring the entropy of pixel intensity histograms.
5. **The Edge-Based Region (EBR) Detector:** An affine invariant region combined operator. While regions are detected using the Harris operator, image edges are detected with a Canny operator.
6. **The Intensity Extrema-Based Region (IBR) Detector:** Applies the image intensity function and its local extremum to detect the affine-invariant regions.

SIFT Detector

SIFT (Scale Invariant Feature Transform) detector was proposed by Lowe in 2004 to solve the viewpoint change, image rotation, scaling, and affine deformation, noise and illumination changes. This SIFT detector has strong robustness. Its main principle is to take an image and transforms it into a group of local feature vectors, to any scaling, rotation or translation of the image. Each of these feature vectors is presumed to be distinctive and invariant. These features can be used to detect distinctive objects in various images.

The SIFT algorithm has four main steps (Panchal et al., 2013):

Scale Space Extrema Detection: To recognize location and scales using scale space extrema in the DoG (Difference-of-Gaussian) functions with different values of σ. The DoG function D is convolved of image in scale space separated by a constant factor *k* as in the following equation:

$$D\left(x,y,\sigma\right)=\left(F\left(x,y,\sigma\right)-F\left(x,y,\sigma\right)\right)\times I\left(x,y\right) \tag{10}$$

Here, *F* is the Gaussian function and I is the image.

2. **Key Point Localization:** By eliminating the key points with the low contrast. The key points are exactly localized by fitting a 3D quadratic function to the scale-space local sample point.

Orientation Assignment: Once the SIFT-feature location is established, a foremost orientation is allocated to each feature based on local image gradients. For each pixel in the region of the feature location, the gradient magnitude and orientation are calculated; respectively as (Alhwarin *et al.*, 2008):

$$m\left(x,y\right)=\left(F\left(x+1,y,\sigma\right)-F\left(x-1,y,\sigma\right)\right)^2+\left(F\left(x,y+1,\sigma\right)-F\left(x,y-1,\sigma\right)\right) \tag{11}$$

$$\Theta\left(x,y\right)=\arctan\left(F\left(x+1,y,\sigma\right)-F\left(x-1,y,\sigma\right)\right)^2/\left(F\left(x,y+1,\sigma\right)-F\left(x,y-1,\sigma\right)\right) \tag{12}$$

4. **Description Generation:** Founded on the image gradient magnitude and orientation at each image sample point in a region centered at key points. This step computes the local image descriptor for each key point.

SURF Detector

The SURF algorithm «Fast hessian detector» is based on the same standards and steps as the SIFT detector algorithm utilizing different scheme to provide better and faster results. It is a local feature detector and descriptor that can be used for tasks such as object recognition or registration or classification or 3D reconstruction. The Speed Up Robust Features (SURF) creates a scale space via convolving rectangular masks of increasing size similar to diverse scales with the input image using an integral image illustration. Consequently, a series of blob response maps at different scales is constructed. Through grouping blob response maps for adjacent scales, a number of octaves is formed to divide the scale space. Usually, four scales per octave are used to be adequate for scale space analysis (Bay et al., 2008).

To diminish computation, the SURF detector can double the spatial sampling interval with increasing octave. Then, a 3D non-maximum suppression followed by 3D quadratic interpolation were used to achieve sub-pixel, sub-scale accuracy. A blob response threshold is normally applied to select high-contrast interest points.

The SURF detector is found on multi-scale space theory along with the feature detector based on Hessian matrix, since Hessian matrix has good performance and accuracy. To facilitate feature point detection in a scale invariant manner, SIFT uses a cascading filtering approach, where the DoG is computed on increasingly downscale images. SURF detector divides the scale space into levels and octaves similar to the SIFT detector. An octave matches to a doubling of σ and the octave is separated into uniformly spaced levels.

Recall equation (1) for the Hessian matrix. The core of the SURF detection is non-maximal-suppression of the Hessian matrix determinants. The convolutions are very costly to calculate. It is approximated and accelerated using integral images and approximated kernels. Here, an integral image $I\left(x\right)$ is an image where each point $X = \left(x, y\right)^T$ stores the sum of all pixels in a rectangular area, as follows:

$$I\left(X\right) = \sum_{i=0}^{i \le x} \sum_{j=0}^{j \le y} I\left(x, y\right) \tag{13}$$

To compute the response in a rectangular area with arbitrary size using 4 lookups, the integral images are used.

The second order Gaussian kernels used for the hessian matrix should be discretized/ cropped before applying to the kernel. The SURF algorithm approximates these kernels with rectangular box filters. Therefore, it is possible to compute the approximated convolution effectively for arbitrarily sized kernel utilizing the integral image.

$$\det\left(He_{approximated}\right) = D_{xx} D_{xy} - \left(wD_{xy}\right)^2 \tag{14}$$

The approximated and discrete kernels are referred to as D_{xx} for $I_{xx}(x,\sigma)$ and D_{xy} for $I_{xy}(x,\sigma)$ with weight w that approximate the kernels to calculate the determinant of the Hessian matrix.

FAST Detector

The detector known as Features from Accelerated Segment Test (FAST) is an algorithm proposed formerly by Rosten and Drummond (2006) for recognizing interest points in an image. This detector is considered as a corner detection method used to extract feature points and builds on the SUSAN detector. Afterward, FAST detector is used to track/ map objects in various computer vision applications. Its computational efficiency is considered the main advantage of the FAST corner detector. In fact, it is faster than many other feature detector methods.

In the FAST detector algorithm, the segment test criterion operates by considering a circle of sixteen pixels around the corner pixel "*A*" (Rosten and Drummond, 2006):

FAST Detector Algorithm

1. Choose a pixel *A* in the image that needed to be identified it as an interest point or not.
2. Let I_A being the intensity of this pixel.
3. Determine a threshold intensity value *T*.
4. Consider a Bresenham circle of radius 3, that has 16 pixels, to surround the pixel *A*.
5. Assume *Q* neighboring pixels out of the 16 should to decide if they are either above or below I_A by the value *T* (if the pixel needs to be detected as an interest point).
6. For fast algorithm:

First compare the intensity of pixels 1, 5, 9 and 13 of the circle with I_A. At least three of these four pixels should satisfy the threshold criterion.

Pixels that satisfy the threshold are considered as an interest point (exist).

If at least three of the four pixel values - I_1, I_5, I_9, I_{13} are not above or below ($I_A + T$), then *A* is not an interest point (corner). Consequently, reject the pixel *A* as a possible interest point.

Else if at least three of the pixels are above or below $I_A + T$,

Then check for all 16 pixels and check if 12 contiguous pixels fall in the criterion.

Repeat the procedure for all the pixels in the image.

End

As mentioned in (Rosten & Drummond, 2006), FAST detector reveals high performance. However, there are some weaknesses:

1. The high-speed test does not generalized well if the number of the adjacent pixels in the circle that are all brighter than the intensity was Q< 12.
2. The selection/ ordering of the fast test pixels contains the implicit hypothesis about the distribution of feature appearance.
3. The information from the first 4 tests is discarded.
4. Numerous features are detected adjacent to one another.
5. FAST is not robust to the presence of noise. Since, high speed is attained by analyzing the few pixels possible. Therefore, the FAST detector's ability to average out noise is reduced.

FEATURE DETECTOR APPLICATIONS

All feature detectors can be used in the same applications, however each detector provides better performance with certain applications. Consequently, the most common applications for each previously mentioned detector are as follows based related literatures.

Hessian Detector Applications

The Hessian detector method discovers the interest objects from a multi-scale image set, where the determinant of the Hessian matrix is at maxima.The Hessian matrix operator is computed using the convolution of the second-order partial derivative of the Gaussian to provide gradient maxima.

The DoH method analyzes the Gaussian partial derivatives very quickly using integral images. Therefore, the performance of the Hessian Matrix calculation is marvelous and its accuracy is better than many methods. Also, the related Hessian-Laplace method manages on local extrema using the determinant of the Hessian at several scales for spatial localization and the Laplacian at multiple scales for scale localization.

Hessian detector is used in the medical field by Gerig et al. (1993) who detected the local direction of the vessel as an extension of their nonlinear detector of 2D and 3D. They used the Hessian matrix to straightforwardly locate the optimal orientation in which to apply the non-linear filtering scheme corresponding to the eigenvector associated to the smallest eigenvalue. For vessel detection and direction estimation, the eigenvalue analysis of the Hessian matrix was first initiated in (Lorenz et al., 1997). According to the values of the eigenvalues of the local Hessian matrix, Frangi et al. (1998) classified the local structure of a 3D image in numerous basic shapes. Using eigen analysis of the Hessian matrix, Jacob and Unser extended a set of precise ridge detectors in the structure of steerable filters subsequent to Canny-like optimization.

A filter based on a linear combination of the Hessian matrix components was derived in vessels feature detection. The trace and determinant of the Hessian matrix were successfully used in scale and affine invariant extensions of interest point detectors when other feature properties became more important.

Azarbayejani et al. (1993) detected points where the image intensity has a large Hessian as features for live face detection. For each frame, a set of features are applied for tracking by gratifying the Hessian criterion, e.g., eyes, pupils, nostrils. Hessian-Affine feature detector and Gillie's key detector are used in biometric applications. These two detectors are employed to identify the feature key areas for embedding. This work proposed a feature-based robust digital image watermarking algorithm to attain the objective of image authentication and protection simultaneously. Depending on the number of feature points detected with and without attacks, a comparison was done between the two feature selections algorithm's performances. Both the algorithms established their efficiency in terms of watermark embedding/ extraction. The authors recommended that these feature detector algorithms can resist diverse image processing and geometric attacks.

Generally, the Hessian matrix base detector has good performance in both computation time and accuracy.

Moravec's Detector Applications

The Moravec operator is considered a corner detector; as it defines interest points like points where there are large intensity variations in all directions. This is the case at corners. Conversely, Moravec was not particularly interested in finding corners. Moravec suggested measuring the intensity variation by placing a small square window centered at a point, then moving this window with one pixel in every of the eight principle directions. The intensity variation for a specified shift is computed by taking the sum of squares of intensity differences of corresponding pixels in these two windows. Intensity variation at this point is the minimum intensity variation determined over the eight principle directions.

The Moravec detector was improved by Harris and Stephens (1988) to make it more repeatable for small image variations and near edges. By 1987, Förstner detector was used the auto-correlation function to categorize the pixels into groups (interest points, edges or region) (Förstner & Gülch, 1987). The detection and localization phases are split into the selection of windows and feature location within the selected windows. Additional statistics are executed locally to tolerate estimating automatically the thresholds for the classification. Compared to other detectors, this operator requires a complicated execution and is usually slower.

Harris Detector Applications

Harris detector can be used in the medical field. A new corner detection method for omnidirectional images was developed (Dey et al., 2011). The method was based on the standard Harris corner detector and a virtual spherical electrostatic model for edge detection. Experimental results affirmed the performance of the proposed approach against several image degradations and confirmed its robustness against noise. Wherever, two conducted sets of the omnidirectional images are

1. Gray level synthetic images,
2. Real catadioptric images.

Harris and Stephens showed that using an appropriate local auto-correlation response function, corners and edge detection can be straightforwardly matchable. Therefore, Harris can be used for Fingerprint Registration as a biometric application. Harris corner detector was applied on a preprocessed fused image using wavelet decomposition as in [69]. Where, Harris corner detector detects fingerprint images, retinal blood vessel and histogram equalized fused image which gave a very effective result. The number of the detected corners was stored in a database to be used for future in tracking or recognition of objects.

Lee et al. (2014) modified Harris corner detection for infrared images to reduce corner clustering, increase contrast, and easier edges detection. The authors use an adaptive radius suppression technique which reduced the corner clustering. In addition, it avoids the loss of valuable corners because of over-suppression and automates the selection of suppression radius. Moreover, corners detected by the proposed algorithm improved the accuracy of infrared image registration for chemotherapy assessment and early detection.

Djara et al. (2013) proposed two fingerprint registration algorithms based on Harris Detector and Minutiae matching for a Contactless Biometric system approach. The performances of the algorithms were evaluated using statistical mean and variance. Despite the fact that Harris feature points are easily matchable, the comparison have proved that the bifurcation point detector based algorithm perform better than the Harris detector. To detect thermal facial features Harris interest point detector was used in (Bhowmik et al., 2013). It is based on intensity and computes the presence of interest point directly from the gray values. The result demonstrated that Harris operator can detect a fine number of interest points even in illumination variation, rotation, and scaling condition.

SUSAN Detector Applications

Generally, SUSAN detects ramp and ridge edges as well as step edges. The ridge edges arise as two similar surfaces intersect in an image. The SUSAN detector recognizes the single dimensional features (edges), in addition to the two-dimensional features (corners) in an image.

Mauricio and Martinez (2004) used SUSAN algorithm to detect and extract facial features such as chin and cheeks border, eye corners/ center, nose corner, and mouth corners/center. Nevertheless, this technique necessitates the use of threshold which increases the detector sensitivity and may adversely influence the performance. The palm print biometric is used for personal authentication (Kumar, & Nagappan, 2012). A variety of feature extraction methods were discussed and compared; Forstner operator, SUSAN operator, Wavelet based salient point detection and Trajkovic and Hedley corner detector.

The face and its regions that contain facial features were extracted from the image in (Wu et al., 2001) using the integral projection method. This method constructed both the color information (the skin color and the hair color) and the edge information (the strength and the orientation). Applying SUSAN detector, the facial feature points were then detected from each region enclosed facial feature. The proposed system was verified to be efficient and robust. It can deal with the face images with a variety of translations, slightly rotations and changes of facial expressions.

SIFT Detector Applications

Lowe et al. (2004) employed the SIFT detector in the spherical coordinates for omnidirectional images. Results verified the promising and accurate performance of these algorithms for object recognition. Along with the authors work, the invariant features extracted from images can be used to execute consistent matching between diverse views of an object or scene. The approach was efficient on feature extraction and had the capacity to identify large numbers of features. Helmer et al. (2004) used SIFT features grouped using a probabilistic model initiated with a few parts of an object. The parts were learnt incrementally and additional possible parts were added to the model during the process of training. Then, the authors use the Expectation Maximization (EM) algorithm to update the model. The results show that this approach is time efficient and can be used with complicated models.

Mikolajczyk and Schmid (2005) found that the SIFT detector is the most effective as they yielded the best matching results compared to the performances of some other presently used local detectors. The SIFT improving detector developed targeted minimization of the computational time.

SURF Detector Applications

Badrinath et al. (2011) introduced an efficient Finger-knuckle-print (FKP) based Recognition System fusing SIFT and SURF. Resultant features of the enrolled and the query FKPs were matched using nearest-neighbor-ratio method. Then, the derived SIFT and SURF matching scores were fused using weighted sum rule. Facial marks are classically nearby in isolated blobs. Consequently, SURF was used (Choudhury & Mehata, 2012) to extract the facial marks emerged on the face such as cosmetic items.

A novel symmetrical SURF algorithm to improve the power of SURF to detect all possible symmetrical matching pairs through a mirroring transformation was proposed in work by Hsieh et al. (2014). A vehicle make and model recognition (MMR) application is assumed to prove the practicability and feasibility of the method for Vehicle Detection. To detect vehicles from the road, the proposed system was first applied to conclude the region of interest of each vehicle from the road without using any motion features.

FAST Detector Applications

Classical feature detectors such as Harris, SUSAN and SIFT, provided high quality features. Nevertheless, there detectors are too computationally exhaustive for real-time applications uses. Due to the computational properties of the FAST detector, it becomes widely used. However, FAST features do not have an orientation component. It deploys the intensity threshold between the center pixel and those in a circular ring about the center.

Rosten et al. (2006) introduced a new heuristic algorithm called FAST for fast feature detection using machine learning. The authors used the proposed algorithm to fully process live PAL video with less processing time. The experimental results proved that neither the Harris detector nor the SIFT detector can operate at full frame rate. This method showed a high level of repeatability for different kinds of feature compared to the classical detectors. However, it is not robust to high levels noise and depend on a threshold value. This generalized detector is optimized for repeatability with minute loss of efficiency, and performs a rigorous comparison of corner detectors based on the repeatability criterion applied to 3D scenes. The comparison reveals that using machine learning creates significant improvements in repeatability, yielding a detector that is both very fast and of very high quality.

The FAST detector was developed to be used in real time frame rate applications such as the SLAM on a mobile robot that have limited computational resources. The FAST corner detector is very suitable for real-time video processing application because of high-speed performance.

Based on the FAST detector in (Koelstra & Patras, 2009), interest region detectors were used. These Spatio temporal interest region detectors are used for event detection in video. Interest points can assist in recognizing local regions of interest that can be tracked over time. Sheu & Hu (1996) extracted edges as a chain code, achieves a polygonal approximation on the chains. Then, searches for the line segment intersections. Polygonal approximations, however, do not preserve many corners because they have a different goal in mind, simplifying curves by polygons.

Generally, due to the effectiveness of the feature detectors, Jego et al. (2012) proposed a low complexity interest point detector written in OpenCL language and executed on different desktop and embedded computing platforms to perform interest point detection in images. The proposed algorithm has been optimized on the targeted platforms.

RECENT FEATURE DETECTORS

Previously mentioned detectors are considered the classical feature detectors. However, due to the increased processing power of the standard computers, several different schemes for edge/ corner detection are identified in various literatures. All endeavor to discover more efficient, reliable and accurate detectors. Some of the recent advanced detectors are addressed as follows.

Robust Fuzzy Rule Corner Detector

For general purposes, fuzzy approaches have exclusively addressed for corner detection. Russo (1999) employed a local structure matrix to support a fuzzy corner detector. The authors established a continuous transient between the localized and not localized corner points to attain good performance method. The proposed algorithm used a fuzzy pre-filter to improve the image quality. The main drawback of this method compared to other classical algorithms such as the Harris method or SUSAN, it its expensive computing load. Conversely, Banerjee and Kundu (2008) suggested an algorithm to extract the significant gray level corner points. The authors used the fuzzy edge strength and the gradient direction to measure the cornerness in each point. By considering different threshold values from the fuzzy edge map, several corner fuzzy-sets are attained.

Adaptive and Generic Corner Detection Based on the Accelerated Segment Test

Mair et al. (2010) employed the accelerated segment test to significantly improve the FAST feature detector to be more generic with increased performance. The authors have suggested an optimal decision tree to yield an adaptive and generic accelerated segment test. This adaptive and generic accelerated segment test (AGAST) afforded high performance for any arbitrary environment and unlike FAST, it does not required adaptation to a specific scene structure.

The experimental results proved increased performance of the accelerated segment test by merging specialized decision trees. The algorithm was dynamically adjusts to an arbitrary scene which led to generic accelerated segment test. Thus, approach become most efficient corner detection algorithm. Moreover, no further need for any decision tree learning to adapt to an environment.

Miscellaneous Feature Detectors

An image mosaic method was proposed by Mahesh (2012) that used the FAST detector followed by the SURF to implement the matching feature points. The authors compared the proposed method to the other detectors such as Harris, SUSAN, SIFT. The results proved that the proposed method is very fast in detecting corners. Also, the time taken for mosaicing 12 images is approximately 8 seconds, which is superior than that obtained using the SIFT detector, which is taking more than 20 secons. This gives good results for the image rotation and image brightness variations. Especially FAST is preferable for video processing.

Alhamzi et al. (2014) proposed a novel corner detection algorithm CLDC. The authors used the Line Detection using Contours (LDC) algorithm that outputs the list of all detected line segments together with their endpoints. Each line segment is expanded in a post-processing step, and then the CLDC (Corners from LDC) detected the corners. The detected corners are linked through line segments that describe them. This algorithm was comparable in time complexity with other algorithms, while providing more information about the line segments in the image. The results proved that, CLDC is robust to image transformations, such as rotation and translations. Figure 1 provides an example of a real house

Figure 1. Various corner detectors with real images
Mokhtarian & Suomela, 1998.

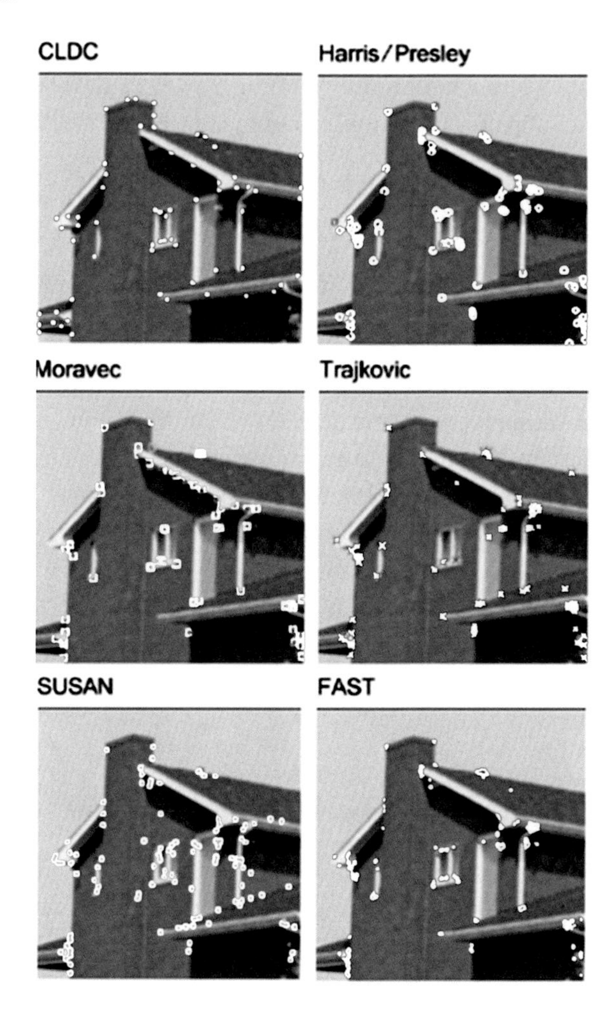

image with the corners detected using various detector method. It illustrates that all methods except the proposed method CLDC have too many corners detected on the camera in the image which are in close proximity to each other, and apparently missing corners along the man's borders and in the background. The house image also has regions saturated with corners, yet some important corners are missing, by all methods except the CLDC.

DISCUSSION

Real-time processing of a video stream attracts the focus of researches to develop efficient methods to deal with large amounts of data and moving targets. Feature detection is considered the first step in the image/video processing chain, followed by matching, tracking and object recognition.

Numerous computer vision techniques for extracting features and perceiving image content are anchored at detecting points of interest. Interest points have numerous local information content and must be superlatively repeatable between different images. An interesting point in an image is known as a pixel that has a well-defined location and need to be robustly detected. Interest point detection has applications in image matching, object recognition, tracking, etc. Based on the extracted image structure type, FDs can be categorized into:

1. Corners,
2. Blobs, or
3. Regions.

According to the image content, several image structures are more widespread than others. Thus, the detected features' number is varied for the same detector, according to the different image dataset used.

Human being can effortlessly identify a corner, but a mathematical detection is compulsory in case of algorithms. Corner detection algorithms facilitate a speedy association between interest points among diverse images of the same object. Where, motion is ambiguous at an edge, but definitely at a corner. Therefore, shapes can be roughly reconstructed from their corners. Detecting corners in images, as exacting points of interest that are frequently used in different domains such as:

- Object detection/recognition,
- Stereo matching, motion tracking,
- Panoramic photographs,
- Image database retrieval, and
- Robot navigation.

According to the detector type used, the corners differentiate will changes consequently. In images, the corners can be described as the intersection of two or more edges. In addition, a corner can be identified as:

- The local minimum/ maximum intensity of a point,
- The point where two central and dissimilar edge directions in a local neighborhood of the point exist,
- Line ending or point on a curve where the curvature is locally maximal.

The efficiency of a corner detector is significant as it determines whether the detector merged with a supplementary processing can operate at an acceptable rate. The quality of a corner detector is often adjudicated based on its capability to detect the same corner in multiple images under different conditions, such as dissimilar translation, lighting, rotation, and other transforms. The Computational time required to detect corners is also imperative, and faster algorithms are desired.

The main aim of this chapter is to introduce detailed survey of a feature detector/ descriptor. There are several well established algorithms for corner detection. These algorithms are almost categorized into corner detects

1. From grayscale images, and
2. From digital curves.

Corner detectors from Grayscale image based algorithms typically detect interest points other than corners. They generally search for lightness and brightness contrasts in images, locate local maxima in discriminate geometry operators, or calculate the determinant of the Hessian matrix and second order derivatives.

The majority trendy algorithms for corner detectors as mentioned most of them through this work. One of the first suggested corner detector was Moravec, who employed it in a stereo matching application. Then, Harris and Stephens recovered its repeatability, addressed the problem of poorly localized points along the edges. They used their original detector in an efficient structure from motion algorithm. Other corner detectors are Beaudet, SUSAN by Smith and Brady (1997), Curvature Scale Space by Mokhtarian and Suomela, (1998), and FAST detector by Rosten et al.(2006). Harris and Stephens in 1988 recovered upon Moravec's corner detector by considering the account the differential of the corner score with respect to direction directly, in place of using shifted patches. Where, Moravec considered only shifts in discrete 45 degree angle, while Harris employed all directions. Harris detector has verified to be more accurate in distinguishing between edges and corners using a circular Gaussian window to reduce noise.

The key step in numerous image processing and computer vision applications is the detecting interest points, means to use corner detectors, to find analogous points across multiple images. Some of the mainly remarkable applications are:

1. Image registration especially in medical imaging,
2. Stereo matching,
3. Biometric identification and tracking applications,
4. Stitching of panoramic photographs,
5. Object detection/recognition,
6. Motion tracking,
7. Robot navigation.

Based on this study a comparison is done between the different detectors mentioned through it with respect to the principal, advantages and drawbacks for each type, as illustrated in Table 1.

Generally, various studies were done to describe and compare between the FDs performance and applications.

Performance Evaluation of the Detectors and Descriptors

Local feature detectors/descriptors performance evaluation is a challenging task. Since, the local feature detectors are based on the intensity changes of pixels in the digital images and the detected features are blobs, corners or regions depending on the used detector technique. Therefore, the performance evaluation can be used to measure how well the same local features could be found in the same spatial locations despite a deformation. It measures the detector's ability to manage the deformations that exist in the real world images.

Table 1. Comparison between the different common feature detectors/descriptors

Edge Detector	Main Principle	Advantages/Uses	Limitations/ Drawbacks
Hessian Detector	• The corners in the image are determined by the local maxima of the corner strength. • The determinant is related to the Gaussian curvature of the image. • Computed the corner strength as the determinant of the Hessian matrix.	• Hessian matrix has good performance and accuracy. • It is invariant to rotation. • Used by the medical and biometric field applications.	
Moravec's Detector	• Give a measure of corners to each pixel in the image. • Corners are local maxima, except those with corners value below a threshold.	It is computationally efficient.	• It is not rotationally invariant. • It is susceptible to reporting false corners along edges and at isolated pixels. • It is sensitive to noise. • It is difficult to find corners in diagonal edges. • It is not isotropic (If an edge is present that is not in the direction of the neighbours, and then it will not be detected as an interest point).
SUSAN Detector	• The pre-determined mask is centered on each pixel of the image. • Its brightness compare to the brightness of each pixel (nucleus) of the mask. • The SUSAN corner detector is recognized by a circular mask. • Then an area of the mask "USAN" can be described which has similar brightness as the nucleus. • Consistent with the area of the USAN, corners can be detected	• No image derivatives are used. • No noise reduction is required.	• It requires the use of threshold, which increase the detector sensitivity and may adversely influence the performance. • A fixed global threshold is not suitable for a general situation. • The corner detector needs an improved adaptive threshold and the shape of the mask can be improved. • The anti-noise ability is weak. • The robustness is not good enough.
SIFT Detector	• It transforms the image into a group of local feature vectors (any scaling, rotation or translation). • Each of these feature vectors is presumed to be distinctive and invariant. • These features can be used to detect distinctive objects in various images.	• Robust to the variations corresponding to typical viewing conditions. • SIFT is good performance compare to other descriptors.	Has strong robustness.

continued on next page

Table 1. Continued

Edge Detector	Main Principle	Advantages/Uses	Limitations/ Drawbacks
Harris /Plessey Detector	• Allows an estimate of the intensity variation calculations in any direction. • Generalize Moravec's operator. • Corner point is determined by the variation of gray value in a small window whose size is determined by the actual situation. • A point is not perceived if the gray value changes in both x and y direction. • The gray value of the edge area varies only in x or y direction, but not both. • The value of smooth area will not be changed in either x nor y direction. • Second image derivatives are convoluted by the Gaussian window. • The resulting matrix contains all the differential operators describing the geometry of the image surface at a given point. • The Eigenvalues of are proportional to the principle curvatures of the image surface and form a rotationally invariant description. • The corners measure is defined via the determinant and trace.	• It is widely used in practice. • On a whole, Harris algorithm is superior to SUSAN algorithm. • Considered the most widely used corner detection algorithm based on intensity. • *It has a good performance on its stability and robustness. • Used in the medical and biometric applications. • It is a stable and robust method.	• It is not invariant to scale and correlation. • It is not rotation invariant. • It requires high computational demand. • It is sensitive to noise. • Has poor localization on many junction types. • For more improvement it needs to choose difference operators and better Gaussian smoothing filter operators. • Limited in the computing speed.
SURF Detector	• Use a circular mask for corner detection, without using derivatives. • It computes the fraction of pixels within a neighborhood which have similar intensity of the center pixel. • Corners can then be localized by thresholding this measure and selecting local minima. • The position of the center of gravity is used to filter out false positives.	Useful for various vision-based applications.	Unable to detect symmetrical objects.
Fast Detector	• A new heuristic algorithm provides fast feature detection using machine learning. • Its basic idea to reduce the number of calculations which are necessary at each pixel in order to decide whether a keypoint is detected at the pixel or not. • This is via placing a circle consisting of 16 pixels centered at the pixel under investigation. • For the corner test, only gray value differences between each of the 16 circle pixels and the center pixel are evaluated.	• Has few losses of efficiency. • Has significant improvements in repeatability. • It is computationally efficient. • Very fast and of very high quality. • It is considered as the only detector that is capable of video rate. • FAST is preferable for video processing.	• Do not have an orientation component. • It is not very robust to the presence of noise. • The FAST algorithm has several weak points as mentioned previously.

Typically, the features evaluation can be performed either at the image-level or using a system-level evaluation framework (Philbin et al., 2007). The first approach is founded only on a few numbers of images. It has the ability to compare several features in a short period of time, independent of a any specific application. However, it cannot reveal the actual performance. The most imperative criteria for evaluating detectors and descriptors are the repeatability and the matching score, respectively. On the other hand, the second approach namely the System-level feature evaluation (Miksik & Mikolajczyk,

2012), involves a large number of images and is performed for just one specific feature or one specific application.

Commonly, the repeatability is the most significant measure for detectors comparison. It can be defined as the frequency with which keypoints detected in one image are found within (x= 1.5) pixels of the corresponding location in a transformed image. It can be defined in terms of the correspondences as:

$$\frac{\#\,of\ \ correspondences}{\min\left(\#of\ \ regions\ \ in\ \ image\ \ A, \#\,of\ \ regions\ \ in\ \ image\ \ B\right)} \cdot 100\% \tag{15}$$

As only regions present in both images are to be included for the region count. Features in both images are projected to each other using a known homography.

While, after detecting the features (keypoints), it is required to match them, i.e., determine which features come from corresponding locations in different images, this is performed via feature descriptors.

Once features and their descriptors are extracted from two or more images, the next step is to establish preliminary feature matches between these images that depends partially on the application. Different strategies may be preferable for matching images that are known to overlap versus images that may have no correspondence whatsoever. Therefore, a matching strategy must be selected that determines which correspondences to be used further, where the simplest matching strategy isto set a threshold value. Generally, the performance of the different combinations of detectors and descriptors can be evaluated on a feature matching problem.

Mikolajczyk and Schmid presented examples of using these performance evaluation methods.The authors examined planar scenes via assessing the repeatability of the Harris-Laplace detector using the evaluation technique. The results proved that Harris-Laplace points surpass both DoG points and Harris points in repeatability.

Alhamzi and Elmogy (2014) discussed the current computer vision literature on 3D object recognition. The authors presented particular challenges in 3D object recognition approach based on the depth map. Results suggested that, the conventional local feature based object recognition methods, such as SIFT, SURF, and ORB, that are used to retrieve the strong features of the object. The Line Detection using Contours (LDC) method used was compared to other classical methods.

Figure 1 compares different corner detectors response with a real image. The figure establishes that:

1. The SAUSAN detector detected much more number of edges, thus it is considered the worst detector,
2. The Fast and Trajkovic methods are superior to all other detectors, except the CLDC, while
3. Harris and Moravec are approximately has the same performance.

Object Recognition Application

Generally, object recognition refers to finding out to the class an object in a given image is belongs to. For example, an object in two different scenes can be recognized. Lowe (1999) used the SIFT descriptors for object recognition. However, the use of local features is realistic only when the objects have at least some textures to form distinctive descriptions. Another example of based on feature descriptions and locations to calculate a signature for a class using local features is as intermediate level of representation in an object class recognition as a very compact representation of an image can be formed. Clustering can be employed to form classes from the training data. After the system training, new samples can be classified using trained clusters.

Various Detectors for Video Recognition Systems

The presented smart feature detection system has low complexity SIFT based key point extraction algorithm and its hardware engine. In addition, it established key point of interest detection algorithm based on spatio-temporal feature. This contribution will increase the potential for a cloud based video recognition system. Its advantages are reduction of:

- Descriptor data communicated with cloud Systems,
- Computational complexity of descriptor calculations,
- Data amount of key points.

From the comparative analysis of various recognition systems in different publications to the proposed algorithm that achieved about 95% reduction of key points and 53% reduction of the computational complexity, the results shown that:

1. **Using Cloud Computing and SIFT Detector:** Provided average classification rate of 84%.
2. **Using SIFT with Hough Transform:** Afforded recognition accuracy for 3D objects rotated in depth by 20 degrees increased from 35% for correlation of gradients to 94%.
3. **Using Harris Corner Detector:** Gave 50% accuracy.
4. **Using Moravec Corner Detector:** Provided only 20% accuracy.

Consequently, FDs and their descriptors are the basic blocks of many computer vision algorithms. Their applications include object detection and classification, image registration, tracking, and motion estimation as well as are used in various domains. Using the FD methods enable better handle for the scale changes, rotation, and occlusion. Such FDs are the FAST, Harris, Hessian, and Shi & Tomasi methods for detecting corner features, and the SURF and maximally stable extremal regions (MSER) methods for detecting blob features. While, the SURF, Fast Retina Keypoint (FREAK), Binary Robust Invariant Scalable Keypoints (BRISK),Scale-Invariant Corner Keypoints (SICK) and Histograms of Oriented Gradient (HOG) descriptors. Therefore, depending on the application requirements, the detectors and the descriptors are to be matched.

Feature Detector and Descriptor for Objects/Face Tracking

Suitable features selection play a vital role in video object tracking. Feature detectors/descriptors can be used for tracking applications due to their high performance. Nevertheless, using only one of the feature detectors for object tracking may lead to inadequate accuracy due to different challenges in tracking, such as:

- Non-rigid object structure,
- Abrupt change in object motion,
- Occlusions in the scene,

- Change in appearance of object, and
- Camera motion.

Patel et al. (2014) measured the tracking speed and accuracy of several feature detectors in real time video for face tracking. The authors used several parameters; namely average detection time of key-point, average number of detected key points and frame per second. The experimental results depicted that in low light condition, the number of detected keypoints and matches were decreased. The SURF detector acheived the lowest distance deviation and mean so it is accurate. But it takes almost double time than other detectors. The authors concluded that combined feature detectors can provide improved objects tracking accuracy.

Feature Detectors for Video Shot Boundary Detection

Video content within a shot is tend to be continuous, due to the continuity of both the the parameters (motion, zoom, focus) of the camera images and the physical scene. Thus, the most essential temporal video segmentation task is shot boundary detection. Li et al. (2010) used local feature (SIFT) based shot boundary detection approach which utilized support vector machine (SVM) to decide the frame similarities. From the preceding sections, it is clear that feature is defined as a descriptive parameter which is extracted from an image or a video sequence. Video content analysis effectiveness depends on the efficiency of features/attributes used for the representation of the content. Features can be classified into:

1. Low-level features (primitive features), such as color, shape, texture, spatial location of image elements and object motion (in video), which can be extracted automatically. Nevertheless, these features are meaningful from the human perception point of view.
2. High-level features (semantic, logical features) which can describe the physical objects in images and action in video, or can concerne with abstract attributes.

Thus, efficient video content analysis can be accomplished by collaboratively using both low-level and high-level features. Low-level features can be used to segment a video sequence into individual shots as well as to generate representative key frames for each shot.

There are different types for FDs and descriptors which can be used efficiently in various applications. However, using a combination of them can realize more efficient approaches.

CONCLUSION

The main concept behind the computer vision and perception is to obtain information about the world and create sense of it. Where, vision is like hearing as it is considered a distance sense, evolved to sense objects without direct contact. Visual features consist of spots and edges, shapes and colors, movements and textures. These are all attributes that are not in themselves objects; however, in combination they can define the observed objects.

This study presents a huge survey on the FDs and their various applications of discrete scale, rotation and illumination features. An overview of the literature pointed out some of the major challenges facing the community and stressed some of the characteristic approaches attempted for selecting the appropriate detector cording to the application was presented. This work started with defining the properties of the feature detectors and their types as well as describing a simple, fast, relatively parameterless and robust algorithm for detecting corners in an image. This is followed by an overview of the literature over the past decades organized in different categories of feature detection methods with detailed analysis of a selection of methods.

Moravec, Hessian, Harris, SUSAN, SIFT, SURF and FAST are the detectors/ descriptors, where this study focuses on as they are the most common detectors. Each detector has different concepts and performed well with different applications not like the other detectors. For example, SUSAN and Harris corner detection algorithms that are both based on intensity were discussed in details. The FAST technique uses the intensities of surrounding pixels to detect features. It shows how machine learning can be used to improve even further the speed of the detector. The FAST corner detector is considered an appealing feature detector due to its efficiency and faster performance and less intermediate memory requirements.

Various applications are discussed in details through the discussion section, which includes also a comparison between different detectors this had a particularly significant impact on the research field.

The addressed advanced feature detectors indicated that combining a novel heuristic for feature detection with machine learning methods provided a robust, high-speed corner detection algorithm that can be use in real-time image processing applications.

REFERENCES

Alhamzi, K., Elmogy, M., & Barakat, S. (2014). 3D Object Recognition Based on Image Features: A Survey. *International Journal of Computer and Information Technology*, *3*(3).

Alhwarin, F., Wang, C., Ristic-Durrant, D., & Gräser, A. (2008, September). Improved SIFT-Features Matching for Object Recognition. In *British Computer Society International Academic Conference* (pp. 178-190).

Azarbayejani, A., Starner, T., Horowitz, B., & Pentland, A. (1993). Visually controlled graphics. *IEEE Transactions on Pattern Analysis and Machine Intelligence*, *15*(6), 602–605. doi:10.1109/34.216730

Badrinath, G. S., Nigam, A., & Gupta, P. (2011, November). An efficient finger-knuckle-print based recognition system fusing sift and surf matching scores. In *International Conference on Information and Communications Security* (pp. 374-387). Springer Berlin Heidelberg. doi:10.1007/978-3-642-25243-3_30

Banerjee, M., & Kundu, M. K. (2008). Handling of impreciseness in gray level corner detection using fuzzy set theoretic approach. *Applied Soft Computing*, *8*(4), 1680–1691. doi:10.1016/j.asoc.2007.09.001

Bay, H., Ess, A., Tuytelaars, T., & Van Gool, L. (2008). Speeded-up robust features (SURF). *Computer Vision and Image Understanding*, *110*(3), 346–359. doi:10.1016/j.cviu.2007.09.014

Beaudet, P. R. (1978, November). Rotationally invariant image operators. In *International Joint Conference on Pattern Recognition* (Vol. 579, p. 583).

Bhowmik, M. K., Shil, S., & Saha, P. (2013). Feature Points Extraction of Thermal Face Using Harris Interest Point Detection. *Procedia Technology*, *10*, 724–730. doi:10.1016/j.protcy.2013.12.415

Bloomenthal, J. (1983, July). Edge inference with applications to antialiasing. *Computer Graphics*, *17*(3), 157–162. doi:10.1145/964967.801145

Cagnoni, S., Mordonini, M., & Sartori, J. (2007, April). Particle swarm optimization for object detection and segmentation. In *Workshops on Applications of Evolutionary Computation* (pp. 241–250). Springer Berlin Heidelberg. doi:10.1007/978-3-540-71805-5_27

Choudhury, Z. H., & Mehata, K. M. (2012). Robust facial Marks detection method Using AAM and SURF. *International Journal Engineering Research and Applications*, *2*(6), 708–715.

Dey, N., Das, S., & Rakshit, P. (2011). A novel approach of obtaining features using wavelet based image fusion and Harris corner detection. *Int J Mod Eng Res*, *1*(2), 396–399.

Dey, N., Nandi, P., Barman, N., Das, D., & Chakraborty, S. (2012). A comparative study between Moravec and Harris corner detection of noisy images using adaptive wavelet thresholding technique. *arXiv preprint arXiv:1209.1558.*

Djara, T., Assogba, M. K., Naït-Ali, A., & Vianou, A. (2013). Comparison of Harris Detector and ridge bifurcation points in the process of fingerprint registration using supervised contactless biometric system. *International Journal of Innovative Technology and Exploring Engineering*, *2*(6).

Förstner, W., & Gülch, E. (1987, June). A fast operator for detection and precise location of distinct points, corners and centres of circular features. In *Proceedings of International Society for Photogrammetry and Remote Sensing intercommission conference on fast processing of photogrammetric data* (pp. 281-305).

Frangi, A. F., Niessen, W. J., Vincken, K. L., & Viergever, M. A. (1998, October). Multiscale vessel enhancement filtering. In *International Conference on Medical Image Computing and Computer-Assisted Intervention* (pp. 130-137). Springer Berlin Heidelberg.

Gerig, G., Koller, T., Székely, G., Brechbühler, C., & Kübler, O. (1993, June). Symbolic description of 3-D structures applied to cerebral vessel tree obtained from MR angiography volume data. In *Biennial International Conference on Information Processing in Medical Imaging* (pp. 94-111). Springer Berlin Heidelberg. doi:10.1007/BFb0013783

Hardie, R. C., & Boncelet, C. G. (1994). Gradient-based edge detection using nonlinear edge enhancing prefilters. *IEEE Transactions on Image Processing, 4*(11), 1572-1577.

Helmer, S., & Lowe, D. G. (2004, June). Object class recognition with many local features. In *Computer Vision and Pattern Recognition Workshop, 2004. CVPRW'04. Conference on* (pp. 187-187). IEEE. doi:10.1109/CVPR.2004.409

Hess, M., & Martinez, G. (2004, December). Facial feature extraction based on the smallest univalue segment assimilating nucleus (susan) algorithm. In *Proceedings of Picture Coding Symposium* (Vol. 1, pp. 261-266).

Hsieh, J. W., Chen, L. C., & Chen, D. Y. (2014). Symmetrical surf and its applications to vehicle detection and vehicle make and model recognition. *IEEE Transactions on Intelligent Transportation Systems*, *15*(1), 6–20. doi:10.1109/TITS.2013.2294646

Jego, B., Robart, M., Saha, K., & Pau, D. P. (2012, September). FAST detector on many-core computers. In *Consumer Electronics-Berlin (ICCE-Berlin), 2012 IEEE International Conference on* (pp. 263-266). IEEE. doi:10.1109/ICCE-Berlin.2012.6336453

Kim, S. (2012). Robust corner detection by image-based direct curvature field estimation for mobile robot navigation. *International Journal of Advanced Robotic Systems*, 9.

Koelstra, S., & Patras, I. (2009, May). The FAST-3D spatio-temporal interest region detector. In *2009 10th Workshop on Image Analysis for Multimedia Interactive Services* (pp. 242-245). IEEE. doi:10.1109/WIAMIS.2009.5031478

Kumar, M., & Saxena, R. (2013). Algorithm and technique on various edge detection: A survey. *Signal & Image Processing*, *4*(3), 65.

Kumar, V., & Nagappan, A. (2012). *Study and comparison of various point based feature extraction methods in palmprint authentication system*. Editorial Committees.

Lee, C. Y., Wang, H. J., Chen, C. M., Chuang, C. C., Chang, Y. C., & Chou, N. S. (2014). A modified harris corner detection for breast ir Image. *Mathematical Problems in Engineering*.

Li, J., Ding, Y., Shi, Y., & Li, W. (2010). A divide-and-rule scheme for shot boundary detection based on SIFT. *JDCTA*, *4*(3), 202–214. doi:10.4156/jdcta.vol4.issue3.20

Lorenz, C., Carlsen, I. C., Buzug, T. M., Fassnacht, C., & Weese, J. (1997). Multi-scale line segmentation with automatic estimation of width, contrast and tangential direction in 2D and 3D medical images. In CVRMed-MRCAS'97 (pp. 233-242). Springer Berlin Heidelberg.

Lowe, D. G. (1999). Object recognition from local scale-invariant features. In *Computer vision, 1999. The proceedings of the seventh IEEE international conference on* (Vol. 2, pp. 1150-1157). IEEE. doi:10.1109/ICCV.1999.790410

Lowe, D. G. (2004). Distinctive image features from scale-invariant keypoints. *International Journal of Computer Vision*, *60*(2), 91–110. doi:10.1023/B:VISI.0000029664.99615.94

Mahesh, S. M. N. (2012). Image mosaic using fast corner detection. *International Journal of Advanced Research in Electronics and Communication Engineering*, *1*(6).

Maini, R., & Aggarwal, H. (2009). Study and comparison of various image edge detection techniques. *International Journal of Image Processing,3*(1), 1-11.

Mair, E., Hager, G. D., Burschka, D., Suppa, M., & Hirzinger, G. (2010, September). Adaptive and generic corner detection based on the accelerated segment test. In *European conference on Computer vision* (pp. 183-196). Springer Berlin Heidelberg. doi:10.1007/978-3-642-15552-9_14

Mikolajczyk, K., & Schmid, C. (2001). Indexing based on scale invariant interest points. In *Computer Vision, 2001. ICCV 2001. Proceedings. Eighth IEEE International Conference on* (Vol. 1, pp. 525-531). IEEE. doi:10.1109/ICCV.2001.937561

Mikolajczyk, K., & Schmid, C. (2004). Scale & affine invariant interest point detectors. *International Journal of Computer Vision*, *60*(1), 63–86. doi:10.1023/B:VISI.0000027790.02288.f2

Mikolajczyk, K., & Schmid, C. (2005). A performance evaluation of local descriptors. *IEEE Transactions on Pattern Analysis and Machine Intelligence*, *27*(10), 1615–1630. doi:10.1109/TPAMI.2005.188 PMID:16237996

Miksik, O., & Mikolajczyk, K. (2012, November). Evaluation of local detectors and descriptors for fast feature matching. In *International Conference on Pattern Recognition (ICPR), 2012 21st International Conference on* (pp. 2681-2684). IEEE.

Mokhtarian, F., & Suomela, R. (1998). Robust image corner detection through curvature scale space. *IEEE Transactions on Pattern Analysis and Machine Intelligence*, *20*(12), 1376–1381. doi:10.1109/34.735812

Morgan, B. (2010). Interest point detection for reconstruction in high granularity tracking detectors. *Journal of Instrumentation*, *5*(07), P07006. doi:10.1088/1748-0221/5/07/P07006

Panchal, P. M., Panchal, S. R., & Shah, S. K. (2013). A comparison of SIFT and SURF. *International Journal of Innovative Research in Computer and Communication Engineering*, *1*(2), 323–327.

Patel, A., Kasat, D. R., Jain, S., & Thakare, V. M. (2014). Performance analysis of various feature detector and descriptor for real-time video based face tracking. *International Journal of Computers and Applications*, *93*(1).

Philbin, J., Chum, O., Isard, M., Sivic, J., & Zisserman, A. (2007, June). Object retrieval with large vocabularies and fast spatial matching. In *2007 IEEE Conference on Computer Vision and Pattern Recognition* (pp. 1-8). IEEE. doi:10.1109/CVPR.2007.383172

Rosenfeld, A., & Thurston, M. (1971). Edge and curve detection for visual scene analysis. *IEEE Transactions on Computers*, *100*(5), 562–569. doi:10.1109/T-C.1971.223290

Rosten, E., & Drummond, T. (2006, May). Machine learning for high-speed corner detection. In *European Conference on Computer Vision* (pp. 430-443). Springer Berlin Heidelberg.

Russo, F. (1999). FIRE operators for image processing. *Fuzzy Sets and Systems*, *103*(2), 265–275. doi:10.1016/S0165-0114(98)00226-7

Sheu, H. T., & Hu, W. C. (1996). A rotationally invariant two-phase scheme for corner detection. *Pattern Recognition*, *29*(5), 819–828. doi:10.1016/0031-3203(95)00121-2

Shin, J., & Kim, D. (2014). Hybrid approach for facial feature detection and tracking under occlusion. *IEEE Signal Processing Letters*, *21*(12), 1486–1490. doi:10.1109/LSP.2014.2338911

Smith, S. M., & Brady, J. M. (1997). SUSAN—A new approach to low level image processing. *International Journal of Computer Vision*, *23*(1), 45–78. doi:10.1023/A:1007963824710

Tien, M. C., Wang, Y. T., Chou, C. W., Hsieh, K. Y., Chu, W. T., & Wu, J. L. (2008, June). Event detection in tennis matches based on video data mining. In *2008 IEEE International Conference on Multimedia and Expo* (pp. 1477-1480). doi:10.1109/ICME.2008.4607725

Tuytelaars, T., & Mikolajczyk, K. (2008). Local invariant feature detectors: A survey. *Foundations and Trends in Computer Graphics and Vision*, *3*(3), 177–280. doi:10.1561/0600000017

Tuytelaars, T., & Van Gool, L. (2004). Matching widely separated views based on affine invariant regions. *International Journal of Computer Vision*, *59*(1), 61–85. doi:10.1023/B:VISI.0000020671.28016.e8

Wu, H., Inada, J., Shioyama, T., Chen, Q., & Simada, T. (2001, June). Automatic facial feature points detection with susan operator. In *Proceedings of the Scandinavian Conference on Image Analysis* (pp. 257-263).

KEY TERMS AND DEFINITIONS

Descriptor: A vector of values, which somehow describes the image patch around an interest point. It could be as simple as the raw pixel values, or it could be more complicated, such as a histogram of gradient orientations.

Feature Extraction: Extract, find out, match the feature or object in one frame of the video. Feature extraction means to find out the "point of interest" or differentiating frames of video. These features are consistent over several video frames of the same scene and after the video scene is changed the value of these features are changed. therefore we can use features to classifies videos.

Feature Tracking: After extracting the feature, we track the same feature or same object in every frames.

Feature: In this context usually means an interest point, also known as key point, interest region, etc. In most cases, the interest point detector tends to extract interest points from edges and corners because this kind of points are more "characteristic" than others. That means, these points can be more reliably tracked over multiple frames.

Harris, Min Eigen, and FAST: Interest point detectors, or more specifically, corner detectors.

Interest Point (Key Point, Salient Point) Detector: An algorithm that chooses points from an image based on some criterion. Typically, an interest point is a local maximum of some function, such as a "cornerness" metric.

Local Feature: Together an interest point and its descriptor is usually called a local feature. Local features are used for many computer vision tasks, such as image registration, 3D reconstruction, object detection, and object recognition.

SIFT: Includes both a detector and a descriptor. The detector is based on the difference-of-Gaussians (DoG), which is an approximation of the Laplacian. The DoG detector detects centers of blob-like structures. The SIFT descriptor is a based on a histogram of gradient orientations.

Chapter 4
Analysis of Different Feature Description Algorithm in object Recognition

Sirshendu Hore
HETC, India

Shouvik Chakraborty
University of Kalyani, India

Sankhadeep Chatterjee
University of Calcutta, India

Rahul Kumar Shaw
HETC, India

ABSTRACT

Object recognition can be done based on local feature description algorithm or through global feature description algorithm. Both types of these descriptors have the efficiency in recognizing an object quickly and accurately. The proposed work judges their performance in different circumstances such as rotational effect scaling effect, illumination effect and blurring effect. Authors also investigate the speed of each algorithm in different situations. The experimental result shows that each one has some advantages as well as some drawbacks. SIFT (Scale Invariant Feature Transformation) and SURF (Speeded Up Robust Features) performs relatively better under scale and rotation change. MSER (Maximally stable extremal regions) performs better under scale change, MinEigen in affine change and illumination change while FAST (Feature from Accelerated segment test) and SURF consume less time.

INTRODUCTION

Digital image processing makes the use of different algorithms to accomplish image processing on digital images. The object detection and extraction of feature from the object plays a vital role in case of digital image processing. To obtain some useful information from different digital media such as photo, video or any form of multimedia content digital Image processing relies heavily on feature extraction and object detection and subsequent object recognition. Successful and efficient object recognition is an important research domain in computer vision and image processing. Though object recognition has started it journey four decade back, it has started making its acceptance rapidly in recent years due to the advances in computational intelligence. It is also influenced by the advancement made in the field

DOI: 10.4018/978-1-5225-1025-3.ch004

of feature extraction techniques Object recognition is the process of determining the distinctiveness of an object being perceived in the image. This is often done using a set of known labels. Significant effort has been made earlier to develop some generic made to overcome the challenges often encountered in the case of object recognition. Recognition of object in cognitive way is much easier then recognizing the same object through computer vision or image processing. Pose of an object relative to a camera, variation in lighting under different condition, and difficulty in generalizing across objects from a set of images causes much difficulties in object recognition process. In the literature different way of recognizing an object is reported.

Feature

The concept of feature is very common and choice of feature to be obtained has been depended heavily on given specific problem. Thus there is no complete or precise definition of what make up a feature and the accurate definition often relay on the given problem; the application type In image processing or in computer vision a portion of information which is substantial for resolving the computational job associated with a certain application, can be coined as feature. On the other words a feature can be defined as an interesting section with in an image; interest points in an image are the area whose position does matter and can be detected under different changing circumstances. The noticeable part found within the interest points is that they store large number of local information and the information they store are rather same between different images; therefore many computer vision algorithms used features as a starting point. This concept of feature in general is also applicable in case of machine learning or pattern recognition . Features of digital image may have some precise structures in the image such as objects, points or edges. Features of any image can also be obtained by applying neighborhood operation in the adjoin region. In case of tracking a motion object it can be sequence of image, boundaries or curves of different image regions.

Feature Detector

In digital Image processing and in computer vision to find the interested point or key point of an object feature detector plays a crucial role. A feature detector is nothing but an algorithm that takes an image as input and produce pixel coordinates or locations as outputs. These locations are significant areas of the targeted image or inputted image. An example of feature detector is a corner detector, template detector, blob etc. that produce locations of corners or templates or regions as an outputs in the observed image but does not provide user any type of information about the features it detected.

Feature Descriptor

It has been observed that detecting interested points or key points in the inputted image most of the researchers in the field of digital image processing or in computer vision uses feature descriptor. Like a feature detector in case of feature descriptor an algorithm is used to take an image as input and produce feature descriptors/feature vectors as outputs. In Feature descriptors interesting information are encoded into a chains of numbers and used as a kind of numerical "template". This template can be used to discriminate one feature from another. Unlike feature detector s information stored inside a feature descriptors are invariant under image transformation. Therefore we can retrieve the feature again even though the image is transformed in some way or other. An example of feature descriptor are SIFT, HOG, SURF etc. In each such cases information related to the local neighborhood image gradients are encode inside the feature descriptors. Since many of the work related to object recognition are based on feature descriptors, therefore there is a huge growth of very large number of feature descriptor. Feature descriptors are divided based on the computational complexity and the repeatability. The function of a descriptor is to give an exclusive as well as vital description of a feature; usually descriptors are generated based on the region adjacent to an interest point. Much work has been done towards this direction to improve the performance of object recognition techniques, and many researchers have proposed different types of algorithms to meet the challenge (Alnihoud, 2008; Rossion et al., 2004; Ravela, 2003; Ullman, 1989).The object recognition algorithms can be broadly divided into two categories (Birinci, Diaz-de-Maria & Abdollahian, 2011).

- Global feature based descriptor algorithm,
- Local feature based descriptor algorithm.

Both types of these descriptors have the efficiency in recognizing an object quickly and accurately. A local feature descriptor detects a feature with in a digital image is considered to be an image pattern that varies from its immediate neighborhood. It is typically related with a change of an image property or several properties at the same time, although it is not inevitably confined exactly on this change. The image belongings normally considered are intensity, color, and texture. Whereas a Global feature descriptor detects a feature with in a whole image is considered to be an image pattern comparing with global feature based descriptor matching algorithms; local feature-based descriptor matching algorithms are more stable. One can use a local feature based descriptor successfully in many real-world applications; they can be applied in recognition of texture, localization of robot, retrieval of image, in data and video mining, panoramas building, and in object category recognition (Lowe, 1999; Mikulka et al., 2012; Moravec, 1980). Two stages are involved in local feature-based matching algorithms they are:

1. Detection of interest point, and
2. Careful extraction of feature description.

All the popular local features descriptor should have the following features:

- Feature detection has a high repeatability rate,
- High speed and time efficient,
- Feature description should have a low feature dimension, which helps us to achieve quick and efficient matching,
- Robust against illumination, occlusion, clutter, blur, noise, rotation, scale, affine transformation and viewpoint change,
- **Distinctiveness:** The region should contain "interesting" structure,
- **Quantity:** There should be enough points to represent the image.

The procedures of obtaining feature descriptor mainly include three steps:

1. Detection of Key point,
2. Establishing proper Descriptor,
3. Extract image feature matching information through descriptor.

Researchers for many years try to improve the performance by adjusting these above mention steps.

Figure 1 illustrates the extraction of features using local and global descriptor under normal condition i.e. no transformation is applied over the image while Table 1 summarizes and classifies different feature detector or descriptor algorithm based on the feature such as key points or region of interest.

Feature extraction is the process by which some sort of computing is performed in the descriptor. This operation has been performed from the pixels around each interest point. Thus feature descriptor are the raw pixel values stored in a small patch around the interest point of the image.

Feature matching is the process of finding corresponding points between two images.

BACKGROUND

Literature Review

A lot of research is being done in the fields of object recognition and for success full object recognition some popular features descriptor has been used. Success of object recognition can be related with

Figure 1. Untransformed features has been extracted using local and global descriptor

Table 1. Classification of different algorithm based on key point or region of interest and the type

Algorithm	Corner	Blob	Type
SIFT	N	Y	Local
SURF	N	Y	Local
FAST	Y	Y	Local
MSER	N	Y	Global
Harris	Y	N	Global
Min Eigen	Y	N	Global

cognitive behavior and science. An important behavior of animals is that it can easily able to recognize any objects which are vital for their existence. Humans can quickly and unsurprisingly recognize these objects under different condition such as change in orientations, scale, rotation, illumination or partially overlapped by other objects under perplexing environment. Obliviously to recognize an object animal consider or store some information that are invariant in nature. To that animals follows some cognitive approach such learning and reasoning (Banks & Krajicek, 1991; Biederman,1987; Pinker, 1985), neurobiology (Gross, 1973; Gross et al., 1993; Miyashita, 1993; Rolls, 1994), neuropsychology (Damasio, Tranel & Damasio, 1990; Davis, 1992; Farah, 1990; Farah et al., 1991; Griisser et al., 1991; Humphreys et al., 1987; Humphreys et al., 1987), and computation and engineering (Aloimonos, 1993; Kim et al., 2010; Koenderink, 1990; Marr et al., 1978; Mian et al., 2010; Ullman, 1989). Traditionally object recognition was based on structural components (Biederman & Ju, 1988; Biederman, 1990; Biederman & Bar, 1999; Grossbarg et al., 1985). Theoretically object recognition is basically motivated by shape. Since a single dimension needed to be attended and there are a finite number of mutually exclusive components. However, recent work provides evidence that surface level information (e.g., object color) is readily used in object recognition (Naor-Raz et al., 2003; Oliva et al., 2000; Tanaka et al., 1999; Tanaka et al., 2001). Beiderman and Ju (1988) first argued that structural (edge-based) properties of objects are theoretically preferred over viewpoint, texture, and color information. It is not the case that these features can't be used, but that they are only useful in certain circumstances when object shape is compromised or extremely variable. Edward et al. (2006) proposed Feature from Accelerated segment test for identifying interest points in an image. In their proposed work author tries to use FAST algorithm detector in real time application to obtain frame rate from a digital image. Since FAST detector algorithm compares pixels at circle of predetermined radius nearby a point. In the proposed work a point is consider as corner if and only if a large section of pixels in the circle of fixed radius near the point are considerably darker or brighter than the point at the central. Harris and Stephens (1988) proposed a combined corner and edge detector algorithm popularly known as Harris corner detection algorithm. In this work the author(s) improved Moravec's corner detector. Here, place of using shifted patches author(s) directly differentiated the corner score with respect to direction. Speeded up Robust Features was proposed by Herbert Bay et al. (2006), is a robust local feature detector, which can be used in computer vision for object recognition or reconstruction. SURF feature descriptor algorithm is based on 2D Haar wavelet. Juan et al. (2009) proposed a comparative study and the performance of three popular robust feature detectors (SIFT, PCA-SIFT, and SURF) under different conditions to prove the robustness of the algorithm. Lowe et al. (1999, 2004) proposed a local feature description algorithm SIFT, stands for Scale-invariant Feature Transform. SIFT is used as an invariance-based feature detection method. SIFT shown remarkable good stability and invariance. It is successfully able to detect large numbers of local key points that stored a huge quantity of information. The unique advantages offered by SIFT has drawn many researchers interest in this topic and thus it become very popular. Matas et al. (2002) proposed maximally stable extremal regions is a feature descriptor algorithm which is very much effective where two images have different viewpoints, It is a better stereo matching and object recognition algorithms to find correspondences between image features. It is based on blob detection. The earliest corner detection algorithm was proposed by Moravec (1980). Here authors define that a corner is actually a point that has low self-similarity. Here the author(s) uses a local window in the digital image, while defining the average changes of image intensity by changing the window by a minor amount in several directions. In their proposed word author(s) considered the following three cases.

1. If the patch of the windowed image is nearly constant in intensity, then only a small change is observed in the result.
2. If the window overlaps at the edge, then small changes are observed in the result, on the other hand if there is a change perpendicular to the edge a large change is observed in result.
3. If the corner or isolated point appears as window parched, then for all changes a large change is noticed.

Ouyang et.al.(2012) proposed to create a benchmark dataset for the effective analysis and comparison of different algorithms used for robust pattern matchings. Ravela (2003) proposed how multi-scale differential features and their representations are helpful for image retrieval and recognition. In this submission author argued that for face recognition multi-scale Gaussian differential features (MGDFs) proves to be a better alternative. For Face recognition distributions of image brightness is used as local features to construct the surface. Here author(s) claimed that on standard benchmark datasets 96% recognition accuracy can be achieved, their works shows better performance against other well-known techniques available. MGDF based technique is very common; and it can be used effectively in retrieving similarity in trademarks, textures, and also for other similar application. Shi and Tomasi (1994) proposed a novel feature selection and tracking algorithm. For selecting features from an object algorithm which was based on affine image change. In their work author(s) avoid adhoc measure of texturedness thereby provides a better optimal solution for object tracking. Use of affine image model helps to discriminate between good feature and bad feature. Though this feature does not provide good solution under all case and suffer from illumination problem. Later Kanade-Tomasi (2004) proposed corner detection method to find corners using the minimum Eigen values.This work demonstrated use of minimum feature point for feature matching. Tuytelaars et al. (2008) carried out a survey on most commonly used local invariant feature detectors .This survey systematically divided the local invariant feature detector into two parts. In the first section authors investigated local invariant feature detectors; in the second part authors investigated corner detectors, blob detectors, and region detectors. Younes et al. (2012) discussed in details about the three implementations of the SIFT algorithm as suggested by Lowe in 2004. In this paper, author(s) survey an artificial visualization of the SIFT algorithm asserting on the awkward steps of its implementation. In this submission they claim that their proposed algorithm is better in comparison to lowes algorithm, Here, they studied and argued the properties of the detected feature points and their equivalent descriptors. They conducted a survey of three implementations of the Lowe's algorithm with parameter as suggested by Lowes. Wang et al. (2014) suggest how appropriate and efficient color feature played an important role in object recognition in Traffic management.

System Component Overview

The operation involved in object recognition process must be kept low time consuming. In this section we describe some popular and well-practiced feature descriptors.

SIFT

SIFT stands for Scale Invariant Feature Transformation algorithm. SIFT uses DoG (Difference of Gaussian) function to achieve scale invariance. SIFT follow two step convolution on an image. It obtains the change in different scale images by the change of σ. Finally the DoG pyramid is obtained by subtracting the images that are adjoining in the same resolution. An improvement of a Gauss-Laplace algorithm is the DoG function.

$$G(x,y,\sigma) = \frac{1}{2\pi\sigma^2}\exp\left[-\frac{x^2+y^2}{2\sigma^2}\right] \quad (1)$$

$$D(x,y,\sigma) = G(x,y,k\sigma) - G(x,y,\sigma)) * I(x,y) = L(x,y,k\sigma) - L(x,y,\sigma) \quad (2)$$

To describe each key point SIFT uses a 128-dimensional vector. As the number of dimension use by SIFT are pretty high therefore image feature matching using SIFT are relatively slow. To detect the feature SIFT follows the following steps:

Step 1 - Scale-Space Extrema Detection: Detect interesting points (invariant to scale and orientation) using DOG has been detected.

Step 2 - Key Point Localization: Determine location and scale at each candidate location, and select them based on stability i.e. reject the low contrast points and points that are located near the edge.

Step 3 - Orientation Estimation: Use local image gradients to assigned orientation to each localized keypoint. Preserve theta, scale and location for each feature.

Step 4 - Keypoint Descriptor: Extract local image gradients at selected scale around keypoint and form a representation invariant to local shape distortion and representation invariant to local shape distortion and illumination them.

FAST

Feature from Accelerated segment test (FAST) algorithm was proposed by Rosten and Drummond to discover interest points within an image. Increasing the computational efficiency is the most striking feature of FAST corner detector. As name stand it is fast and undeniably it is faster than many other renowned feature extractions. FAST corner detector uses a Bresenham circle of radius 3 where each circle consists of 16 pixels. Whether a candidate point (p) is actually a corner can be decided based on the combination of 16 pixels. Integer number 1 to 16 is used clockwise to label each pixel of the circle. To be called a corner point a candidate point must satisfy following two criteria.

Criteria 1: A set of N contiguous pixels S, ∀ x ∀ S, the intensity of x (Ix) > Ip + threshold t.
Criteria 2: A set of N contiguous pixels S, ∀ x ∀ S, Ix < Ip – t.

SURF

SURF was proposed by Herbert Bay et al. in 2006. It is a robust local feature detector, which can be used in computer vision for object recognition or reconstruction. The experimental results shows that SURF is reasonably faster than SIFT, detection and description of 1529 interest points takes round about 610ms and in some situation it out perform SIFT. The foundation of SURF is sums of 2D Haar wavelet response. SURF makes an efficient use of integral images. Computational efficiency of box type convolution filters can be increased with the use of Integral images. The entry of an integral image I(x) at a location x = (x; y) T represents the sum of all pixels in the input image 'I' within a rectangular region formed by the origin and x.

$$I\sum(x) = \sum_{i=0}^{i<=x} \sum_{j=0}^{9<=y} I(i,j) \tag{3}$$

Interest points can be detected using a hessian based blob detector. The determinants of a hessian matrix express the local change around the area and the extent of the response.

Figure 2. Features has been extracted from a digital image using FAST descriptor algorithm under normal condition

$$H(x,\sigma)=\begin{bmatrix}L_{xx}(X,\sigma) & L_{xy}(X,\sigma)\\ L_{xy}(X,\sigma) & L_{yy}(X,\sigma)\end{bmatrix} \tag{4}$$

where

$$L_{xx}(X,\sigma)=I(X)*\frac{\delta^2}{\delta x^2}g(\sigma) \tag{5}$$

$$L_{xy}(X,\sigma)=I(X)*\frac{\delta^2}{\delta xy}g(\sigma) \tag{6}$$

$$L_{xy}\left(X,\sigma\right)=I(x)*\frac{\partial^2}{\partial_{xy}}g(\sigma)$$

Figure 3. Interested points of a digital image has been detected using SURF feature descriptor algorithm under normal condition.

MIN-EIGEN

The algorithm was developed by Shi and Tomasi (1994) is also known as Kanade-Tomasi (1990) corner detector to find corners using the minimum eigenvalues min ($\lambda1$, $\lambda2$). The algorithm returns a corner point's object within a 2-D grayscale image. The Corner points contain information about the feature points. A corner point is described by a large variation of pixel in all the directions of the vector(X, Y). It can be expressed in the following way:

An interest point should have two large eigenvalues. The magnitudes of the eigenvalues helps us to drawn the following inferences.

- The pixel (X, Y) has no corner point or features of interest iff $\lambda1<>0$ and $\lambda2<>0$.
- An edge is found iff $\lambda1<>0$ and $\lambda2$ has some large positive value.
- A corner is found iff $\lambda2$ and $\lambda2$ have large positive values.

Noble's corner measure M'C is used to avoid the setting of the parameter K which amounts to the harmonic mean of the eigenvalues:

$$M_c' = 2\frac{\det(A)}{trace(A) + \varepsilon'} \tag{7}$$

where ε being a small positive constant. The covariance matrix for the corner position is A^{-1} i.e.

Figure 4. Interested points of a digital image has been detected using MIN-EIGEN feature descriptor algorithm under normal condition.

$$\frac{1}{\langle I_x^2\rangle\langle I_y^2\rangle - \langle I_x I_y\rangle^2}\begin{bmatrix}\langle I_y^2\rangle & -\langle I_x I_y\rangle \\ -\langle I_x I_y\rangle & \langle I_x^2\rangle\end{bmatrix} \tag{8}$$

HARRIS Corner

Corner detection is one of the most practiced methods used to infer the contents and to extract certain kinds of features from an image. Corner detection is very often used in computer vision systems, image registration, motion detection and object recognition. Different researchers proposed different algorithm for detecting corner. One of the most practice corner detection algorithms was Harris and Stephens by improving Moravec's corner detector. Let an image patch over the area is (u, v) and (x, y) after shifting the image patch. The weighted sum of squared differences (SSD) between (u, v) and (x, y) can be obtained using:

$$S(x,y) = \sum_u \sum_v w(u,v)(I(u+x,v+y) - I(u,v))^2 \tag{9}$$

I (u + x, v+ y) can be expressed by a Taylor expansion. If I_x and I_y be the partial derivatives of I, such that

$$I\left(u+x,v+y\right) \approx I\left(u,v\right) + I_x\left(u,v\right)x + I_y\left(u,v\right)y \tag{10}$$

This produces the approximation

$$S(x,y) \approx \sum_u \sum_v w(u,v)(I_x(u,v)x + I_y(u,v)y)^2 \tag{11}$$

This can be written in matrix form

$$S(x,y) \approx (xy)A\binom{x}{y} \tag{12}$$

where A is the structure tensor,

$$A = \sum_u \sum_v w(u,v)\begin{bmatrix}I_x^2 & I_x I_y \\ I_x I_y & I_y^2\end{bmatrix} = \begin{bmatrix}\langle I_x^2\rangle & \langle I_x I_y\rangle \\ \langle I_x I_y\rangle & \langle I_y^2\rangle\end{bmatrix} \tag{13}$$

This matrix is known as Harris matrix, and angle brackets denote averaging.

Figure 5. Interested points of a digital image has been detected using HARRIS feature descriptor algorithm under normal condition.

MSER

Maximally stable extremal regions a better stereo matching and object recognition algorithms was proposed by Matas et al. MSER is used to find correspondences between image features. It proves to be very effective where two images have different viewpoints. It is based on blob detection. The use of blob helps in extracting a comprehensive number of feature between corresponding image elements. Thus it contributes wide-baseline matching and makes MSER quite popular among researchers. Region Q is a contiguous subset of D. (For each p, q belongs to Q there is a sequence p, a1, a2, an, q and p Aa1, ai, Aai+1, an Aq). Maximally Stable Extremal Region can be defined by the following equation. Let Q1… Qi-1. Qi… is a sequence of nested extremal regions where Qi belongs to Qi+1. Extremal region Qi* is maximally stable if and only if q (i) =| Qi+$ \ Qi-$|/|Qi| has a local minimum at i*. (Here |.| denotes cardinality). € Si is a parameter of the method.

Experimental Results

Experiments were performed on a on an Intel Dual core processor running Windows XP with 250 GB of storage. The programs have been implemented in MATLAB (2013). In this part, we perform experiments to observe the performance of both local feature descriptor and global feature descriptor and its deviation under different condition such as change in scale and rotation, effect of blur, change of behavior when illumination varies, and effect of affine change. We evaluated our method using 124 images. The images are captured from different viewpoints and under different scales and orientations. They represent

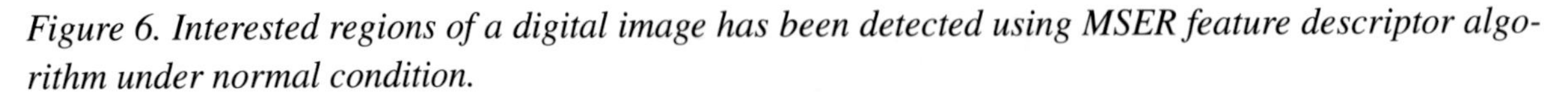

Figure 6. Interested regions of a digital image has been detected using MSER feature descriptor algorithm under normal condition.

moreover different themes such as landscapes, animals, monuments and people. We also study the speed i.e. time taken by each algorithm for feature extraction and feature matching under different conditions. Figure 7 to 19 shows the performance of different descriptor in detecting features under different conditions. Figure 20 to 22 summarize the performance of descriptors i.e. feature description rate, average feature extracted, average time taken in feature extraction. Figure 23, 25, 27, 29, 31 and 33 shows the features matching performance of descriptors. Figure 24, 26,28,30,32 and 34 analyze the performance of different descriptor i.e. time taken (speed of the descriptor) in feature extraction and feature matching. Figure 35-38 methodically analyze the performance of descriptors in feature matching, average time taken by the descriptor while performing feature matching, average feature matched, average feature extracted vs. average feature matched and finally average time taken by descriptor algorithm in feature extraction and feature matching.

CONCLUSION

Object recognition in the fields of computer vision or digital image processing is considered as an active area in research. It has a large amount of different algorithms for features detection and description. The proposed work methodically analyzes the foremost feature detector/descriptor algorithm. In this work, we systematically assessed the strength and weakness of each descriptor under different state and condition such as change in rotation and scale, effect of blur, change in image intensity as well as change in affine transformation on different benchmark dataset. In this study the speed of each descriptor algorithm has

Figure 7. a) Blurred image, b) illuminated, c) 100 rotated, d) affine transformation, e) scale change, f) original image

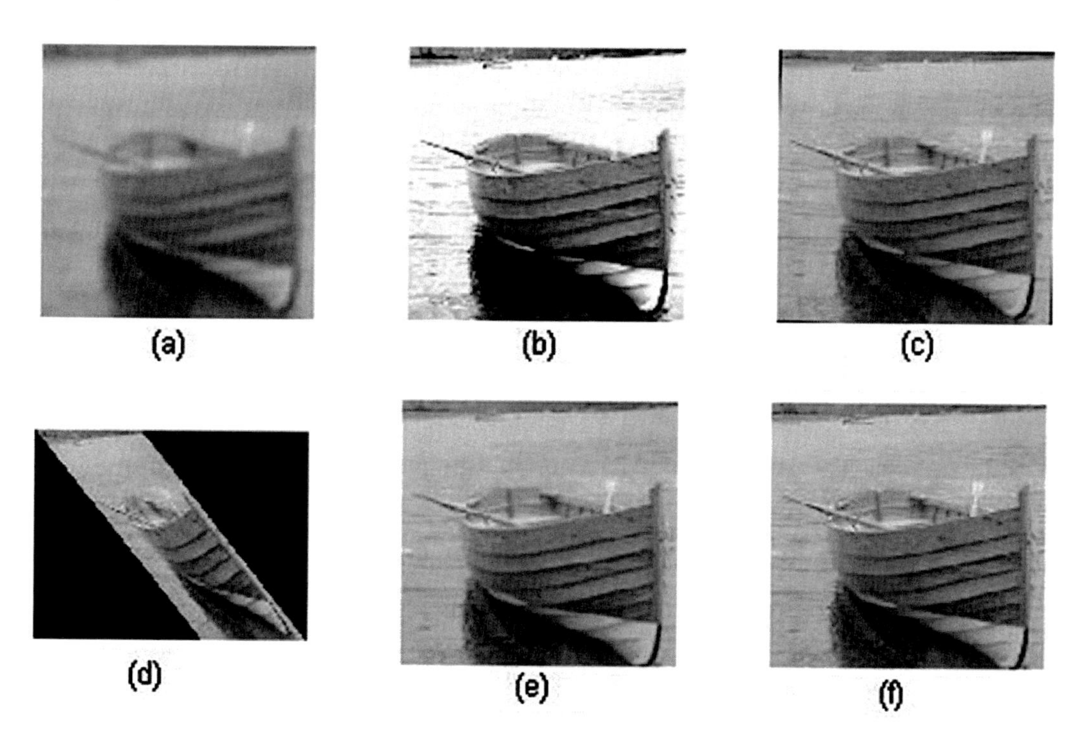

Figure 8. Feature detected using fast descriptor against different condition: (a) blurred, (b) illuminated, (c) 100 rotated, (d) affine transformation, (e) scale change, (f) original image

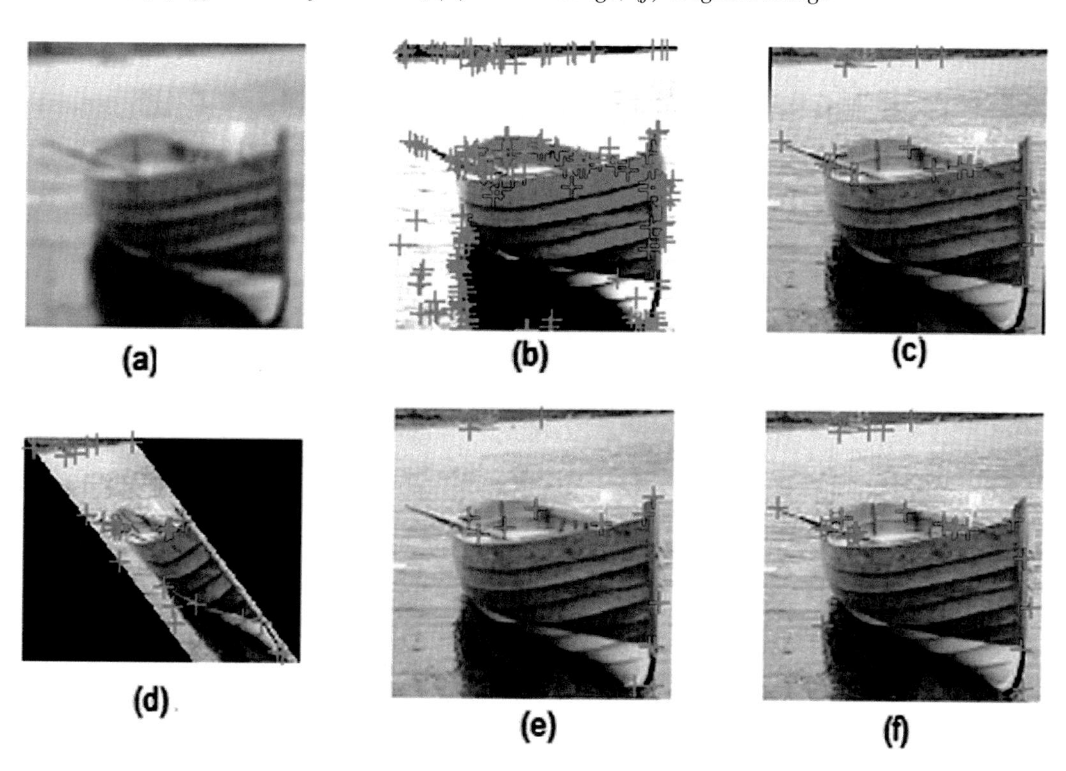

Figure 9. Time (in ms) taken by FAST descriptor in detecting features under different condition

Figure 10. Feature detected using surf descriptor against different condition: (a) blurred, (b) illuminated, (c) 100 rotated, (d) affine transformation, (e) scale change, (f) original image

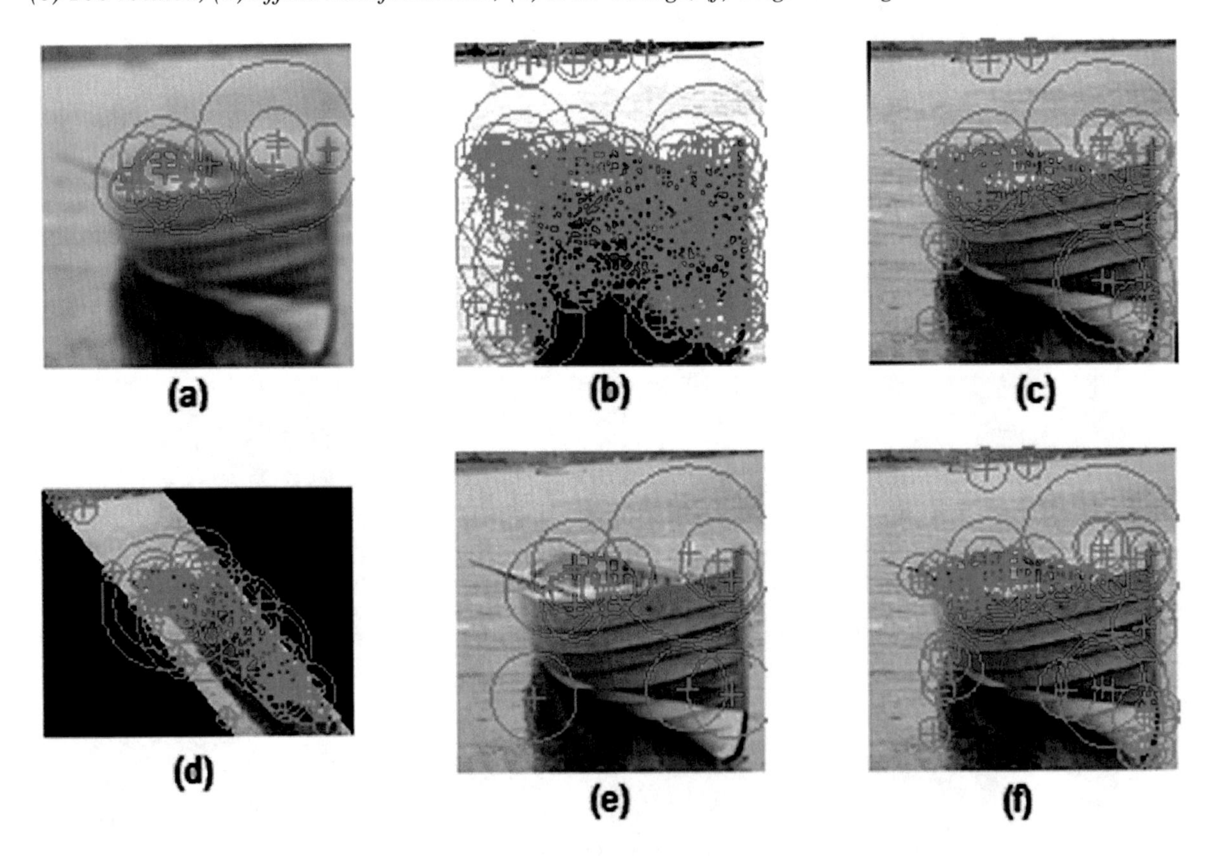

Figure 11. Time (in ms) taken by SURF descriptor in detecting features under different condition

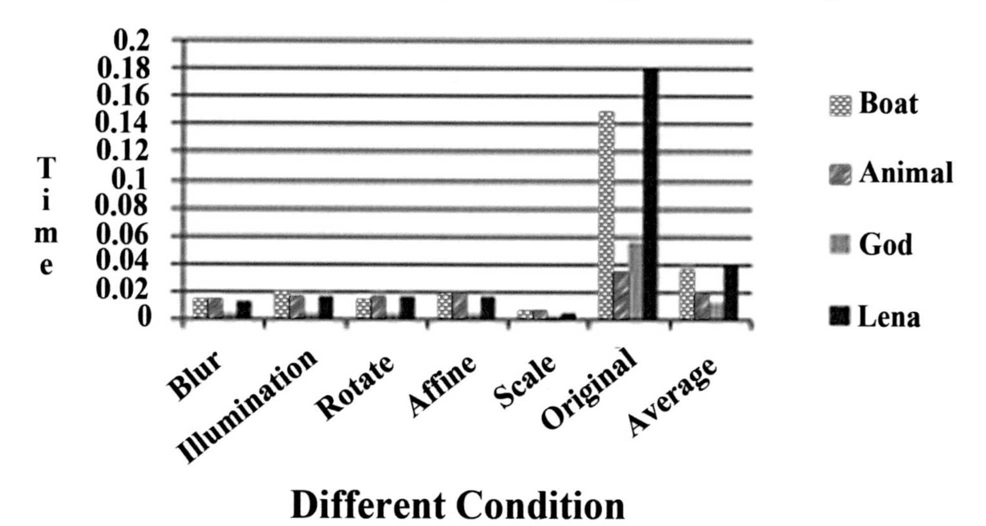

Figure 12. Feature detected using MSER descriptor in different condition: (a) blurred, (b) illuminated, (c) 100 rotated, (d) affine transformation, (e) scale change, (f) original image

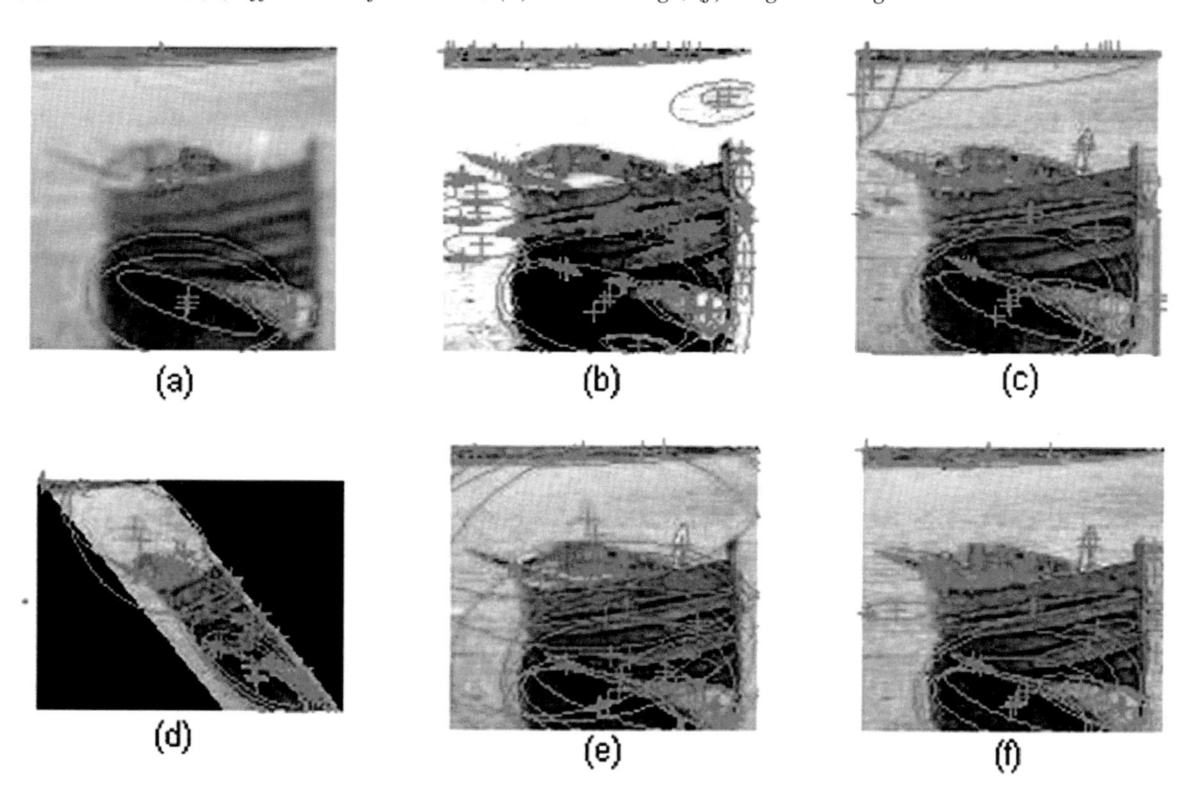

Figure 13. Speed (in ms) of MSER descriptor in detecting features under different condition

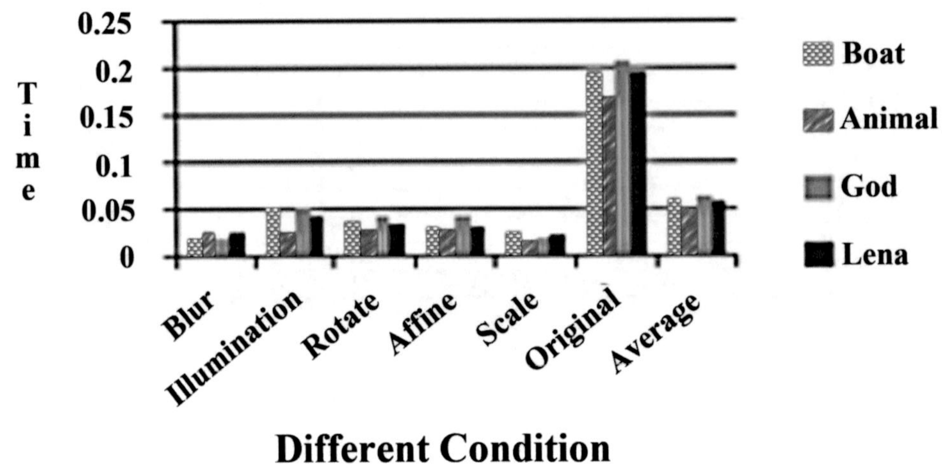

Figure 14. Feature detected using HARRIS descriptor against different condition: (a) blurred, (b) illuminated, (c) 100 rotated, (d) affine transformation, (e) scale change, (f) original image

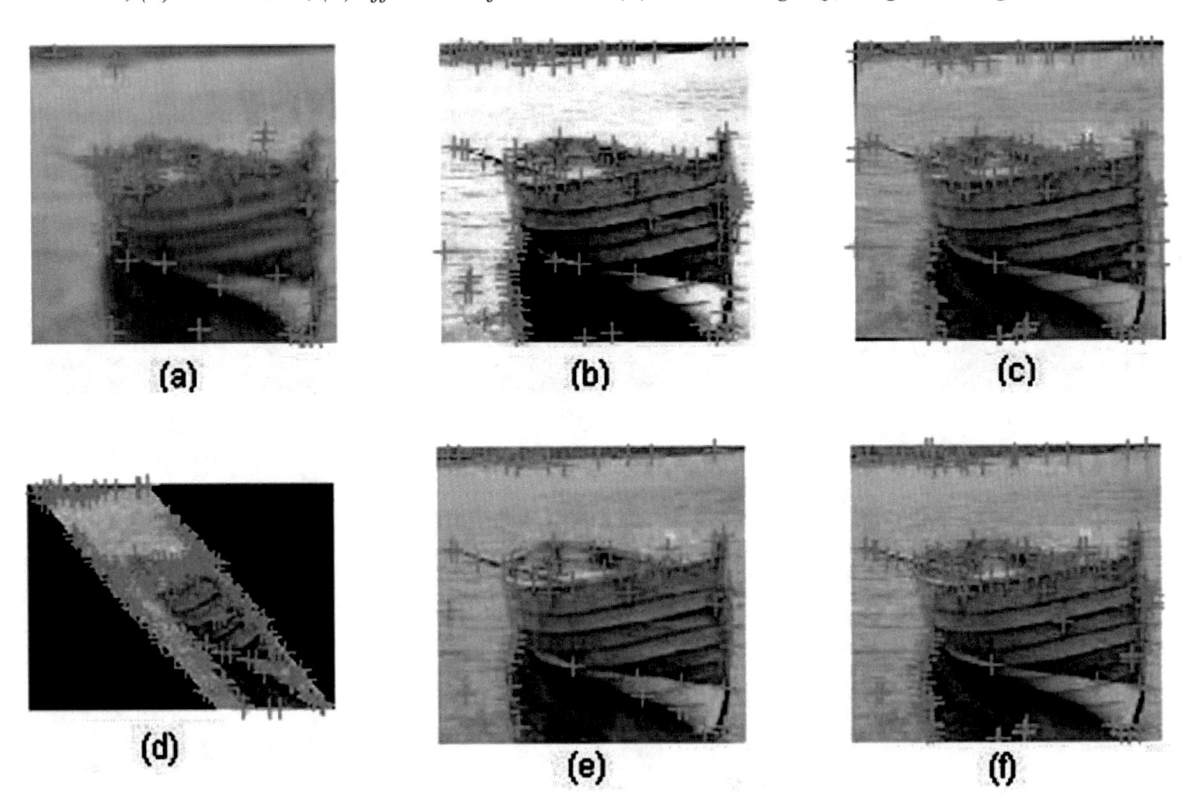

Figure 15. Time (in ms) taken by HARRIS descriptor in detecting features under different condition

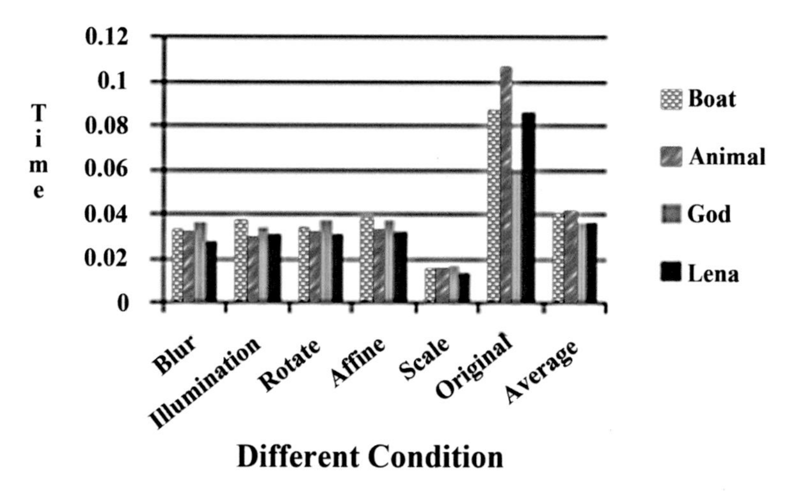

Figure 16. Feature detected using MIN-EIGEN descriptor against different condition: (a) blurred, (b) illuminated, (c) 100 rotated, (d) affine transformation, (e) scale change, (f) original image

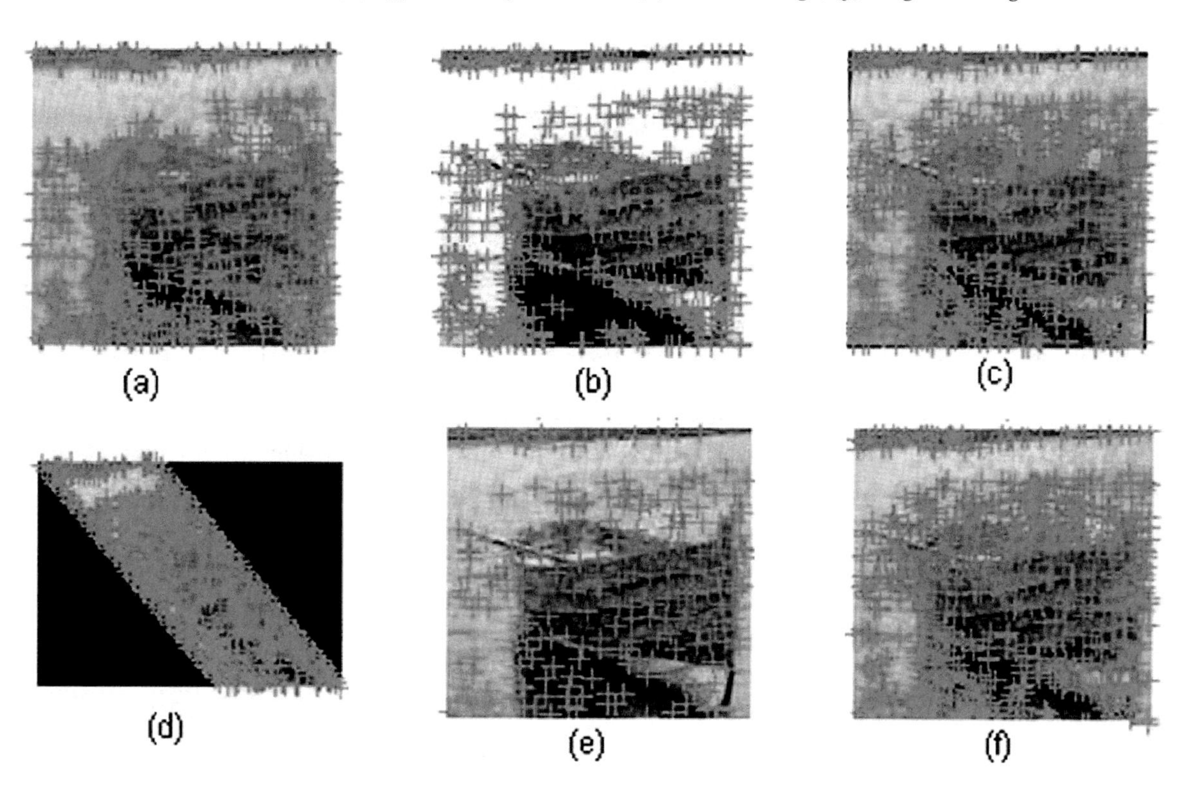

Figure 17. Speed (in ms) of Min-Eigen descriptor in detecting features under different condition

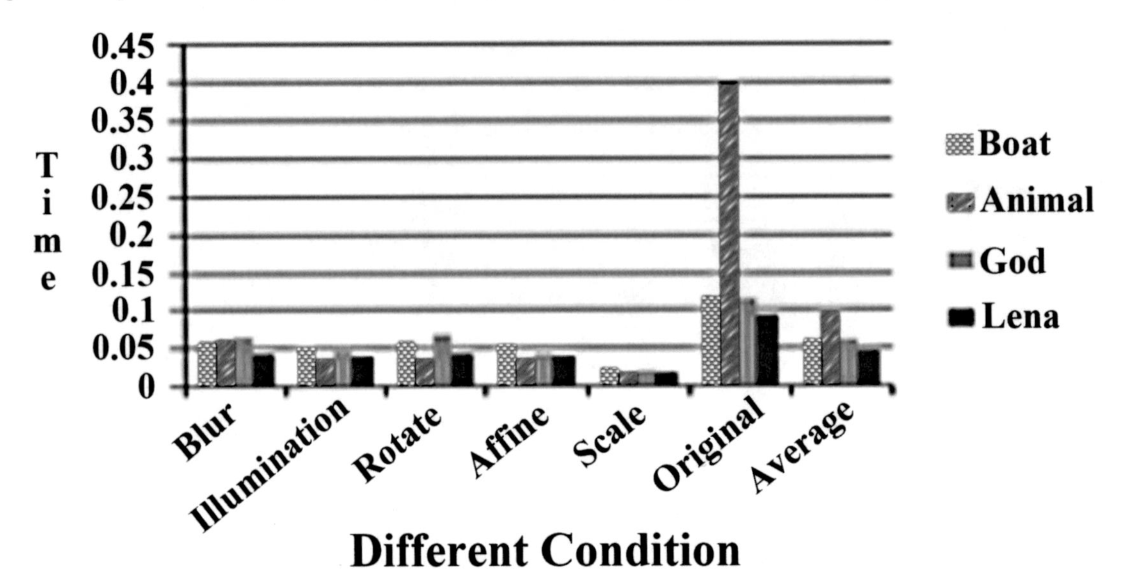

Figure 18. Feature detected using SIFT descriptor against different condition: (a) blurred, (b) illuminated, (c) 100 rotated, (d) affine transformation, (e) scale change, (f) original image

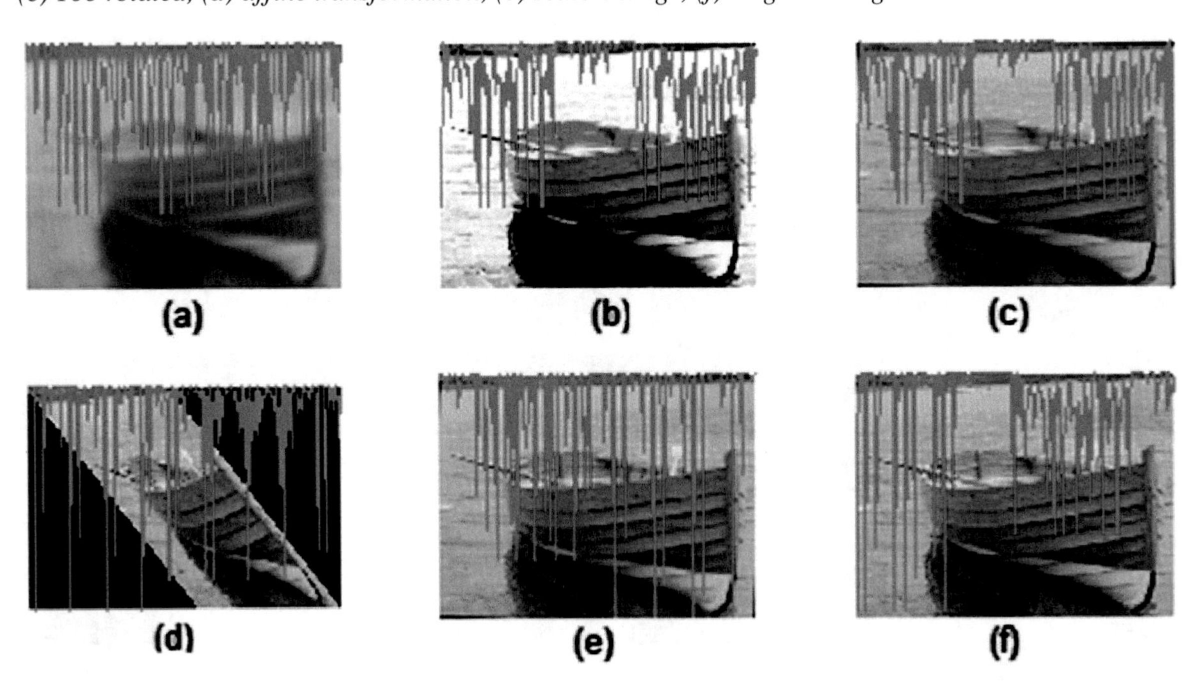

Figure 19. Time (in ms) taken by SIFT descriptor in detecting features under different condition

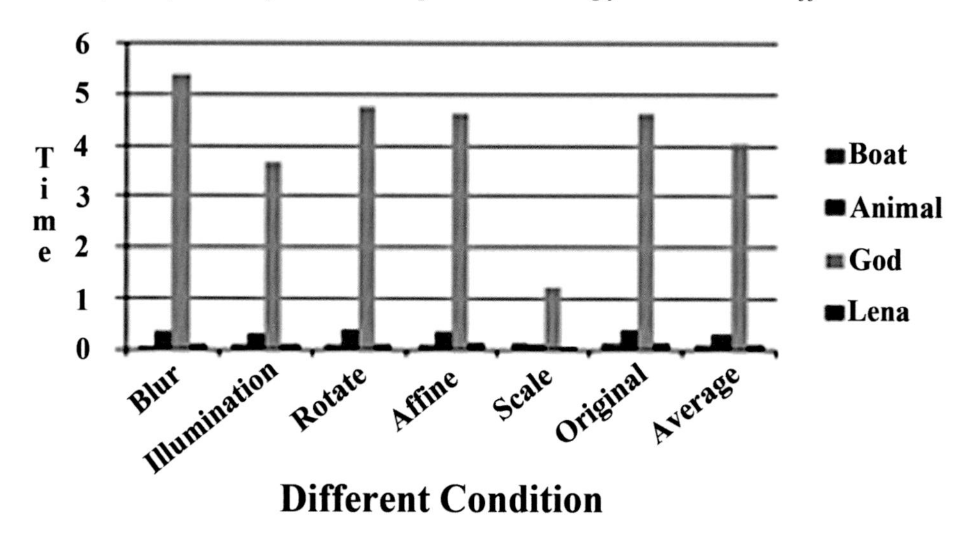

Figure 20. Performance i.e. time (in ms) taken by different descriptors in feature extraction under different condition

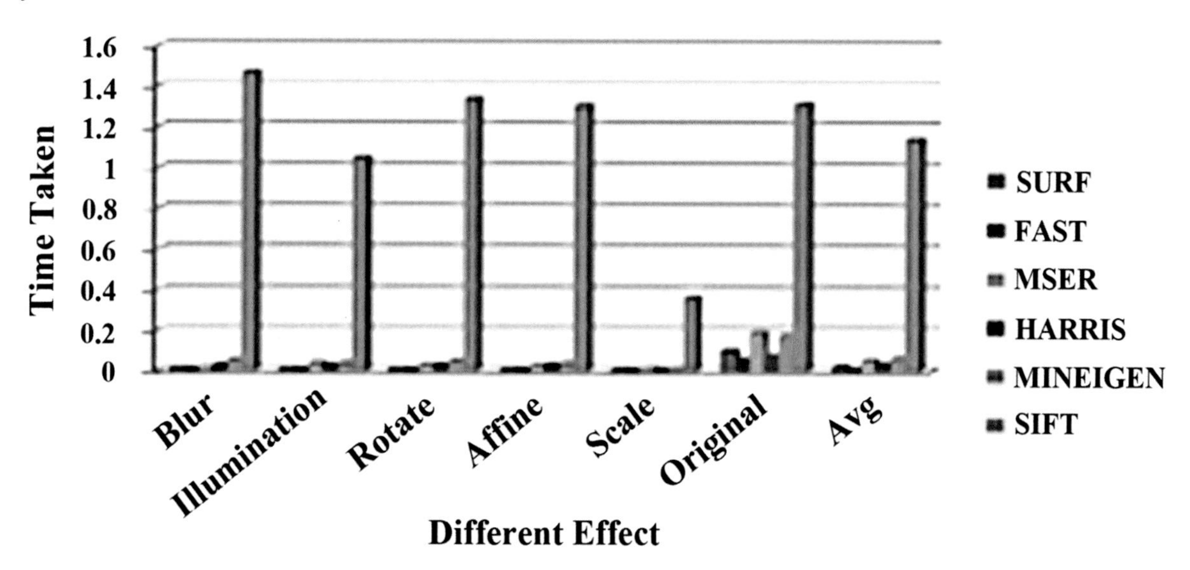

Figure 21. Average features extracted by different descriptors under different condition

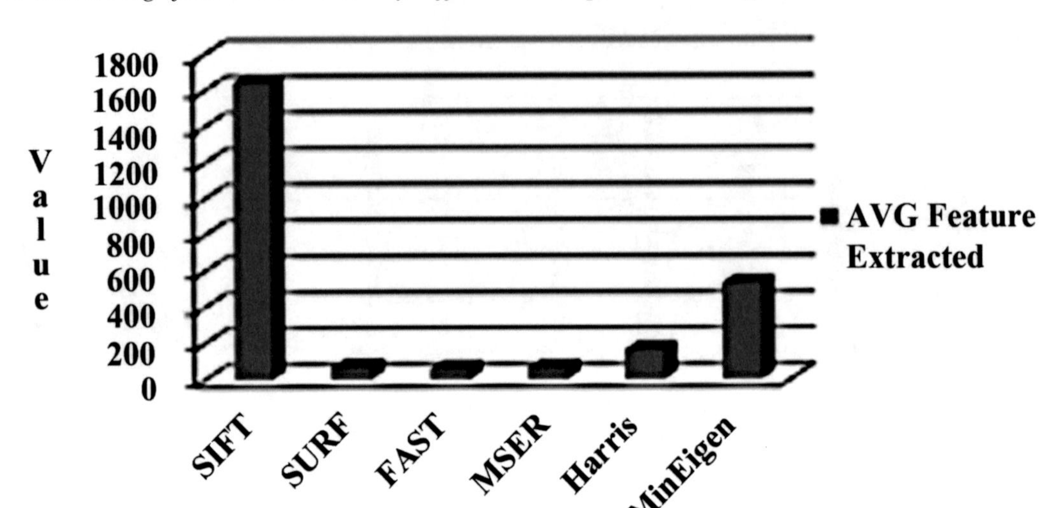

Figure 22. Average speed (in ms) v/s average features extracted by different descriptors under different condition

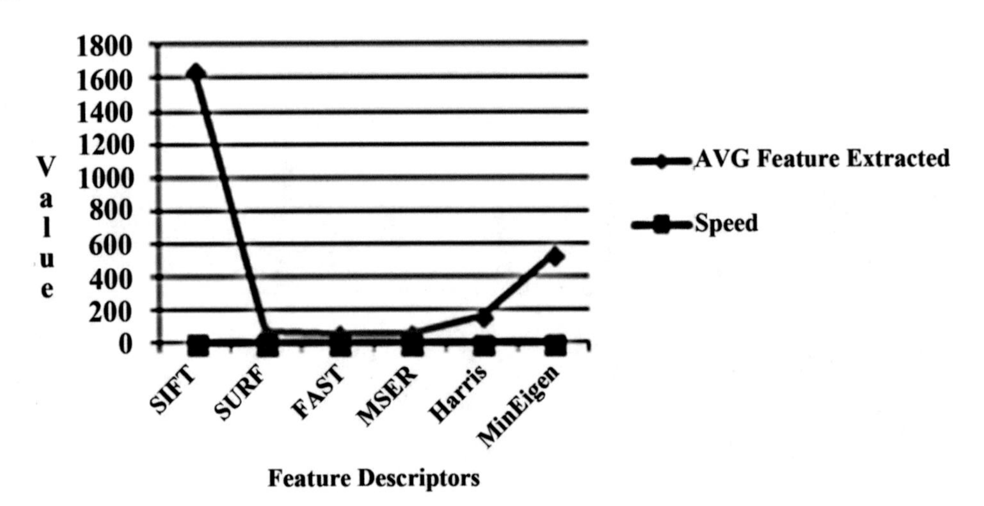

Figure 23. SURF descriptor based feature matching under different condition

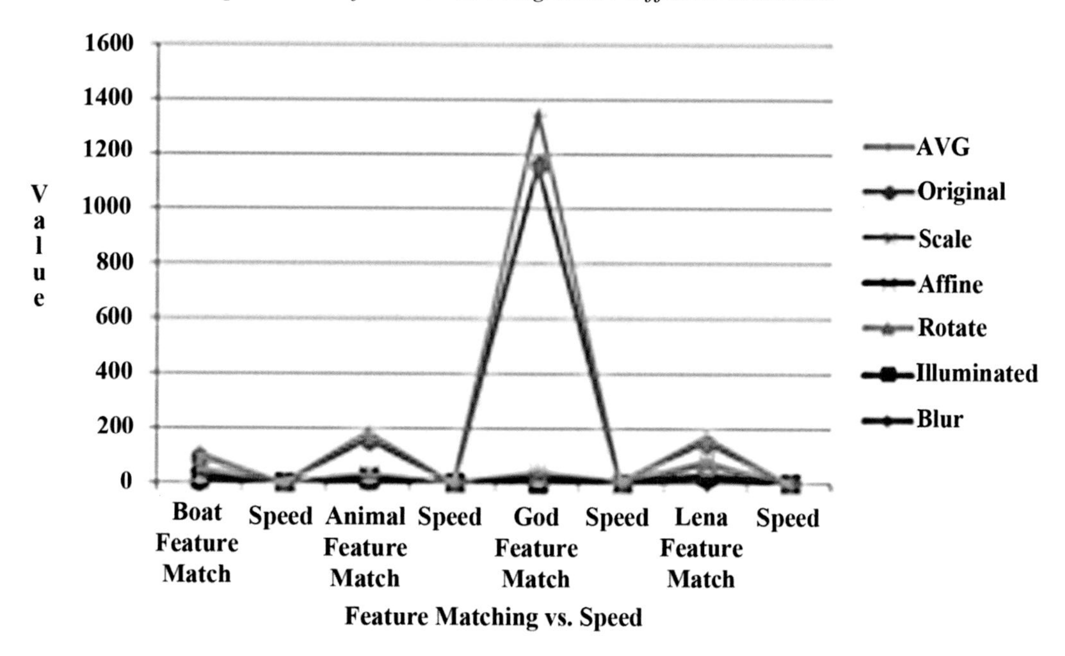

Figure 24. SURF performance: time (in ms) taken in feature extraction and feature matching

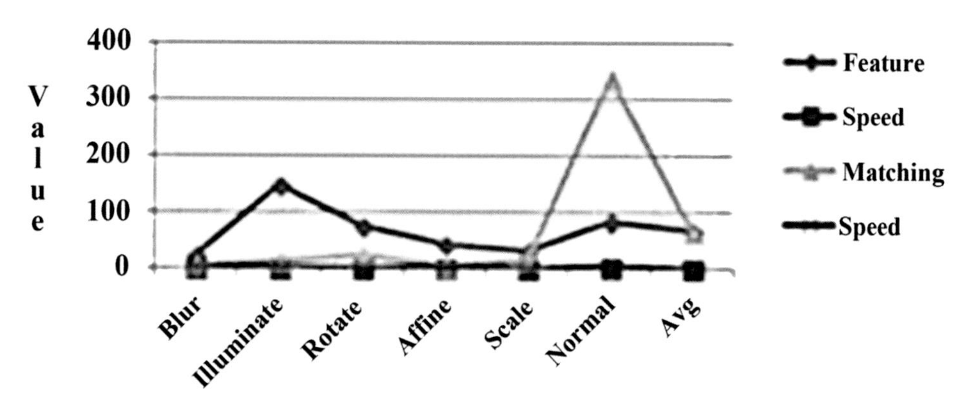

Figure 25. FAST descriptor based feature matching under different condition

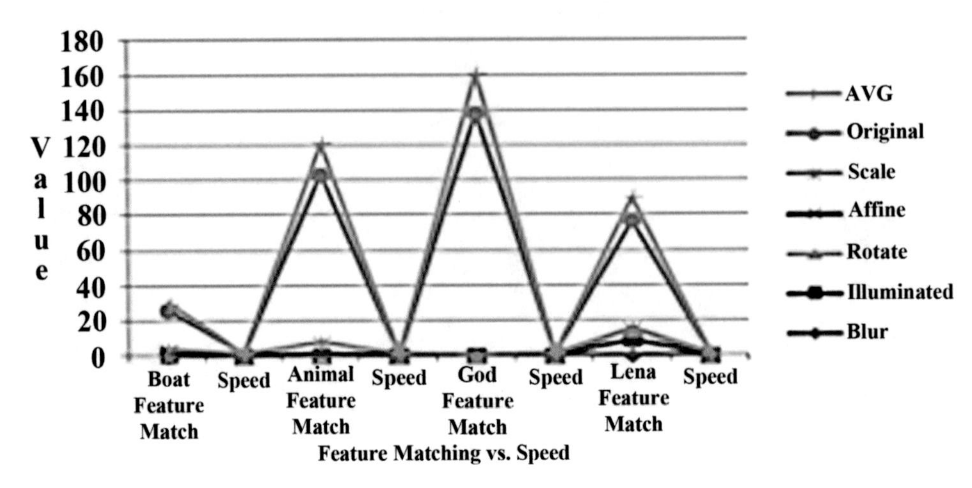

Figure 26. FAST performance: time (in ms) taken in feature extraction and feature matching

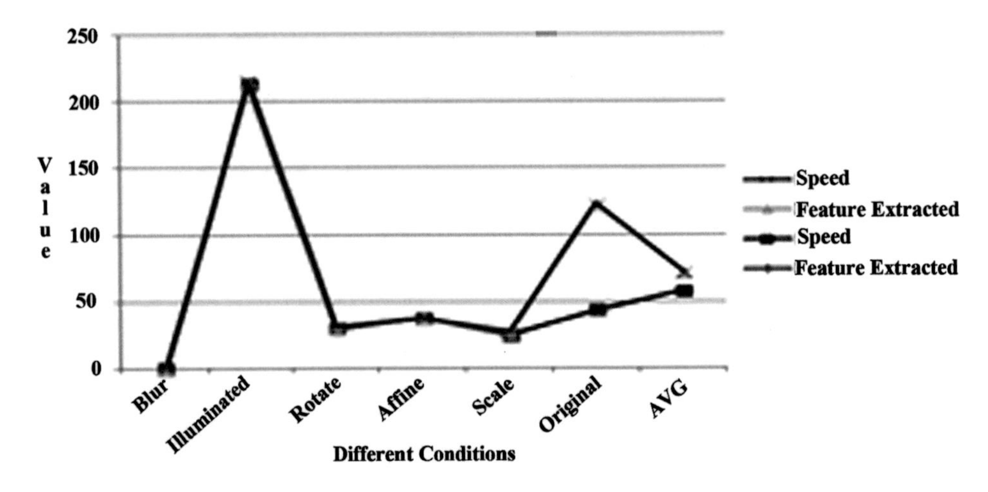

Figure 27. MSER descriptor based feature matching under different condition

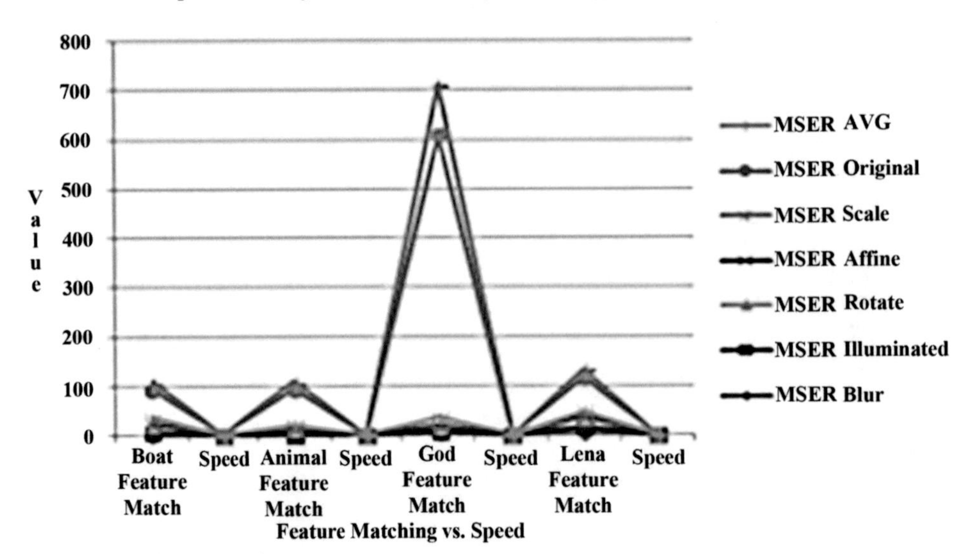

Figure 28. MSER performance: time (in ms) taken in feature extraction and feature matching

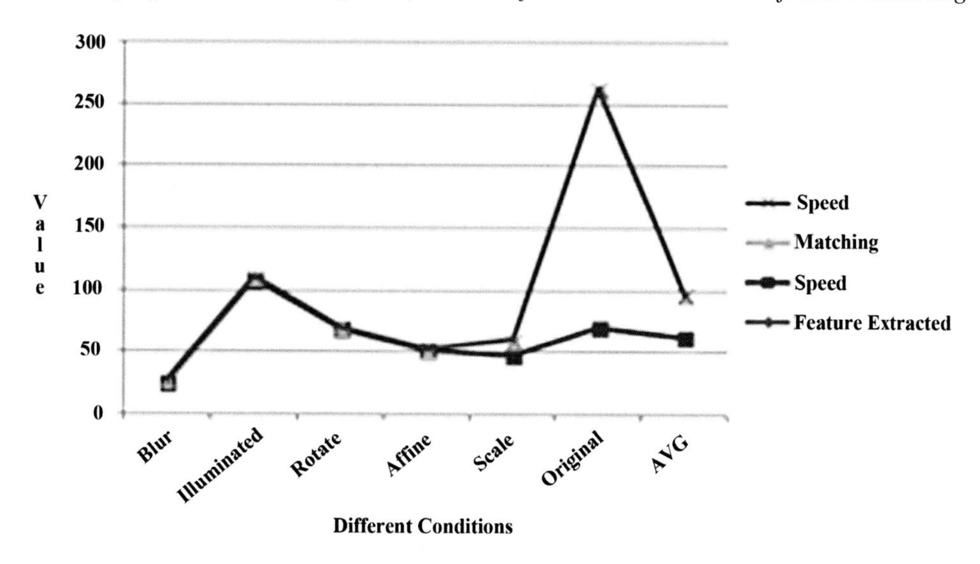

Figure 29. HARRIS descriptor based Feature Matching under different condition

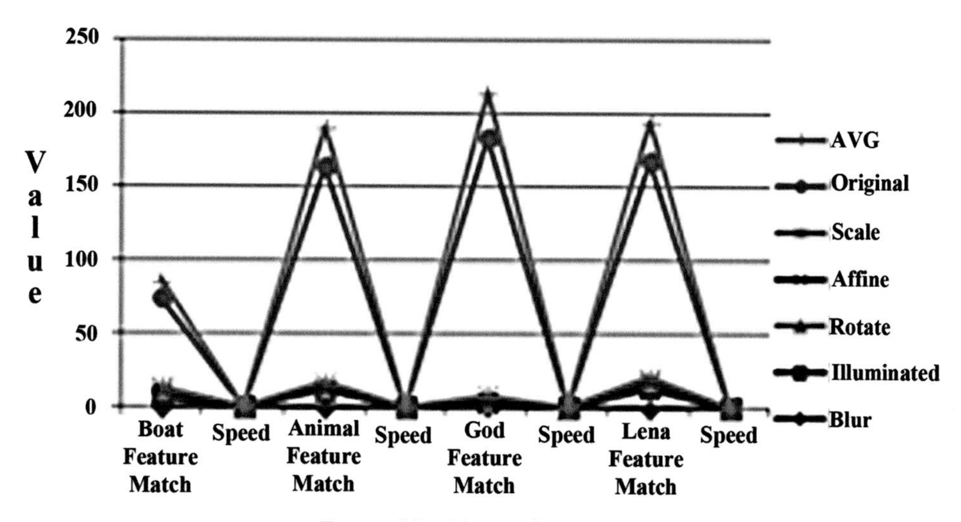

Figure 30. Performance of HARRIS feature descriptor: time (in ms) taken in feature extraction and feature matching

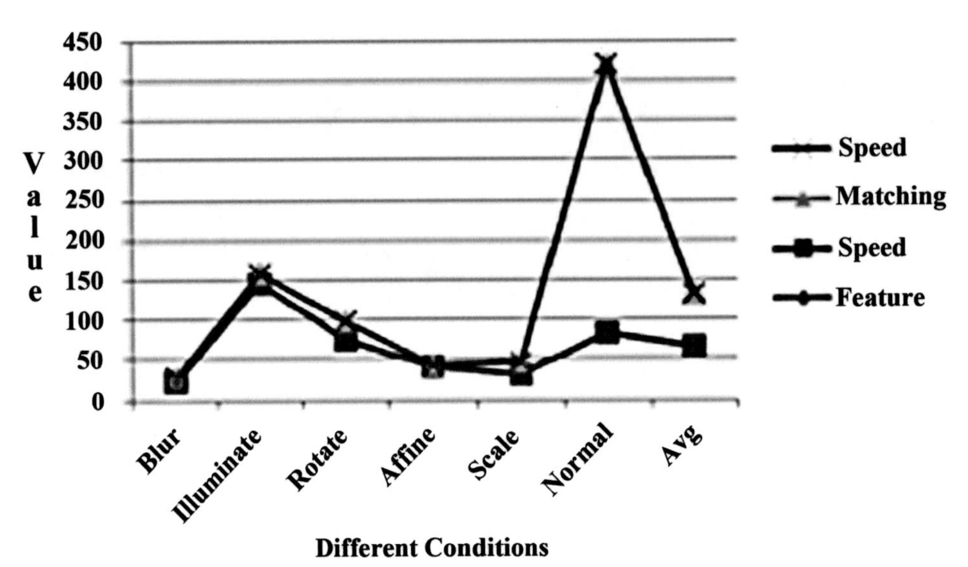

Figure 31. MINEIGEN descriptor algorithm based feature matching under different condition

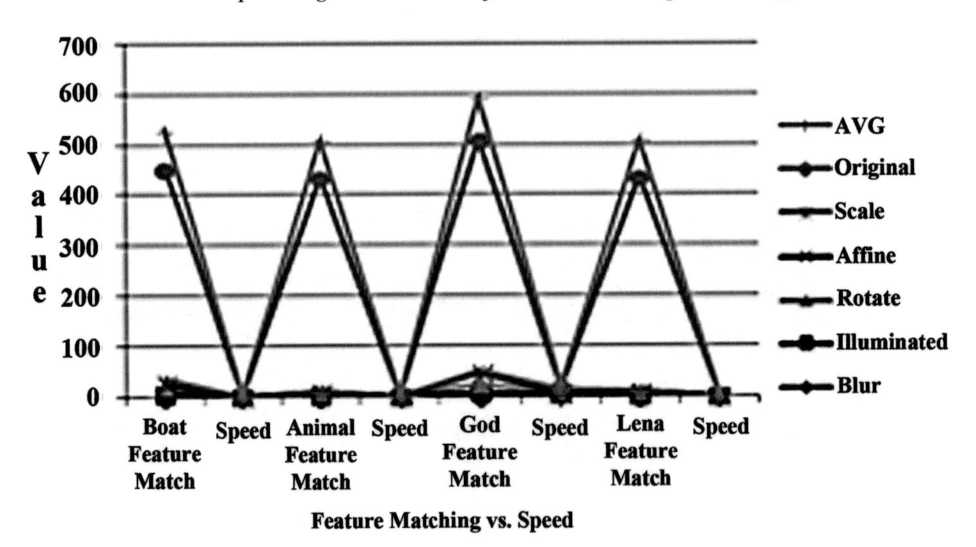

Figure 32. Performance of MINEIGEN feature descriptor has been shown: time (ms) taken in feature extraction v/s feature matching

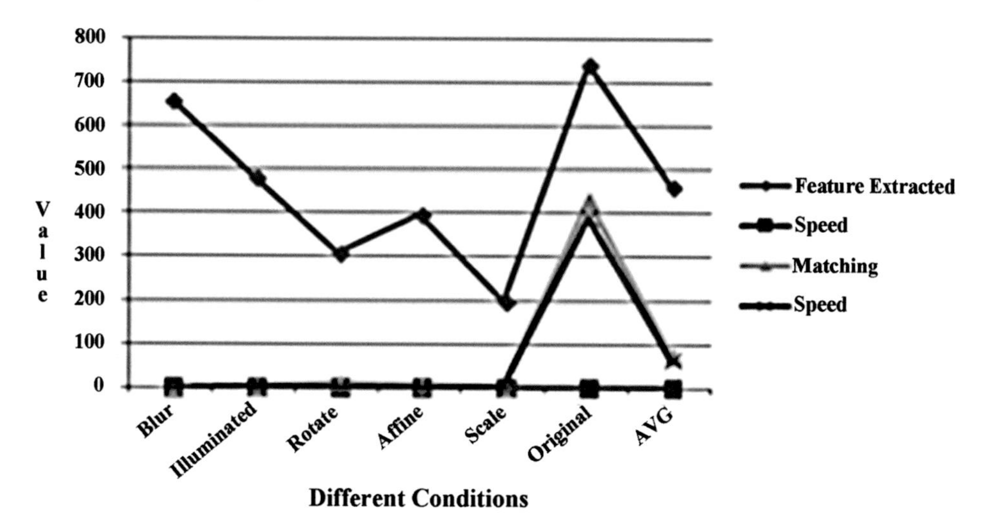

Figure 33. SIFT descriptor algorithm based feature matching under different condition

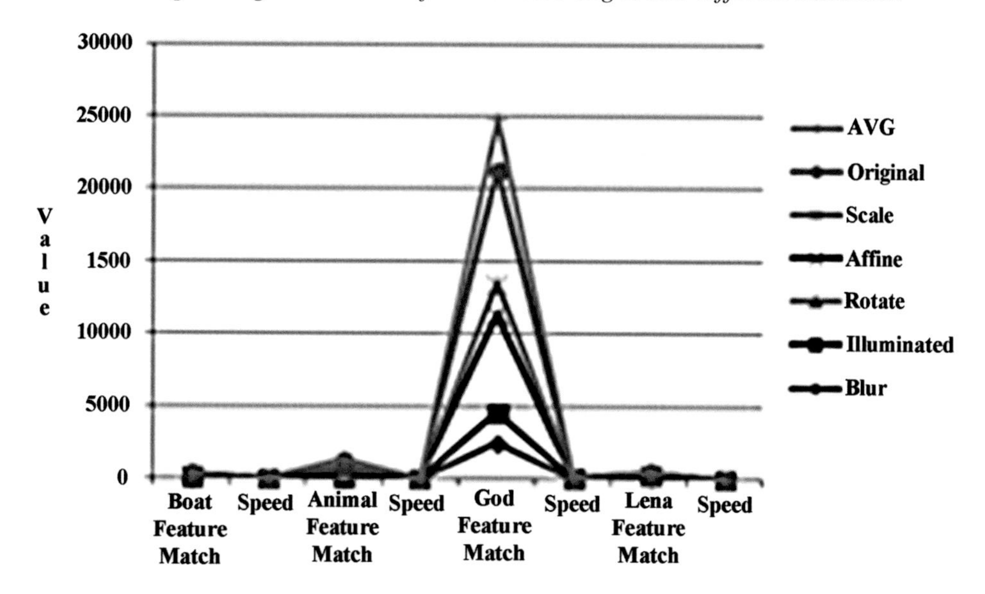

Figure 34. Showing the performance of SIFT algorithm: time (in ms) taken in feature extraction and feature matching

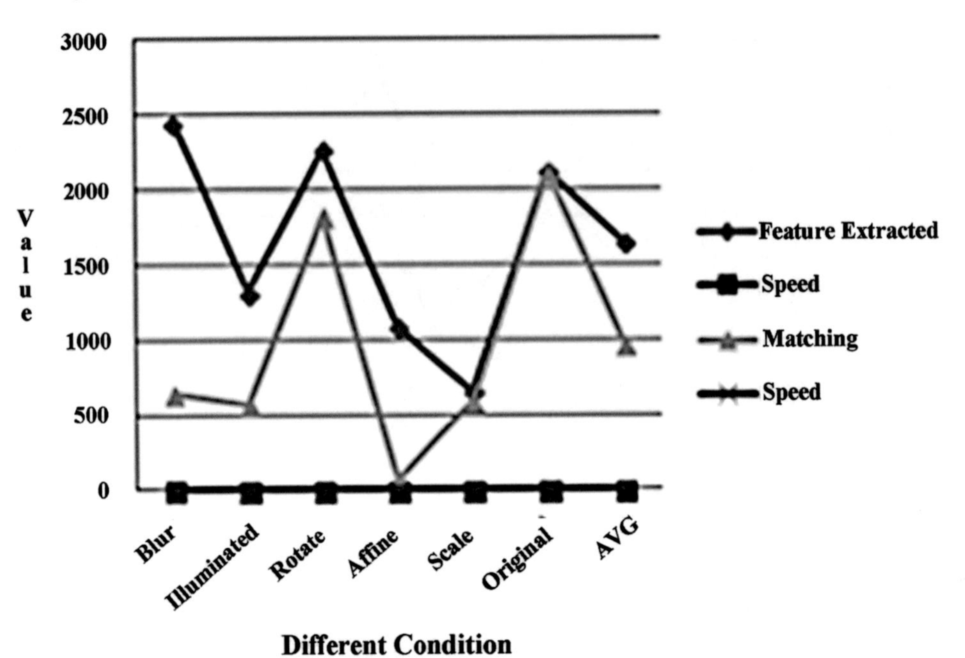

Table 2. Comparative study of feature descriptor algorithm under different condition

Algo	Criteria	Blur	Illuminate	Rotate	Affine	Scale	Normal	Average
SIFT	Avg. Feature Extracted	2437.5	1313.75	2265.5	1084	651.5	2118.25	1645
	Avg. Speed	1.4672355	1.052058	1.34045	1.309	0.36634	1.3156283	1.1417963
	Avg. Matching	648	573.25	1820	70.25	589.25	2114.75	969.25
	Avg. Speed	5.5253244	3.220589	1.372	1.372	1.33223	6.0250331	3.4127838
SURF	Avg. Feature Extracted	23.5	146	75.25	42.25	31	83.5	66.916667
	Avg. Speed	0.011231	0.014097	0.01276	0.014	0.00507	0.105279	0.027151
	Avg. Matching	4.25	11	22	1.5	14.75	336	64.916667
	Avg. Speed	0.06931	0.064606	0.06512	0.063	0.09582	0.3468055	0.1174551
FAST	Avg. Feature Extracted	0	213.5	29.75	37.25	24.75	43.5	58.125
	Avg. Speed	0.0042665	0.004025	0.00404	0.004	0.00288	0.0602215	0.0133053
	Avg. Matching	0	2.5	1.75	0.5	2.25	79.5	14.416667
	Avg. Speed	0.1647783	0.161051	0.16245	0.16	0.16249	0.18332	0.1656228
MSER	Avg. Feature Extracted	25	108.5	69.25	52.5	48	70.5	62.291667
	Avg. Speed	0.0213898	0.041856	0.03427	0.033	0.02056	0.1914331	0.0571239
	Avg. Matching	2.4	2.5	1.75	0.3	12.5	191.75	35.2
	Avg. Speed	0.1647783	0.161051	0.16245	0.16	0.16249	0.18332	0.1656228
Harris	Avg. Feature Extracted	151.25	177	193.5	170	84.5	200.5	162.79167
	Avg. Speed	0.032025	0.032971	0.03349	0.035	0.0156	0.0844713	0.0389548
	Avg. Matching	0.25	9	3	0	4	131.5	24.625
	Avg. Speed	0.157101	0.161051	0.15397	0.161	0.15944	0.1658083	0.1596656

continued on next page

Table 2. Continued

Algo	Criteria	Blur	Illuminate	Rotate	Affine	Scale	Normal	Average
Min Eigen	Avg. Feature Extracted	655.75	479.5	310	395.3	199	741.25	463.45833
	Avg. Speed	0.0566563	0.043492	0.17102	0.044	0.01982	0.1814821	0.0860273
	Avg. Matching	0	4.25	11	7.25	4.25	431.75	76.416667
	Avg. Speed	0.2552858	1.112791	0.45077	0.7	1.22396	293.33682	49.513248

Figure 35. Average feature matched v/s average speed (time taken in ms) by different descriptor algorithm under different condition

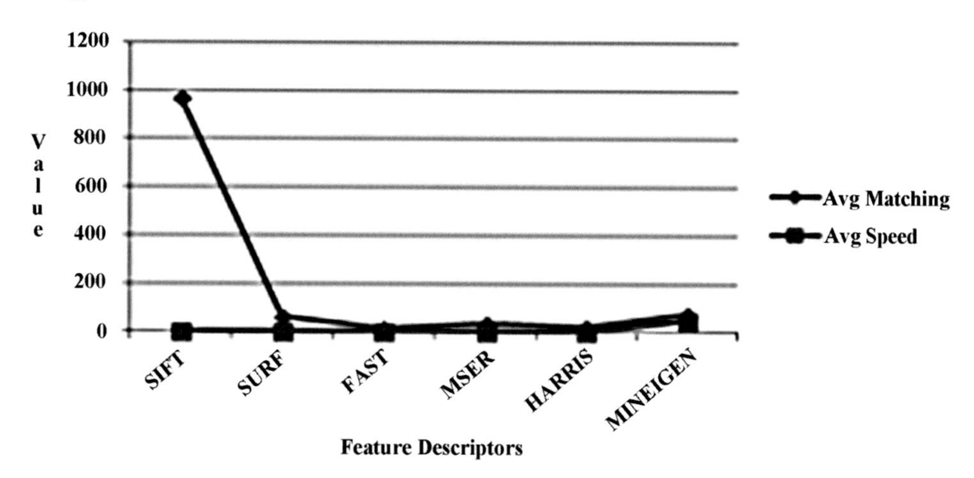

Figure 36. Average feature matched by different descriptors under different condition

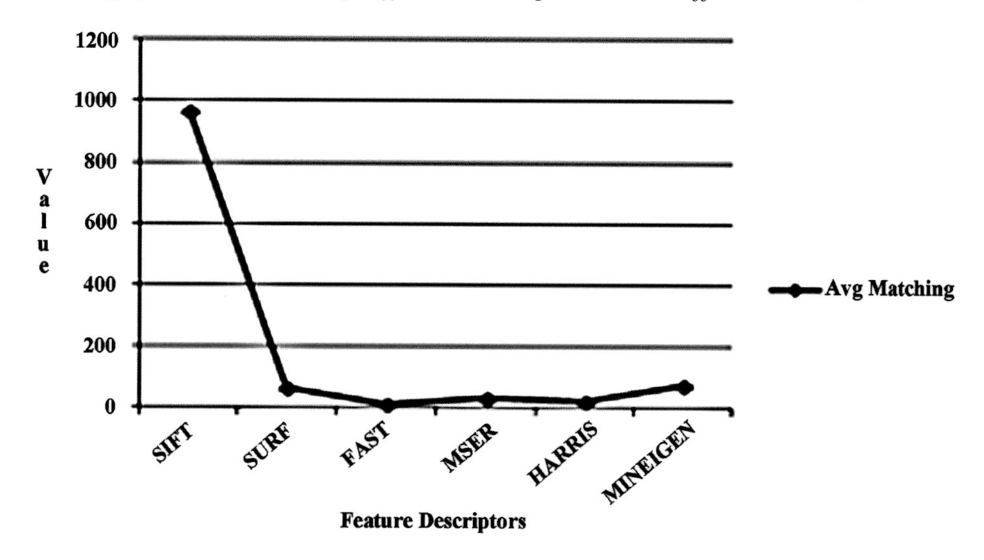

Figure 37. Average feature extracted v/s average feature matched by different descriptors algorithm under different condition

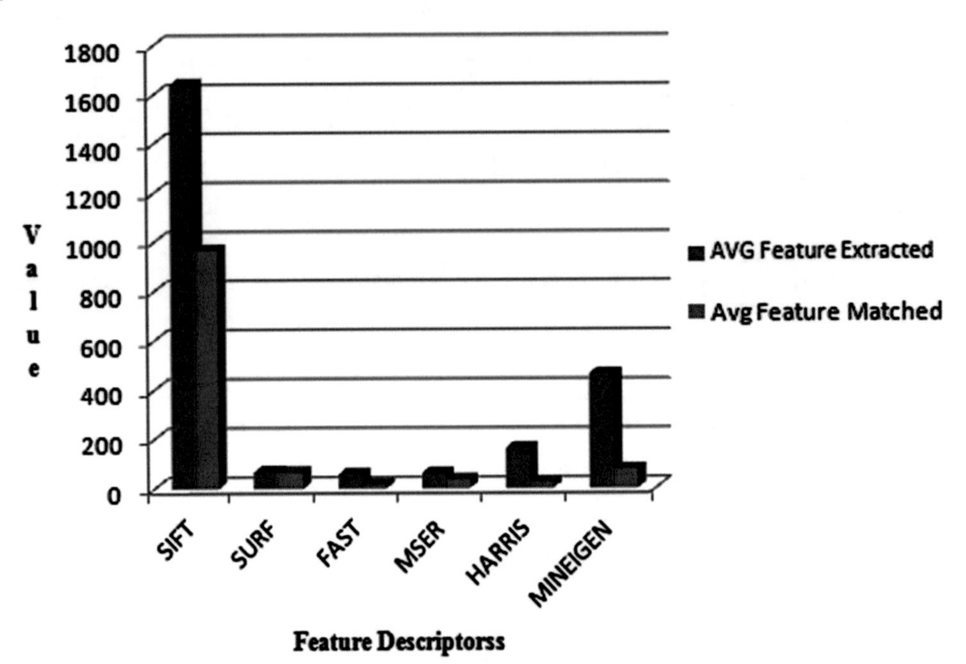

Figure 38. Average feature extracted speed (in ms) v/s. average feature matching speed by different descriptor algorithm under different condition

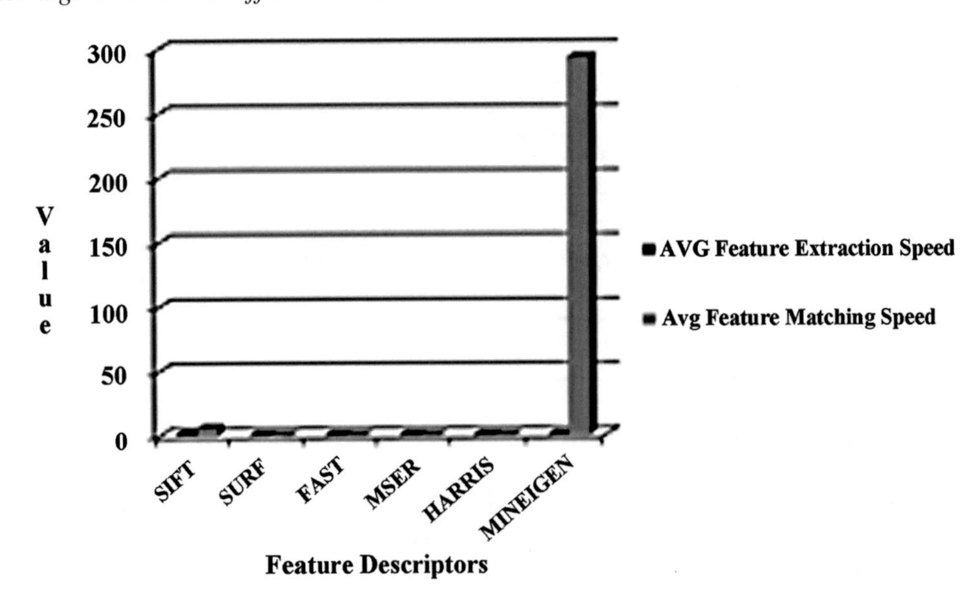

Table 3. Qualitative summarization of descriptors performance

Descriptor	Blur	Illuminated	Rotate	Affine	Scale	Speed
SIFT	Bs	Bs	Bs	Bs	Bs	P
SURF	G	G	B	B	G	B
FAST	P	A	A	P	A	BS
MSER	A	A	A	A	G	A
HARRIS	G	G	A	P	G	A
Min Eigen	P	A	A	A	G	P

G=Good; A=Average; B=Better; BS=Best; P=Poor

been carried out in detailed under different condition to check the robustness of each descriptor or detector algorithm. Based on the investigational results, we carefully analyzed and observed the performance of each algorithm, that shows that how effective are this algorithm for a specific application. The work can be further enhanced and can be more robust with the use of machine Intelligence.

Future Direction

This proposed work systematically analyze the strength and weakens of different feature descriptor algorithm for the purpose of successful object recognition under some adverse condition. The object recognition system can be applied effectively in the area of surveillance system, face recognition, fault detection, character recognition etc. The performance of the object recognition system depends on the features used. The object recognition system developed in this research was tested with the benchmark datasets. This dataset provides both 2D and 3D view of the data. The proposed work can be further extended if we used some other feature detector/descriptor for the purpose of feature extraction such as PCA-SIFT, GSIFT, CSIFT, and ASIFT, BRISK etc. In Future for the purpose of object recognition machine intelligence can be used. The feature can be recognize using k-Nearest Neighbor classifier, Support vector machine(SVM),Fuzzy Logic or using Fuzzy c means or Fuzzy K means. The work can be further extended if we can combine the classifier such as K-Nearest Neighbor can be clubbed with support vector machine.

REFERENCES

Alnihoud, J. (2008). An efficient region-based approach for object recognition and retrieval based on mathematical morphology and correlation coefficient. *International Arab Journal Information Technology, 5*(2).

Aloimonos, Y. (Ed.). (1993). *Active Perception*. Hillsdale, NJ: Erlbaum.

Banks, W.P., & Krajicek, D. (1991). Perception. *Annual Review of Psychology*. DOI: 10.1146/annurev.ps.42.020191.001513

Biederman, I. (1987). Recognition-by components: A theory of human image understanding. *Psychological Review, 94*(2), 115–147. doi:10.1037/0033-295X.94.2.115 PMID:3575582

Biederman, I. (1990). Higher-level vision. In *Visual Cognition and Action*. The MIT Press.

Biederman, I., & Bar, M. (1999). One-shot viewpoint invariance in matching novel objects. *Vision Research, 39*(17), 2885–2899. doi:10.1016/S0042-6989(98)00309-5 PMID:10492817

Biederman, I., & Ju, G. (1988). Surface vs. edge-based determinants of visual recognition. *Cognitive Psychology, 20*(1), 38–64. doi:10.1016/0010-0285(88)90024-2 PMID:3338267

Birinci, M., Diaz-de-Maria, F., & Abdollahian, G. (2011). Neighborhood matching for object recognition algorithms based on local image features. In *IEEE Digital Signal Processing Workshop and IEEE Signal Processing Education Workshop* (pp. 157–162). DSP/SPE. doi:10.1109/DSP-SPE.2011.5739204

Damasio, A. R., Tranel, D., & Damasio, H. (1990). Face agnosia and the neural substrate of memory. *Annual Review of Neuroscience, 13*(1), 89–109. doi:10.1146/annurev.ne.13.030190.000513 PMID:2183687

Davis, M. (1992). The role of amygdala in fear and anxiety. *Annual Review of Neuroscience, 15*(1), 352–375. doi:10.1146/annurev.ne.15.030192.002033 PMID:1575447

Edward, R., & Drummond, T. (2006). Machine learning for high-speed corner detection. In *Proceedings European Conference on Computer Vision*. Cambridge, MAMIT Press.

Farah, M. J. (1990). *Visual agnosia*. Cambridge, MA: MIT Press.

Farah, M. J., McMullen, P. A., & Meyer, M. M. (1991). Can recognition of living things be selectively impaired? *Neuropsychologia, 29*(2), 185–193. doi:10.1016/0028-3932(91)90020-9 PMID:2027434

Griisser, O. J., & Landis, T. (1991). *Visual agnosias and other disturbances of visual perception and cognition*. London: Macmillan.

Gross, C. G. (1973). Visual functions of infero temporal cortex. In R. Jung (Ed.), Handbook of Sensory Physiology. Berlin: Springer-Verlag.

Gross, C. G., Rodman, H. R., & Cochin, P. M. (1993). Inferior temporal cortex as a pattern recognition device. In *Proceedings of 3rd Necrotizing Enterocolitis Research Symposium*. Slam: NEC Res.

Grossberg, S., & Mingolla, E. (1985). Neural dynamics of form perception: Boundary completion, illusory figures and neon color spreading. *Psychological Review*, 2(2), 173–211. doi:10.1037/0033-295X.92.2.173 PMID:3887450

Harris, C., & Stephens, M. (1988). A combined corner and edge detector. In *Proceedings of the 4th Alvey Vision Conference*, (pp. 147–151).

Herbert, B., Andreas, E., Tuytelaars, T., & Gool, L. (2008). SURF: Speeded up robust features. *Computer Vision and Image Understanding, 110*(3), 346–359. doi:10.1016/j.cviu.2007.09.014

Humphreys, G. W. & Riddoch, M. J. (Eds.). (1987). Visual object processing: A cognitive neuropsychological approach. Hillsdale, NJ: Lawrence Erlbaum Associates.

Humphreys, G. W., & Riddoch, M. J. (1987). *To See Bur Not To See: A Case Study of Visual Agnosia.* Hillsdale, NJ: Lawrence Erlbaum Associates.

Juan, L., & Gwun, O. (2009). A comparison of SIFT, PCA-SIFT and SURF. *International Journal of Image Processing*, *3*(4), 143–152.

Kim, D., Rho, S., & Hwang, E. (2012). Local feature based multi-object recognition scheme for surveillance. *Engineering Applications of Artificial Intelligence*, *25*(7), 1373–1380. doi:10.1016/j.engappai.2012.03.005

Koenderink, J. J. (1990). *Solid shape*. Cambridge, MA: MIT Press.

Lowe, D. (1999). Object recognition from local scale invariant features. In *Proceedings of the 7th IEEE International Conference on Computer Vision* (Vol. 2, pp. 1150-1157). doi:10.1109/ICCV.1999.790410

Lowe, D. (2004). Distinctive image features from scale-invariant key points. *International Journal of Computer Vision*, *60*(2), 91–110. doi:10.1023/B:VISI.0000029664.99615.94

Marr, D., & Nishihara, H. K. (1978). Representation and recognition of the spatial organization of three-dimensional shapes. In *Proceedings of the Royal Society of London: Series B,* (pp. 200269-94). doi:10.1098/rspb.1978.0020

Matas, J., Chum, O., Urban, M., & Pajdla, T. (2002). Robust wide baseline stereo from maximally stable extremal regions. In *Proceedings of British Machine Vision Conference*, (pp. 384-396). doi:10.5244/C.16.36

Mian, A., Bennamoun, M., & Owens, R. (2010). On the repeatability and quality of key points for local featurebased 3D object retrieval from cluttered scenes. *International Journal of Computer Vision*, *89*(2-3), 348–361. doi:10.1007/s11263-009-0296-z

Mikulka, J., Gescheidtova, E., & Bartusek, K. (2012). Soft-tissues image processing: Comparison of traditional segmentation methods with 2D active contour methods. *Measurement Science Review, 12*(4).

Miyashita, Y. (1993). Inferior temporal cortex: Where visual perception meets memory. *Annual Review of Neuroscience*, *16*(1), 245–263. doi:10.1146/annurev.ne.16.030193.001333 PMID:8460893

Moravec, H. (1980). *Obstacle Avoidance and Navigation in the Real World by a Seeing Robot Rover.* Tech Report CMU-RI-TR-3, Carnegie-Mellon University, Robotics Institute.

Naor-Raz, G., Tarr, M. J., & Kersten, D. (2003). Is color an intrinsic property of object representation? *Perception*, *32*(6), 667–680. doi:10.1068/p5050 PMID:12892428

Oliva, A., & Schyns, P. (2000). Diagnostic colors mediate scene recognition. *Cognitive Psychology*, *41*(2), 176–210. doi:10.1006/cogp.1999.0728 PMID:10968925

Ouyang, W., Tombari, F., Mattoccia, S., & Di Stefano, L. (2012). Performance evaluation of full search equivalent pattern matching algorithms. *IEEE Transactions on Pattern Analysis and Machine Intelligence*, *34*(1), 127–143. doi:10.1109/TPAMI.2011.106 PMID:21576734

Pinker, S. (1985). Visual cognition: An introduction. In S. Pinker (Ed.), *Visual Cognition* (pp. 1–63). Cambridge, MA: MIT Press.

Ravela, S. (2003). On multi-scale differential features and their representations for image retrieval and recognition. University of Massachusetts Amherst.

Rolls, E. T. (1994). Brain mechanisms for invariant visual recognition and learning. *Behavioural Processes*, *22*(1-2), 113–138. doi:10.1016/0376-6357(94)90062-0 PMID:24925242

Rossion, B., & Pourtois, G. (2004). Revisiting Snodgrass and Vander warts object pictorial set: The role of surface detail in basic-level object recognition. *Perception*, *33*(2), 217–236. doi:10.1068/p5117 PMID:15109163

Shi, J., & Tomasi, C. (1994). Good features to track. In *Proceedings of 9th IEEE Conference on Computer vision and Pattern Recognition*. Springer.

Tanaka, J. W., & Presnell, L. M. (1999). Color diagnosticity in object recognition. *Perception & Psychophysics*, *61*(6), 1140–1153. doi:10.3758/BF03207619 PMID:10497433

Tanaka, J. W., Weiskopf, D., & Williams, P. (2001). Of color and objects: The role of color in high-level vision. *Trends in Cognitive Sciences*, *5*(5), 211–215. doi:10.1016/S1364-6613(00)01626-0 PMID:11323266

Tomasi, C., & Kanade, T. (2004). Detection and tracking of point features. *Pattern Recognition*, *37*, 165–168.

Tuytelaars, T., & Mikolajczyk, K. (2008). Local invariant feature detectors: A survey. *Foundations and Trends in Computer Graphics and Vision*, *3*(3), 177–280. doi:10.1561/0600000017

Ullman, S. (1989). Aligning pictorial descriptions: An approach to object recognition. *Cognition*, *32*(3), 193–254. doi:10.1016/0010-0277(89)90036-X PMID:2752709

Wang, H., Mohamad, D., & Ismail. (2014). An efficient parameters selection for object recognition based colour features in traffic. *International Arab Journal of Information Technology*, *11*(3).

Younes, L., Romaniuk, B., & Bittar, E. (2012). A comprehensive and comparative survey of the SIFT algorithm - Feature detection, description, and characterization. In *Proceedings of the International Conference on Computer Vision Theory and Applications (VISAPP)*. SciTePress.

KEY TERMS AND DEFINITIONS

BLOB: In computer vision and image processing, blob detection methods are used for detecting regions in a digital image. It detects the different area in properties, such as color or brightness compared to neighborhoods regions.

DOG: An image gradient is defined as the directional change in the intensity in an image.it used two step process to detect the interesting points. It is somehow similar to LOG and DOB filter used in digital image processing.

FAST: It is proposed by Researcher Rosten et.al. To identify the interesting point with in an image which can be detected robustly under different conditions.

Global Feature Detector/Descriptor: Global feature detector or descriptor detects a feature with in a whole image is considered to be an image pattern.

Harris Corner: One of the most practice corner detection algorithms was developed by Harris and Stephens by improving Moravec's corner detector. Corner detection is one of the most practiced methods used to infer the contents and to extract certain kinds of features from an image. Corner detection is very often used in computer vision systems, image registration, motion detection and object recognition. Different researchers proposed different algorithm for detecting corner.

Local Feature Detector/Descriptor: A local feature detector or descriptor detects a feature with in a digital image is considered to be an image pattern that varies from its immediate neighborhood. It is typically related with a change of an image property or several properties at the same time, although it is not inevitably confined exactly on this change.

MIN-EIGEN: It was proposed by Shi and Tomasi, which is also known as Kanade-Tomasi corner detector to find corners using the minimum eigenvalues min (λ1, λ2). The algorithm returns a corner point's object within a 2-D grayscale image.

MSER: Matas et al. Proposed one feature descriptor algorithm which is very much effective where two images have different viewpoints, known as maximally stable extremal regions. It is a better stereo matching and object recognition algorithms to find correspondences between image features. It is based on blob detection.

SIFT: It is used to detect and to extract feature along with their associated descriptor from a selected image region. These interested points are popularly known as key points.

SURF: An advanced version of SIFT feature detector, used to speed up feature extraction from the selected image area.

Chapter 5
A Study on Different Edge Detection Techniques in Digital Image Processing

Shouvik Chakraborty
University of Kalyani, India

Mousomi Roy
University of Kalyani, India

Sirshendu Hore
HETC, India

ABSTRACT

Image segmentation is one of the fundamental problems in image processing. In digital image processing, there are many image segmentation techniques. One of the most important techniques is Edge detection techniques for natural image segmentation. Edge is a one of the basic feature of an image. Edge detection can be used as a fundamental tool for image segmentation. Edge detection methods transform original images into edge images benefits from the changes of grey tones in the image. The image edges include a good number of rich information that is very significant for obtaining the image characteristic by object recognition and analyzing the image. In a gray scale image, the edge is a local feature that, within a neighborhood, separates two regions, in each of which the gray level is more or less uniform with different values on the two sides of the edge. In this paper, the main objective is to study the theory of edge detection for image segmentation using various computing approaches.

INTRODUCTION

A digital image is a numeric representation (normally binary) of a two-dimensional image. There are two types of digital images. Depending on whether the image resolution is fixed, it may be of vector or raster type. Without qualifications, the term “digital image” usually refers to raster images also called bitmap images. Digital images can be classified according to the number and nature of those values. A

DOI: 10.4018/978-1-5225-1025-3.ch005

binary image is a digital image that has only two possible values (i.e. 0 or 1) for each pixel. Typically, the two colors used for a binary image are black and white though any two colors can be used. A greyscale is an image, in which the value of each pixel is a single value that carries color intensity information. Greyscale or so-called monochromatic images are composed exclusively of shades of gray, varying from black (lowest intensity) to white (highest intensity). Color image contains color information for each pixel. For visually acceptable results, it is necessary to provide three values (color channels, typically, Red, Green, and Blue in RGB format) for each pixel. The RGB color space is commonly used in computer displays, but other spaces such as HSV are often used in other contexts. A true-color image of a subject is an image that appears to the human eye just like the original, while a false-color image is an image that depicts a subject in colors that differ from reality. Range image represents the depth in the value of each pixel. It can be produced by range finder devices, such as a laser scanner, and makes a 3D volume by inserting third dimension (i.e., depth) into the 2D array of pixels. The processing of digital images is called digital image processing. The most common digital image processing tasks include:

- Resizing,
- Zooming,
- Image segmentation,
- Edge detection and
- Color enhancement.

Image segmentation is one of the fundamental problems in digital image processing. Image segmentation is also an essential step in image analysis. In digital image processing and computer vision, image segmentation is the process of partitioning a digital image into multiple segments (sets of pixels, also known as super pixels). Segmentation separates an image into its component parts or objects. The goal of segmentation is to simplify and/or change the representation of an image into something that is more meaningful and easier to analyze. Image segmentation is used to locate objects and boundaries (lines, curves, edges etc.) in images. More precisely, image segmentation is the process of assigning a label to every pixel in an image such that pixels with the same label share certain visual characteristics. Analytically, a digital image (composed of elements called pixels or picture elements), is defined as a two dimensional function f(x, y), where x and y are spatial (plane) co-ordinates. The value of 'f' at any pair of co-ordinate (x, y) is called the intensity 'or' gray level of the image at that point. Various image segmentation techniques are practiced in case of digital image processing.

One of the most natural technique used for image segmentation is edge detection techniques. Edge is a one of the basic feature of an image. Edge detection can be used as a primary tool for image segmentation. The image edges include a good number of information that is very important for obtaining the image characteristic by analyzing the image. In a grayscale image, the edge is a local feature that, within a neighborhood, separates two regions, in each of which the gray level is more or less uniform with different values on the two sides of the edge. In this paper, the main objective is to study the method of edge detection for image segmentation using different computing approaches. In edge detection techniques, we are basically using 2-D filters to detects the edges depending upon the level of the intensity difference of pixels and the level of discontinuity.

In digital image, edges consist of major properties and a good number significant characteristic of an image. Main aim of edge detection is to locate the borders of homogeneous region in an image depending on characteristics such as texture and intensity (Masud, Keshtkar, & Gueaieb, 2012). Generally, an edge detection procedure can be divided into three steps. In the first stage, a noise reduction operation is performed. In order to achieve better performance, image noise should be removed as much as possible. It is usually obtained by performing a low-pass filter. However, the edges can be accidentally removed because they are high-frequency signals. Therefore, a parameter is used to make the best trade-off. In the second step, a high-pass filter (it can be a differential operator) is usually used to find the edges. In the final step, an edge localization operation is performed to identify the real edges, which are separated from those similar responses generated by noise (Yu & Chang, 2006). Figure 1 shows these steps.

BACKGROUND

In the previous research many algorithms were developed to extract the regions in a distinguished manner. Ullrich Kothe, proposed the idea for primary image segmentation which perform as a well interpreted link between low and high level image interpretation. A common algorithmic structure depending on priority queues is introduced that allows for the incorporation of a variety of several segmentation algorithms (Kothe, 1995). Ye et. al. (2003) introduces a technique by incorporating the color feature of the pixels and spatial connectivity. They have considered that an image can be observed like a dataset in which each and every pixel has a spatial position and value of a color. In this connection we can achieve color image segmentation by clustering the pixels into different set of consistent spatial connectivity and color. Density based clustering technique is introduced to find out the spatial connectivity of the pixels (Ye, Gao & Zeng, 2003). Edge detection is a well-developed field on its own within image processing. Region boundaries and edges are closely related, since there is often a sharp adjustment in intensity at the region boundaries. Edge detection techniques have therefore been used as the base of another segmentation technique. Many scientists have given different kinds of filters (will be discussed later) by which we can find the edges. Researches are being done from many years. Masks become powerful as year passes. Monteiro and Campilho (2008) proposed a new image segmentation method that uses edge and region based information with the help of spectral method and morphological algorithm of watershed. Firstly, they reduce the noise from image using bilateral filter as a pre-processing step, secondly, region merging is used to perform preliminary segmentation, region similarity is generated and then graph based region grouping is perform using Multi-class Normalized Cut method (Hameed, Sharif, Raza, Haider, & Iqbal, 2012; Schunck, 1987). Patil et. al. (Patil & Jondhale, 2010) claims that if the number of clusters is estimated in accurate manner, K-means image segmentation will provide better results. They proposed a new method based on edge detection to estimate number of clusters. Phase congruency is used to detect the edges. Then these edges are used to find clusters. Kehtarnavaz and Monaco sort

Figure 1. Basic steps to detect edges

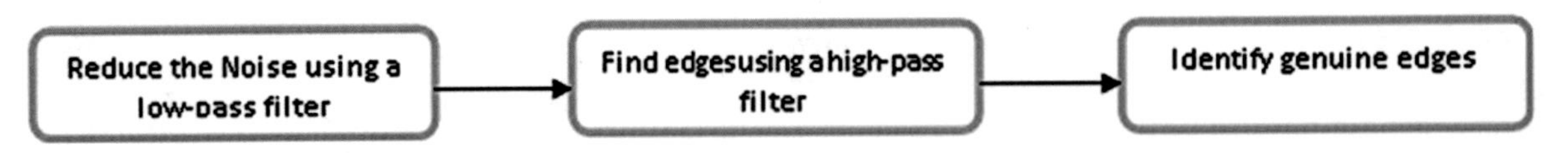

out two main problems in the image segmentation. First, the number of color clumps should be prearranged and the second through the whole of a color space, equal gaps may not be recognized equally by the human visual system. They introduce a clustering algorithm, which find out the eminent numbers of color clusters through an objective measure known as lifetime (Kehtarnavaz, Monaco, Nimtschek, & Weeks, 1998). Bai, Krishna and Sree Devi (2010) described the new morphological approach for noise removal cum edge detection for both binary and grayscale images. Lee, Taipei, Taiwan, and ROC (2012) described the various edge detection algorithms and detector design methods. Mitra Basu, Senior Member (2002) presented a survey of Gaussian-based edge detection techniques. This described in a gray level image of an edge. Edge detection is the process which detects the presence and locations of these intensity transitions. Sujaritha and Annadurai (2009) described a novel level set method for color image segmentation which using Binary Level-set Partitioning Approach. It eliminates the need of the reinitialization, calculation of number of regions procedure which is very costly.

BASICS OF AN EDGE

Edge is a part of an image that has important variation. The most common types of edges are:

- Steps,
- Lines, and
- Junctions.

It is more practical to observe a step edge as a combination of many inflection points. It has been observed that most of the edge detection model used by the different researchers is of double step edge. In the literature two categories of double edges such as the pulse and the staircase are mostly reported. Edges can be of various types. Some types are given in Figure 2.

Figure 3 shows the four types of one- dimensional continuous domain edges. Ramp edge is modeled as a ramp increase in image amplitude from low to high level, or vice versa. The edge is characterized by its height, slope angle and horizontal coordinate of the slope midpoint. A spot, which can only be defined in two dimensions, consists of plateau of high amplitude against a lower amplitude background, or vice versa (Doychev, 1985). Figure 4 shows different steps of a step edge. Here Line edges are local maxima of the grey level variance of the smoothed image or localized as zero-crossings of the first derivative. Edges of this types are typically used for remote sensing based application (Ziou & Tabbone, 1998). Finally, the junction edge is developed where two or more edges are merging to form a new edge. In Figure 6 profiles of line and junction edges are shown. The junction can be a point with high variation in gradient direction or can be a localized point with high curvature.

The edges obtained from a 2D image of a 3D image can be classified based on the basis of whether it is dependent of viewpoint or independent of viewpoint. An edge, which is viewpoint independent usually reflect inherent features of the 3D objects. Surface markings and surface shape are the two inherent features of 3D object obtain form a view independent edge. As the view point changes a viewpoint dependent edge also changes (Yu & Chang, 2006).

Figure 2. Different types of edges

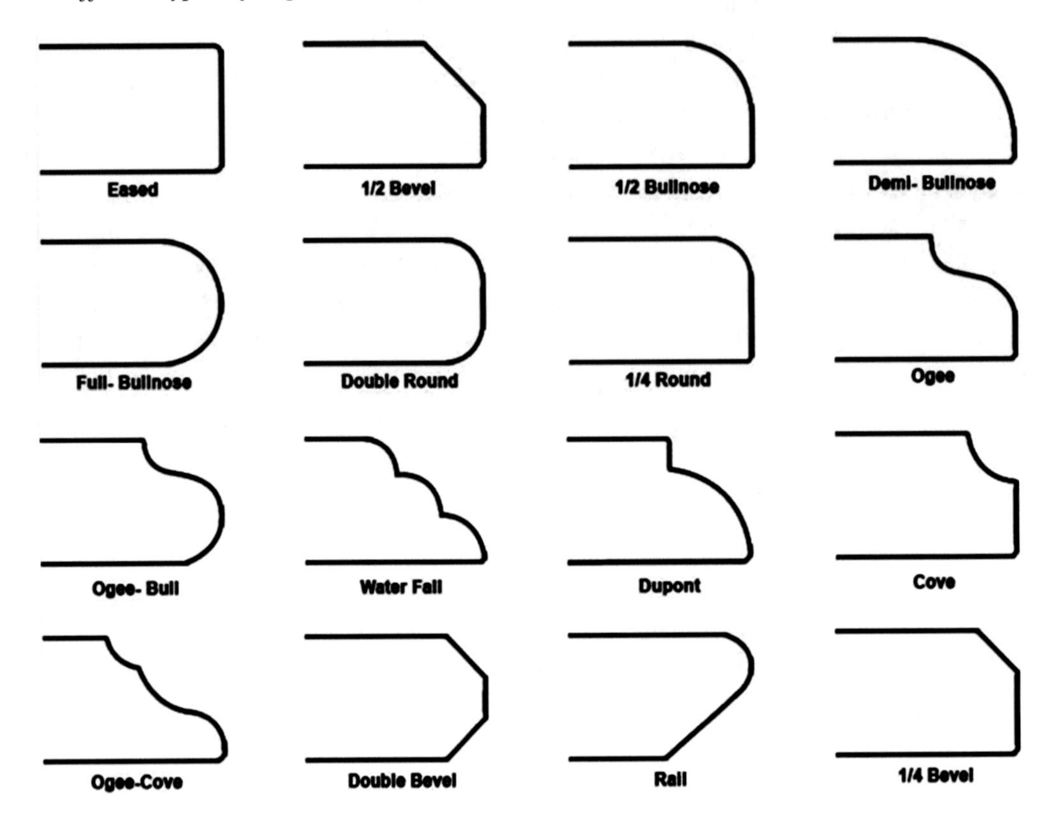

Figure 3. Four type of one-dimensional continuous domain edges

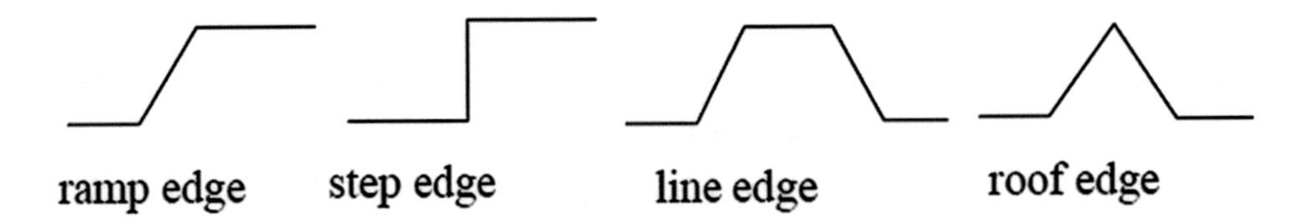

Figure 4. (a) Proper or perfect step edge, (b) smooth type step edge with the presence of noise, (c) first-order derivative step edge, (d) second-order derivative smoothed step edge tarnished by noise

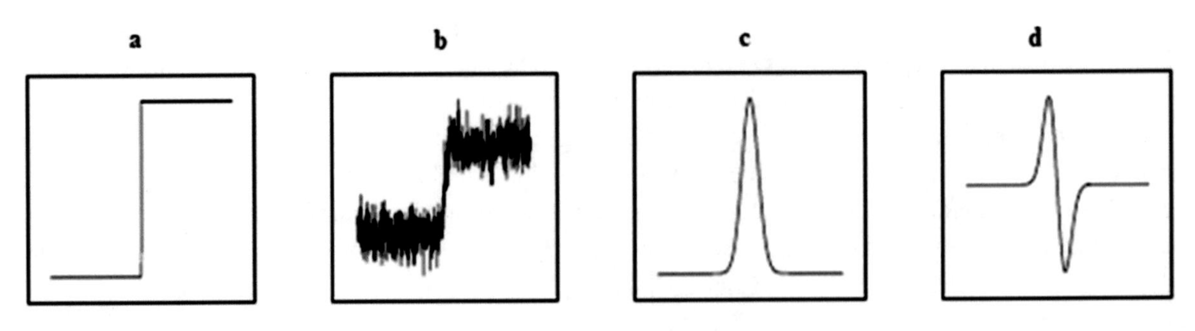

Figure 5. Profile of step edges staircase (right) and pulse (left)

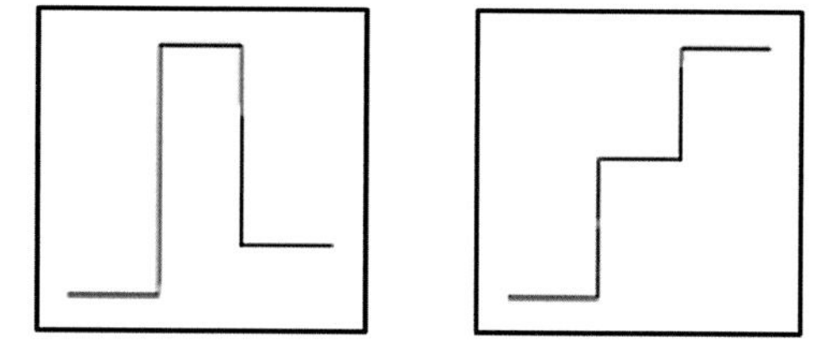

Figure 6. (a) Line and (b) junction edges

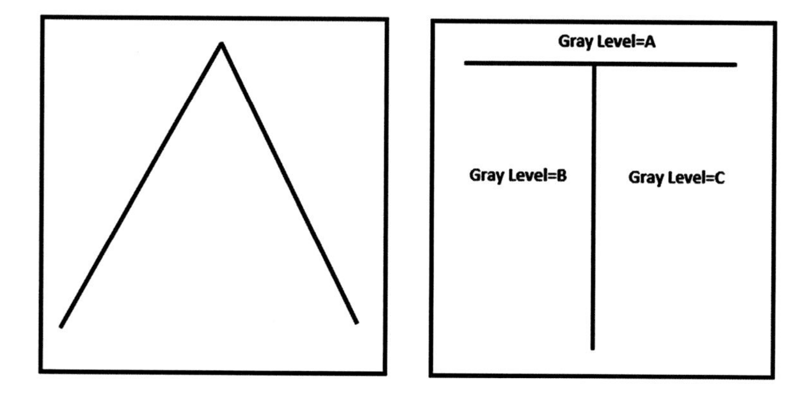

BASICS OF EDGE DETECTION

First we enhance the presence of edges in the original intensity image f(x, y), thus creating a new image g(x,y) edges are more conspicuous. Greater values indicate the likelihood of the presence of an edge. Then we threshold g(x, y) to make an edge/no edge decision, yielding binary edge map. This is shown in Figure 7.

Most of the edge detectors use simple discrete approximations of continuous derivatives. The partial derivatives of the intensity functions can be approximated with finite differences and it is given in equation 1 to equation 3.

$$\frac{\partial f}{\partial x} \approx \Delta_x f\left(x, y\right) = f\left(x+1, y\right) - f\left(x, y\right) \tag{1}$$

Figure 7. Basic steps of edge detection

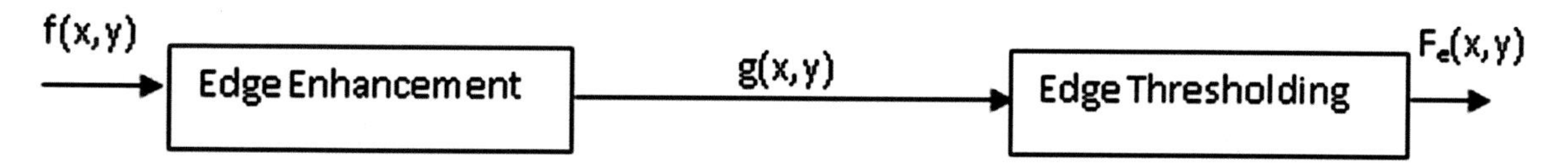

$$\frac{\partial f}{\partial x} \approx \Delta_y f(x,y) = f(x,y+1) - f(x,y) \tag{2}$$

$$\frac{\partial^2 f}{\partial x} \approx \Delta^2{}_x f(x,y) = f(x+2,y) - 2f(x+1,y) + f(x,y) \tag{3}$$

Filtering is generally used to smooth the data before differentiating it. They are number of possible filters.

- Band limited filters can eliminate noise with no frequency content.
- We can also use support-limited filters or finite impulse response (FIR) filters. They are interesting from a computational standpoint, but they have infinite frequency support.
- Another class of filters that can be considered as a compromise between the first two classes is the class of filters that minimize uncertainty. Such class of real functions that minimizes it is the class of Gaussian functions (Doychev, 1985).

Here is a list of some common mathematical tool used in edge detection.

1. Image Differentiation,
2. Discrete Differentiation,
3. Convolution,
4. Image Smoothing,
5. Non-Maximum Suppression,
6. Hysteresis Algorithm,
7. Sub-pixel Accuracy.

DIFFERENT TECHNIQUES TO DETECT DISCONTINUITIES

In a Digital image, Points, Lines, and Edges are the three types of gray-level discontinuities. The most common way to look for such discontinuities is the use of mask. A mask runs through the image to find such discontinuity. For a mask 3 x 3 this procedure involves computing the sum of products of the coefficients with the gray levels contained in the region encompassed by the mask. As usual, the response of the mask is defined with respect to its center location.

- **Detection of Point:** Interest point detection is a recent terminology in computer vision that refers to the detection of interest points for subsequent processing. Figure 8 shows a point detection mask.
- **Detection of Lines:** The next level of complexity is Line detection. Figure 9 shows the line detection masks. If we are interested in detecting all the lines in an image in the direction defined be a given mask, we simply run the mask through the image and threshold the absolute value of the result (Doychev, 1985).

- **Detection of Edges:**
  - **First Order Derivatives:** There two basic methods for generating first-order derivative edge gradients. One method includes generation of gradients in two orthogonal directions in an image; the second utilizes a set of directional derivatives.
    - **Gradient Generation:** An image gradient is a directional change in the intensity or color in an image. An edge in a continuous domain edge segment F(x, y) can be detected by forming the continuous one-dimensional gradient G(x, y). The gradient along the line normal to the edge slope can be computed in terms of the derivatives along orthogonal axes according to the equation 4.

$$G(x,y) = \frac{\partial F(x,y)}{\partial x} + \frac{\partial F(x,y)}{\partial y} \tag{4}$$

    - **Threshold Selection:** After the edge gradient is formed for the differential edge detection methods, the gradient is compared to a threshold to determine if an edge exists. The threshold value determines the sensitivity of the edge detector. For noise free images, the threshold can be chosen such that all amplitude discontinuities of a minimum contrast level are detected as edges, and all others are called non-edges.
  - **Second Order Derivatives:** Second-order derivative edge detection techniques employ some form of spatial second-order differentiation to accentuate edges. For the case of a finite-dimensional graph (having a finite number of edges and vertices), the discrete Laplace operator is more commonly called the Laplacian matrix.

The edge Laplacian (▼^{2}) of an image F(x, y) in the continuous domain is defined in equation 5.

Figure 8. Point detection mask

-1	-1	-1
-1	8	-1
-1	-1	-1

Figure 9. Line detection masks: (a) horizontal, (b) +45o, (c) vertical, (d) -45o

a

-1	-1	2
-1	2	-1
2	-1	-1

b

-1	2	-1
-1	2	-1
-1	2	-1

c

2	-1	-1
-1	2	-1
-1	-1	2

d

-1	-1	-1
2	2	2
-1	-1	-1

$$G(x,y) = -\nabla^2\{F(x,y)\} \Rightarrow \nabla^2 = \frac{\partial^2}{\partial x^2} + \frac{\partial^2}{\partial y^2} \quad (5)$$

The Laplacian G(x, y) is zero if F(x, y) is constant or changing linearly in amplitude. If the rate of change of F(x, y) is greater than linear, G(x, y) exhibits a sign change at the point of inflection of F(x, y). The zero crossing of G(x, y) indicates the presence of an edge. The negative sign in the definition is present so that the zero crossing of G(x, y) has positive slope for an edge whose amplitude increases from left to right or bottom to top in an image.

- **Kernel:** A kernel is used in Digital image for the purpose of image convolutions. A kernel is a nothing but a small matrix of numbers. For the purpose of image convolution, different sized of kernels are used. This kernels contains different patterns of numbers. Thus it helps to get different results under convolution. For example, Figure 10 shows a 3×3 kernel that implements a mean filter.

Figure 10. A 3 X 3 kernel

1	1	1
1	1	1
1	1	1

Set of Coordinate Points=

{ {-1 , -1}, {0, -1}, { 1 , -1 },

{ -1 ,0}, { 0, 0} , {1 ,0},

{-1 , 1} , { 0 , 1} , {1 , 1 } }

CLASSIFICATION OF EDGE DETECTION TECHNIQUES

Edge detection techniques can be classified in eleven ways and it is shown in Figure 11.

Classical Methods

Classical edge detectors have no smoothing filter, and they are only based on a discrete differential operator. The earliest popular works in this category include the algorithms developed by Sobel (1970), Prewitt (1970), Kirsch (1971), Robinson (1977), and Frei-Chen (1977) etc. They compute an estimation of gradient for the pixels, and look for local maxima to localize step edges. Typically, they are simple in computation and capable to detect the edges and their orientation, but due to lack of smoothing stage, they are very sensitive to noise and inaccurate.

Figure 11. Different edge detection techniques

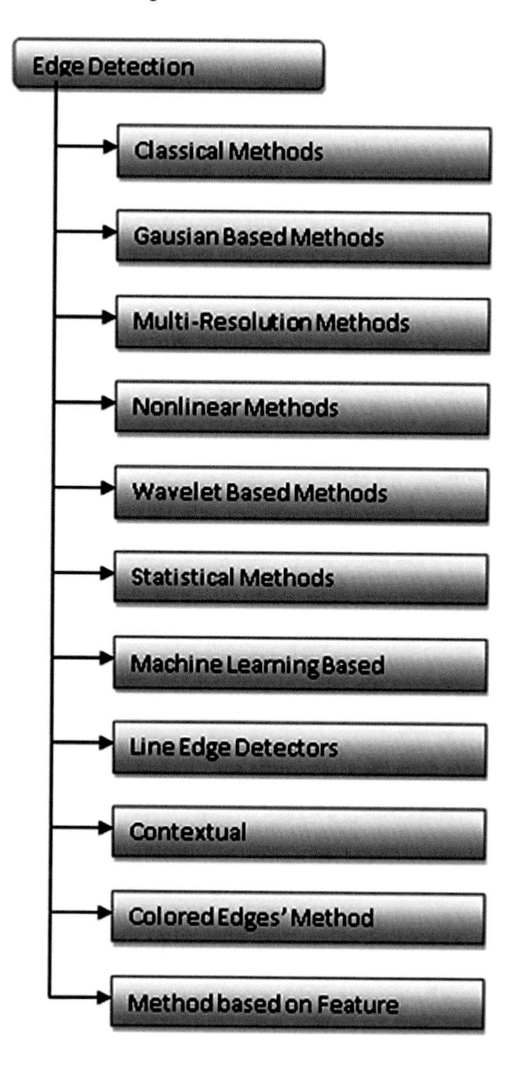

- **Pixel Difference:** This is one of the simplest operators. The kernel of this operator is given in Figure 12.
- **Separated Pixel Difference:** This is the modified version of the pixel difference operator where values are placed in the kernel leaving one gap in between. The kernel of this operator is given in Figure 13.
- **Robert's Cross Operator:** The first edge detector was initially proposed by Lawrence Roberts in 1963, popularly known as Roberts filter in digital image processing. The nature of the operator proposed by Robert is cross operator which is a kind of differential operator, and used widely for the purpose of edge detection in the field of digital image processing or in computer vision. The Roberts cross operator finds the edge with in an image through approximation of the gradient of an image. The gradient is obtained by calculating the sum of the square of the differences between diagonally adjacent pixels. The kernel of the Robert's operator is given in Figure 14.
- **Sobel Operator:** The Sobel operators used in image processing and computer vision, mainly to found the edges in a digital image. The Sobel operator produces an image that emphasizes edges and its transitions. It was Irwin Sobel, who proposed the idea stated above in 1968 (Sobel, 1967). Sobel operator is based on 3x3 Isotropic Image Gradient. The Sobel operator uses a small integer type filter for the purpose of convolution. The Filter operates both in horizontal and vertical direction. It is therefore computationally inexpensive, though the gradient approximation that it has produces is relatively rough. This phenomena is particular observed for those image which has high frequency variations. The kernel of the Sobel's operator is given in Figure 15. Sobel's 5x5 kernel is given in Figure 16.
- **Kayyali Operator:** This operator was developed by Kayyali Selim Mohamed, PhD & CSci FBCS at the USMP Labs in 2000. Kayyali Operator for edge detection is generated from Sobel operator. Generally, we use Kayyali Operator to detect the edges in alignment to directions of (south east -north west (S.E.N.W)) or (north east – south west (N.E.S.W)). The kernels in Kayyali operator are designed to ignore the respond maximally to edges running vertically and horizontally (the horizontal and vertical pixel will not be detected) and it is given in Figure 17.
- **Prewitt Operator:** The Prewitt operator is used in image processing, particularly within edge detection algorithms. The Prewitt operator is based on convolving the image with a small, separable, and integer valued filter in horizontal and vertical directions and is therefore relatively inexpensive in terms of computations. The Prewitt operator was developed by Judith M. S. Prewitt and the kernel of the 3 X 3 Prewitt operator is given in Figure 18 and 7 X 7 Prewitt operator is given in Figure 19.
- **Kirsch Operator:** The Kirsch operator or Kirsch compass kernel is a non-linear edge detector that finds the maximum edge strength in a few predetermined directions. It is named after the computer scientist Russell A. Kirsch.

The operator takes a single kernel mask and rotates it in 45 degree increments through all 8 compass directions: N, NW, W, SW, S, SE, E, and NE. The edge magnitude of the Kirsch operator is calculated as the maximum magnitude across all directions using equation 6.

$$h_{n,m} = \max_{z=1,2,\ldots\ldots,8} \sum_{i=-1}^{1} \sum_{j=-1}^{1} {g^{(z)}}_{ij} \cdot f_{n+i,m+j} \tag{6}$$

Figure 12. Pixel difference operator

Row Gradient

0	0	0
0	1	-1
0	0	0

Column Gradient

0	-1	0
0	1	0
0	0	0

Figure 13. Separated pixel difference operator

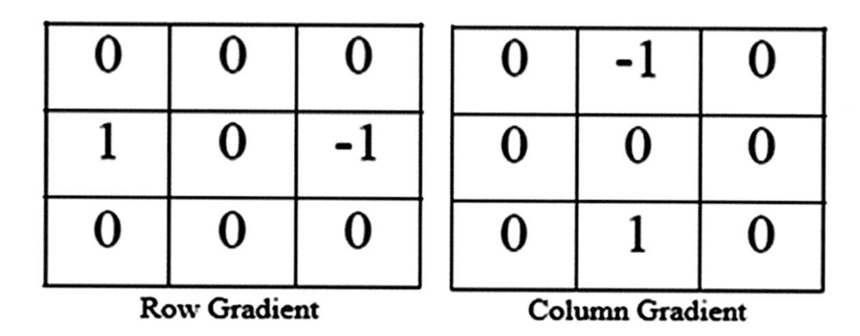

Row Gradient

0	0	0
1	0	-1
0	0	0

Column Gradient

0	-1	0
0	0	0
0	1	0

Figure 14. Robert's cross operator

Row Gradient

0	0	-1
0	1	0
0	0	0

Column Gradient

-1	0	0
0	1	0
0	0	0

Figure 15. Sobel's 3 X 3 operator

Row Gradient: $\frac{1}{4}*$

1	0	-1
2	0	-2
1	0	-1

Column Gradient: $\frac{1}{4}*$

-1	-2	-1
0	0	0
1	2	1

Figure 16. Sobel's 5 X 5 operator

Row Gradient

1	2	0	−2	−1
4	8	0	−8	−4
6	12	0	−12	−6
4	8	0	−8	−4
1	2	0	−2	−1

Column Gradient

−1	−4	−6	−4	−1
−2	−8	−12	−8	−2
0	0	0	0	0
2	8	12	8	2
1	4	6	4	1

Figure 17. Kayyali operator

Kayyali SENW direction

6	0	-6
0	0	0
-6	0	6

Kayyali NESW direction

-6	0	6
0	0	0
6	0	-6

where z enumerates the compass direction kernels (Kirsch, 1971). The kernels are shown in Figure 20.

- **Frei-Chen Operator:** The Frei-Chen edge detector also works on a 3×3 pixel footprint but applies a total of nine convolution masks to the image. Frei-Chen masks are unique masks, which contain all of the basis vectors. The kernels are shown in Figure 21. A 3×3 image area is represented with the weighted sum of nine Frei-Chen masks that can be seen in Figure 22.
- **Pyramid Operator:** The kernels of this operator are shown in Figure 23.
- **Compass Operator:** When using compass edge detection the image is convolved with a set of (in general 8) convolution kernels, each of which is sensitive to edges in a different orientation. Various kernels can be used for this operation; the Prewitt kernel is shown in Figure 24.

Gaussian Based Methods

Gaussian filters are the most widely used filters in image processing and extremely useful as detectors for edge detection. Marr and Hildreth (Zeng, Liu & Tian, 2004; Bhandarkar, Zang & Potter, 1994) were the pioneers that proposed an edge detector based on Gaussian filter. Their method had been a very popular one, before Canny released his detector. They suggested the 2D Gaussian function, defined in equation 7, as the smoothing operator.

Figure 18. Prewitt 3 X 3 operator

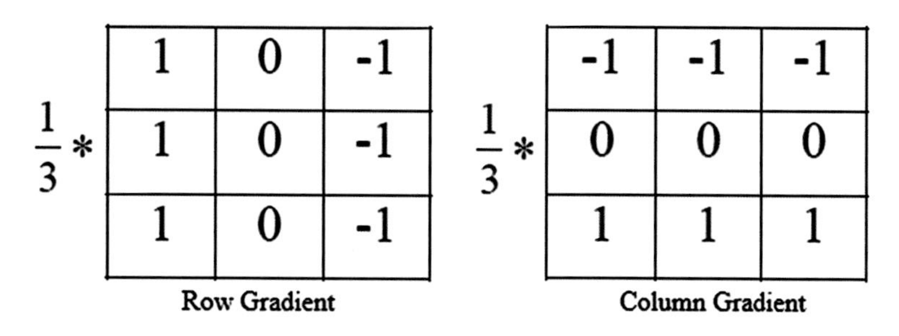

$\frac{1}{3}*$

1	0	-1
1	0	-1
1	0	-1

Row Gradient

$\frac{1}{3}*$

-1	-1	-1
0	0	0
1	1	1

Column Gradient

Figure 19. Prewitt 7 X 7 operator

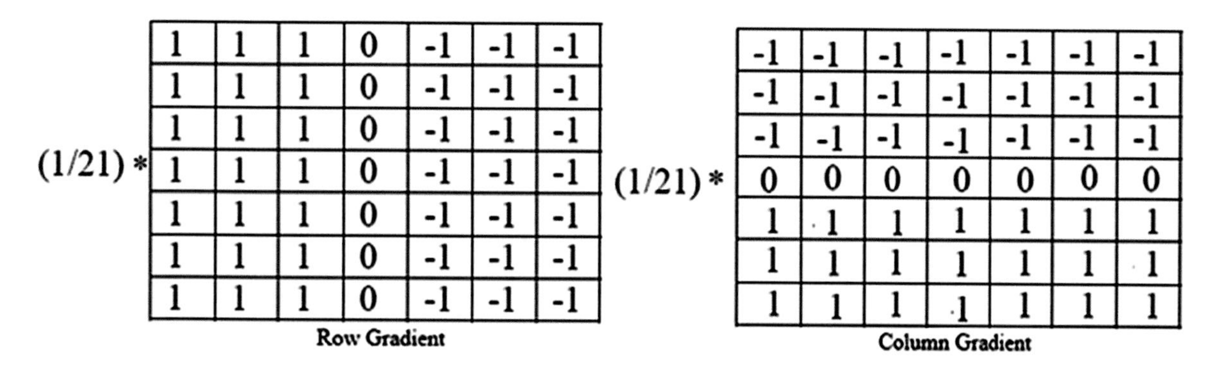

(1/21) *

1	1	1	0	-1	-1	-1
1	1	1	0	-1	-1	-1
1	1	1	0	-1	-1	-1
1	1	1	0	-1	-1	-1
1	1	1	0	-1	-1	-1
1	1	1	0	-1	-1	-1
1	1	1	0	-1	-1	-1

Row Gradient

(1/21) *

-1	-1	-1	-1	-1	-1	-1
-1	-1	-1	-1	-1	-1	-1
-1	-1	-1	-1	-1	-1	-1
0	0	0	0	0	0	0
1	1	1	1	1	1	1
1	1	1	1	1	1	1
1	1	1	1	1	1	1

Column Gradient

Figure 20. Kirsch operator

+5	+5	+5
-3	0	-3
-3	-3	-3

+5	+5	-3
+5	0	-3
-3	-3	-3

+5	-3	-3
+5	0	-3
+5	-3	-3

-3	-3	-3
+5	0	-3
+5	+5	-3

-3	-3	-3
-3	0	-3
+5	+5	+5

-3	-3	-3
-3	0	+5
-3	+5	+5

-3	-3	+5
-3	0	+5
-3	-3	+5

-3	+5	+5
-3	0	+5
-3	-3	-3

Figure 21. Frei-Chen operator

Row Gradient:

$$\frac{1}{2+\sqrt{2}} * \begin{bmatrix} 1 & 0 & -1 \\ \sqrt{2} & 0 & -\sqrt{2} \\ 1 & 0 & -1 \end{bmatrix}$$

Column Gradient:

$$\frac{1}{2+\sqrt{2}} * \begin{bmatrix} -1 & -\sqrt{2} & -1 \\ 0 & 0 & 0 \\ 1 & \sqrt{2} & 1 \end{bmatrix}$$

Figure 22. Frei-Chen operator

$$G_1 = \frac{1}{2\sqrt{2}}\begin{bmatrix} 1 & \sqrt{2} & 1 \\ 0 & 0 & 0 \\ -1 & -\sqrt{2} & -1 \end{bmatrix} \quad G_2 = \frac{1}{2\sqrt{2}}\begin{bmatrix} 1 & 0 & -1 \\ \sqrt{2} & 0 & -\sqrt{2} \\ 1 & 0 & -1 \end{bmatrix} \quad G_3 = \frac{1}{2\sqrt{2}}\begin{bmatrix} 0 & -1 & \sqrt{2} \\ 1 & 0 & -1 \\ -\sqrt{2} & 1 & 0 \end{bmatrix}$$

$$G_4 = \frac{1}{2\sqrt{2}}\begin{bmatrix} \sqrt{2} & -1 & 0 \\ -1 & 0 & 1 \\ 0 & 1 & -\sqrt{2} \end{bmatrix} \quad G_5 = \frac{1}{2}\begin{bmatrix} 0 & 1 & 0 \\ -1 & 0 & -1 \\ 0 & 1 & 0 \end{bmatrix} \quad G_6 = \frac{1}{2}\begin{bmatrix} -1 & 0 & 1 \\ 0 & 0 & 0 \\ 1 & 0 & -1 \end{bmatrix}$$

$$G_7 = \frac{1}{6}\begin{bmatrix} 1 & -2 & 1 \\ -2 & 4 & -2 \\ 1 & -2 & 1 \end{bmatrix} \quad G_8 = \frac{1}{6}\begin{bmatrix} -2 & 1 & -2 \\ 1 & 4 & 1 \\ -2 & 1 & -2 \end{bmatrix} \quad G_9 = \frac{1}{3}\begin{bmatrix} 1 & 1 & 1 \\ 1 & 1 & 1 \\ 1 & 1 & 1 \end{bmatrix}$$

Figure 23. Pyramid operator

(1/34) *

1	1	1	0	-1	-1	-1
1	2	2	0	-2	-2	-1
1	2	3	0	-3	-2	-1
1	2	3	0	-3	-2	-1
1	2	3	0	-3	-2	-1
1	2	2	0	-2	-2	-1
1	1	1	0	-1	-1	-1

Row Gradient

(1/34) *

-1	-1	-1	-1	-1	-1	-1
-1	-2	-2	-2	-2	-2	-1
-1	-2	-3	-3	-3	-2	-1
0	0	0	0	0	0	0
1	2	3	3	3	2	1
1	2	2	2	2	2	1
1	1	1	1	1	1	1

Coloumn Gradient

Figure 24. Compass operator applied on Prewitt kernel

1	0	-1	0	-1	-1	-1	-1	-1	-1	-1	0
1	0	-1	1	0	-1	0	0	0	-1	0	1
1	0	-1	1	1	0	1	1	1	0	1	1

-1	0	1	0	1	1	1	1	1	1	1	0
-1	0	1	-1	0	1	0	0	0	1	0	-1
-1	0	1	-1	-1	0	-1	-1	-1	0	-1	-1

$$G_\sigma(x,y) = \frac{1}{2\pi\sigma^2}\exp\left(-\frac{x^2+y^2}{2\sigma^2}\right) \tag{7}$$

where σ is the standard deviation, and (x, y) are the Cartesian coordinates of the image pixels. They showed that by applying Gaussian filters of different scales (i.e. σ) to an image, a set of images with different levels of smoothing will be obtained using (8).

$$\hat{g}(x,y,\sigma) = G_\sigma(x,y) * g(x,y) \tag{8}$$

Then, to detect the edges in these images they proposed to find the zero-crossings of their second derivatives. Marr and Hildreth achieved this by using Laplacian of Gaussian (LOG) function as filter. Since Laplacian is a scalar estimation of the second derivative, LOG is an orientation-independent filter (i.e. no information about the orientation) that breaks down at corners, curves, and at locations where image intensity function varies in a nonlinear manner along an edge. As a result, it cannot detect edges at such positions. According to Zeng, Liu and Tian (2004), the smoothing and differentiation operations can be implemented by a single operator consisting on the convolution of the image with the Laplacian of the Gaussian function. Different LoG filter kernel is shown in Figure 25. Gauss FDOG operator kernel is given in Figure 26.

In spite of unique features of the Gaussian function, the filter proposed by Marr and Hildreth had some deficiencies related to using the zero-crossing approach. Haralick (1983) proposed the use of zero-crossing of the second directional derivative of the image intensity function. This is theoretically the same as using the maxima in the first directional derivatives, and in one dimension is the same as the LOG filter.

Figure 25. Different LoG filter kernel

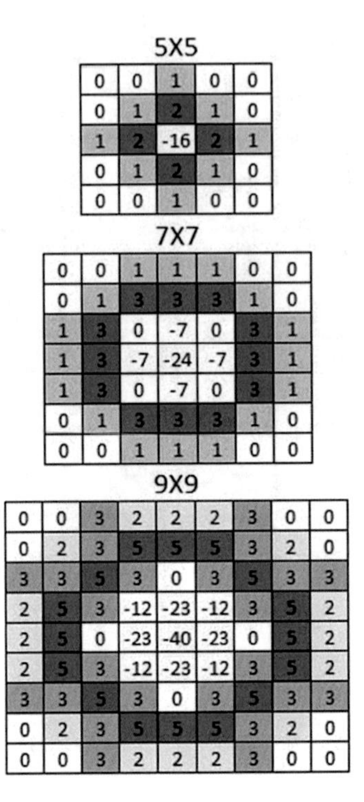

5X5

0	0	1	0	0
0	1	2	1	0
1	2	-16	2	1
0	1	2	1	0
0	0	1	0	0

7X7

0	0	1	1	1	0	0
0	1	3	3	3	1	0
1	3	0	-7	0	3	1
1	3	-7	-24	-7	3	1
1	3	0	-7	0	3	1
0	1	3	3	3	1	0
0	0	1	1	1	0	0

9X9

0	0	3	2	2	2	3	0	0
0	2	3	5	5	5	3	2	0
3	3	5	3	0	3	5	3	3
2	5	3	-12	-23	-12	3	5	2
2	5	0	-23	-40	-23	0	5	2
2	5	3	-12	-23	-12	3	5	2
3	3	5	3	0	3	5	3	3
0	2	3	5	5	5	3	2	0
0	0	3	2	2	2	3	0	0

Figure 26. Gauss FDOG operator

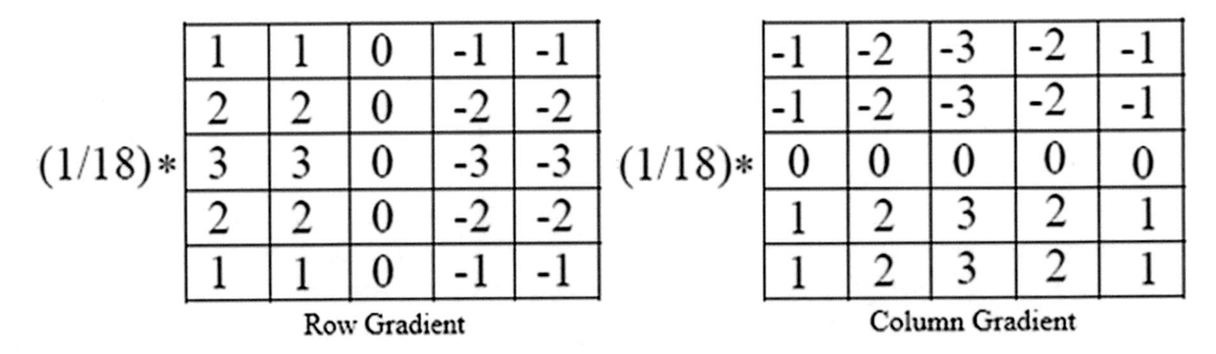

(1/18)*

1	1	0	-1	-1
2	2	0	-2	-2
3	3	0	-3	-3
2	2	0	-2	-2
1	1	0	-1	-1

Row Gradient

(1/18)*

-1	-2	-3	-2	-1
-1	-2	-3	-2	-1
0	0	0	0	0
1	2	3	2	1
1	2	3	2	1

Column Gradient

- **Multi Resolution Methods:** Multi-resolution methods incorporate repeating edge detection for several scales of the Gaussian filter to achieve a quality performance. The main challenges in these methods includes selection a proper range for the scales, combination of the outputs corresponding to different scales, and adaptation to level of Gaussian filters at multiple scales. The initial steps of Schunck's algorithm are based on Canny's method. Schunck's algorithm chooses the width of the smallest Gaussian filter, and the filters that are used differ in width by a factor of two. However, he did not discuss how to determine the number of filters to use. In addition, by choosing such a large size for the smallest filter, Schunck's technique loses a lot of important details which may exist at smaller scales (Basu, 2002). Witkin (1983) studied the property of zero-crossings across scales for

1D signal. He marked the zero-crossings of second derivative of a signal smoothed by Gaussian function in a range of scale, and then presented them versus scales.

- **Non-Linear Methods:** This sub-Section looks into edge-detectors that leave the linear territory in search of better performance. Nonlinear methods based on the Gaussian filter evolved as researchers discovered the relationship between the solution to the heat equation (in physics) and images convolved with Gaussian filter for a smoothing purpose. Consider a set of derived images, g(x, y, σ), by convolving the original image with a Gaussian filter Gσ(x, y) of variance σ (Bergholm, 1987). The parameter σ corresponds to time in the heat equation, whereas in the context of image it refers to the scale. This one parameter family of derived images can be viewed as the solution of the heat equation. However, in the case of linear heat equation as diffusion eradicates noise, it also blurs the edges isotropically (i.e. invariant with respect to direction). To overcome this problem, Perona and Malik (1990) proposed a scale space representation of an image based on anisotropic diffusion. In the mathematical context, this calls for nonlinear partial differential equations rather than the linear heat equation.
- **Wavelet Based Methods:** As it was mentioned, analyzing an image at different scales increases the accuracy and reliability of edge detection. Focusing on localized signal structures, e.g., edges, with a zooming procedure enables simultaneous analysis from a rough to a fine shape. Progressing between scales also simplifies the discrimination of edges versus textures.

Heric and Zazula (2007) presented an edge detection algorithm using Haar wavelet transform. They chose Haar wavelet as the mother wavelet function, because it was orthogonal, compact and without spatial shifting in the transform space. By applying WT, they presented the intensity magnitude variation between adjacent intervals on a time-scale plane. Positive or negative peaks in time-scale representations were called modulus maxima. Their values indicated the edge slope and width.

- **Statistical Methods:** Konishi et al. (2003) shows the edge detection as a statistical inference. Bezdek et al. (1998) suggested that edges of a digital image can be detectedusing a combination of four steps. The steps suggested by the authors are conditioning, extraction of feature, blending, and scaling. In the proposed work authors studied the role of geometry in finding good features for edge. Santis and Sinisgalli (1999) developed an algorithm based on statistical edge detection. In their proposed work to detect the edge a linear stochastic signal model was used. A physical image descriptor was used to derive the model. The mean value of the gray-level image was used a sharp local variation to detect the presence of an edge in the digital image.
- **Machine Learning Based Methods:** Wu et al. (2007) developed fast multilevel fuzzy edge detection. At first, the algorithm enhances the image contrast by means of the fast multilevel fuzzy enhancement (FMFE) algorithm. Second, the edges are extracted from the enhanced image by the two-stage edge detection operator that identifies the edge candidates based on the local characteristics of the image.

A fuzzy neural network based system for edge detection and subsequent enhancement by restoring missing edges and removing false edges generated due to noise was proposed by Lu et al. (Lu, Wang & Shen, 2003).

Zheng et al. (2004) presented an edge detection algorithm that employs estimations of image intensity derivatives produced by least square support vector machine (LS-SVM).

- **Contextual Methods:** Yu and Chang (2006) suggested an adaptive edge detection approach based on context analysis. The proposed approach uses the information from predictive error values produced by the gradient-adjusted predictor (GAP) to detect edges. GAP uses a context, which is a combination of the intensity values of already processed neighboring pixels defined by a template, to produce the predictive values. The context in the casual template of GAP is used to analyze whether the current pixel is an edge point or not.
- **Line Edge Detectors:** As mentioned earlier, line edges correspond to local maxima of intensity function in the image and are of great use in the identification of image features, such as roads and rivers in remote sensing images, as well as the contactless paper counting. Haralick (1984) proposed an algorithm based on polynomial fitting. Toobtain a line that exists in the negative local maximum of the second derivative of the image was proposed by Giraudon (1985). Koundinya and Chanda (1994) give an algorithm based on combinatorial search.
- **Colored Edges Method:** Koschan and Abidi (2005) presented a survey of techniques for the edge detection and classification in color images. The basic difference between color and gray-level images is that, in a color image, a color vector is assigned to each pixel.
- **Feature Detection Based Methods:**
    - **Canny Edge Detector:** The Canny edge detector is an operator that uses a multiple stage process to find a wide range of edges in images (Canny, 1986). It was proposed by John F. Canny in 1986. Canny also developed a computational theory of edge detection. Below are the proposed stages:
        - Noise reduction,
        - Finding the intensity gradient of the image,
        - Non-maximum suppression,
        - Tracing edges through the image and hysteresis thresholding.
    - **Harris Edge Detector:** Harris corner detector is based on the Moravec operator. Moravec's corner detector functions by considering local windows in the image, and determining the average changes of image intensity that result from shifting the window by a small amount in various directions.

Three cases need to be considered (Doychev, 1985):

1. If the windowed image patch is flat (i.e. approximately constant in intensity), then all shifts will result in only a small change.;
2. If the window straddles an edge, then a shift along the edge will result in a small change, but a shift perpendicular to the edge will result in a large change;
3. If the windowed patch is a corner or isolated point, then all shifts will result in a large change. A corner can thus be detected by finding when the minimum change produce by any of the shifts is large.

CONCLUSION

This paper is a review on the various published articles on edge detection method. This article gives a theoretical base, and then studies several of techniques of edge detection in different categories. The manuscript also studies the relationship among categories, and shows evaluations regarding to their application and performance. It was stated that the edge detection techniques are a combination of image smoothing and differentiation plus a post-processing for edge labeling.

Future Direction

The main issues of edge detection, objective of edge detection and different methods of edge detection, presence of noise and their effect on edge detection was presented. Further, the difficulties in edge detection and condition for good edge detection were discussed. Edge detection can be used for image segmentation, extraction of feature etc. Edges provide the geometric representation and structure information of objects present in an image. For example, different buildings can be easily recognized from their body shape as well as structure. The highway and forests from aerial images can be detected in terms of their distribution patterns that are illustrated by edges. By using edge detection method, machine vision and image processing systems can be developed for a variety of applications. For example, edge detection can be applied in assembly line inspection to detect faults of mechanical parts and in semiconductors, for finding the road and recognizing obstacles in automatic vehicle navigation. For medical imaging fields, edge detection and segmentation can be used for positioning tumors and blood vessels, and rigid bony structures. So, various edge detectors were discussed in details to find their performance on noise free and noisy images. Various experiments were performed on different images. From the quantitative as well as qualitative results of edge detectors on noise free images as well as on images corrupted with salt and pepper noise, it was concluded that, Canny and Laplacian of Gaussian are better choice for edge detection in case of noise free images. Sobel and Prewitt edge detectors are good edge detectors for images with sharp edges at low and medium noise levels, whereas LoG and Canny are edge detectors for noisy Lena image (smooth). It was also observed that all the edge detectors in study shows miserable performance at higher noise densities ($d = 0.25$ and $d = 0.5$). Next, the results of various edge detectors on noise free images as well as on images corrupted with Gaussian noise were presented both qualitatively as well as quantitatively. It was observed that Sobel and Prewitt are good edge detectors at noise variance (0.05 and 0.1) on synthetic images whereas their performance is poor on smooth natural gray scale images at all noise levels. All edge detectors show poor performance quantitatively on all noisy images at noise variance (0.25 and 0.5). 248 From the analysis of various edge detectors on noise free as well as noisy images, it was concluded that edge detection is difficult from noisy images, since both the noise and the edges contain high frequency content. Attempts to reduce the noise result in blurred and distorted edges. Operators used on noisy images are typically larger in scope, so they can average enough data to discount localized noisy pixels. After the comprehensive analysis of various edge detectors, it was found that there are problems of false edge detection, missing true edges, edge localization, high computational time and problems due to noise etc. To address some of the issues of existing edge detectors, an improved edge detector has been proposed. Future Directions There is sufficient scope of improvement in the edge detector methods. The following directions may be explored:

- Some edge detector is not able to detect roof type of edges.
- In case of medium and high noise levels, though the accuracy of some edge detector is high but noise is still present in the images
- Various edge detector can be further extended to test its performance on images corrupted with Speckle or Poisson noise.
- Edge detectors can be extended to cover the edge detection of colored images.
- A good edge detector can be designed for to three dimensional images.
- Different edge detectors can be implemented on FPGA for real time application.

REFERENCES

Bai, M. R., Krishna, V. V., & Sree Devi, J. (2010). A new morphological approach for noise removal cum edge detection. *International Journal of Computer Science Issues*, *7*(6), 187–190.

Basu, M. (2002). Gaussian based edge-detection methods: A survey. *IEEE Transactions on System, Man, and Cybernetics Part C: Application and Reviews, 32*(3).

Bergholm, F. (1987). Edge focusing. *IEEE Transactions on Pattern Analysis and Machine Intelligence*, *9*(6), 726–741. doi:10.1109/TPAMI.1987.4767980 PMID:21869435

Bezdek, J. C., Chandrasekhar, R., & Attikiouzel, Y. (1998). A geometric approach to edge detection. *IEEE Transactions on Fuzzy Systems*, *6*(1), 52–75. doi:10.1109/91.660808

Bhandarkar, S. M., Zhang, Y., & Potter, W. D. (1994). An edge detection technique using genetic algorithm-based optimization. *Pattern Recognition*, *27*(9), 1159–1180. doi:10.1016/0031-3203(94)90003-5

Canny, J. (1986). A computational approach to edge detection. *IEEE Transactions on Pattern Analysis and Machine Intelligence*, *8*(6), 679–698. doi:10.1109/TPAMI.1986.4767851 PMID:21869365

Doychev, Z. (1985). *Edge detection and feature extraction.* Springer Verlag.

Giraudon, G. (1985). Edge detection from local negative maximum of second derivative. In *Proceedings of IEEE, International Conference on Computer Vision and Pattern Recognition*, (pp. 643-645).

Hameed, M., Sharif, M., Raza, M., Haider, S. W., & Iqbal, M. (2012). Framework for the comparison of classifiers for medical image segmentation with transform and moment based features. *Research Journal of Recent Sciences*, 2277-2502.

Haralick, R. M. (1983). Ridge and valley on digital images. *Computer Vision Graphics and Image Processing*, *22*(1), 28–38. doi:10.1016/0734-189X(83)90094-4

Haralick, R. M. (1984). Digital step edges from zero-crossing of second directional derivatives. *IEEE Transactions on Pattern Analysis and Machine Intelligence*, *6*(1), 58–68. doi:10.1109/TPAMI.1984.4767475 PMID:21869165

Heric, D., & Zazulam, D. (2007). Combined edge detection using wavelet transform and signal registration. *Elsevier Journal of Image and Vision Computing*, *25*(5), 652–662. doi:10.1016/j.imavis.2006.05.008

Kehtarnavaz, N., Monaco, J., Nimtschek, J., & Weeks, A. (1998). Color image segmentation using multi-scale clustering. In *Image Analysis and Interpretation,IEEE Southwest Symposium*, (pp. 142 – 147). Doi:10.1109/IAI.1998.666875

Kirsch, R. (1971). Computer determination of the constituent structure of biological images. *Computers and Biomedical Research, an International Journal*, *4*(3), 315–328. doi:10.1016/0010-4809(71)90034-6 PMID:5562571

Konishi, S., Yuille, A. L., Coughlan, J. M., & Zhu, S. C. (2003). Statistical edge detection: Learning and evaluating edge cues. *IEEE Transactions on Pattern Analysis and Machine Intelligence, 25*(1), 57–74. doi:10.1109/TPAMI.2003.1159946

Koschan, & Abidi, M. (2005). Detection and classification of edges in color images. *IEEE Signal Processing Magazine*, 64-73.

Kothe, U. (1995). Primary Image Segmentation. Springer. Doi:10.1007/978-3-642-79980-8_65

Koundinya, K., & Chanda, B. (1994). Detecting lines in gray level images using search techniques. *Signal Processing*, *37*(2), 287–299. doi:10.1016/0165-1684(94)90110-4

Lee. (2012). Edge Detection Analysis. *International Journal of Computer Science, 1*(5-6).

Lu, S., Wang, Z., & Shen, J. (2003). Neuro-fuzzy synergism to the intelligent system for edge detection and enhancement. *Elsevier Journal of Pattern Recognition*, *36*(10), 2395–2409. doi:10.1016/S0031-3203(03)00083-9

Masud, M. D. (2012). Knowledge-based Image segmentation using swarm intelligence techniques. *International Journal of Innovative Computing andApplications*, *4*(2), 75–99. doi:10.1504/IJICA.2012.046779

Monteiro, F. C., & Campilho, A. (2008). Watershed framework to region-based image segmentation. In *Proceedings of International Conference on Pattern Recognition, ICPR 19th*, (pp. 1-4).

Patil, R., & Jondhale, K. (2010). Edge based technique to estimate number of clusters in k-means color image segmentation. In *Proceedings.3rd IEEE International Conference on Computer Science and Information Technology (ICCSIT)*, (pp. 117-121). doi:10.1109/ICCSIT.2010.5563647

Perona, P., & Malik, J. (1990). Scale-space and edge detection using anisotropic diffusion. *IEEE Transactions on Pattern Analysis and Machine Intelligence*, *12*(7), 629–639. doi:10.1109/34.56205

Santis, A. D., & Sinisgalli, C. (1999). A Bayesian approach to edge detection in noisy images. *IEEE Transactions on Circuits and Systems. I, Fundamental Theory and Applications*, *46*(6), 686–699. doi:10.1109/81.768825

Schunck, B. G. (1987). Edge detection with Gaussian filters at multiple scales. In *Proceedings IEEE Comp. Soc. Work. Comp. Vis.*

Sobel, I. (2014). *History and definition of the sobel operator, pattern classification and scene analysis* (Vol. 73). John Wiley and Sons.

Sujaritha, M., & Annadurai, S. (2009). Color image segmentation using binary level-set partitioning approach. *International Journal of Soft Computing*, *4*, 76–84.

Witkin, A. P. (1983). Scale-space filtering. In *Proceedings.International Joint Conference on Artificial Intelligence*.

Wu, J., Yin, Z., & Xiong, Y. (2007). The fast multilevel fuzzy edge detection of blurry images. *IEEE Signal Processing Letters*, *14*(5), 344–347. doi:10.1109/LSP.2006.888087

Ye, Q., Gao, W., & Zeng, A. W. (2003). Color image segmentation using density based clustering.*International Conference on Acoustics, Speech and Signal Processing* (Vol. 3, pp. 345-348). Doi:10.1109/ICASSP.2003.1199480

Yu, Y., & Chang, C. (2006). A new edge detection approach based on image context analysis. *Elsevier Journal of Image and Vision Computing*, *24*(10), 1090–1102. doi:10.1016/j.imavis.2006.03.006

Zheng, S., Liu, J., & Tian, J. W. (2004). A new efficient SVM-based edge detection method. *Elsevier Journal of Pattern Recognition Letters*, *25*(10), 1143–1154. doi:10.1016/j.patrec.2004.03.009

Ziou, D., & Tabbone, S. (1998). *Edge detection techniques – An overview, pattern recognition and image analysis* (Vol. 8). Nauka: Interperiodica Publishing.

KEY TERMS AND DEFINITIONS

Convolution: Convolution of two functions involves flipping (rotating by 180°) one function about its origin and sliding it past the other.

Discontinuities: There are three types of gray-level discontinuities in a digital image: points, lines, and edges.

Edge Detection: The approach for image segmentation based on local changes in intensity.

Edge Labeling: Edge labelling process includes edge localization and suppress the false edges.

Image Differentiation: The first and second order derivatives are the most useful in edge detection methods.

Image Gradient: The directional change in the intensity in an image.

Kernel: A small matrix of numbers and it is used in image convolutions.

Chapter 6
A Nearest Opposite Contour Pixel Based Thinning Strategy for Character Images

Soumen Bag
Indian Institute of Technology (Indian School of Mines), Dhanbad, India

ABSTRACT

Thinning of character images is a big challenge. Removal of strokes or deformities in thinning is a difficult problem. In this paper, we have proposed a nearest opposite contour pixel based thinning strategy used for performing skeletonization of printed and handwritten character images. In this method, we have used shape characteristics of text to get skeleton of nearly same as the true character shape. This approach helps to preserve the local features and true shape of the character images. The proposed algorithm produces one pixel-width thin skeleton. As a by-product of our thinning approach, the skeleton also gets segmented into strokes in vector form. Hence further stroke segmentation is not required. Experiment is done on printed English and Bengali characters and we obtain less spurious branches comparing with other thinning methods without any post-processing.

INTRODUCTION

Thinning is the process that reduces the amount of foreground shape information in images to facilitate efficient recognition and faster regeneration. One advantage of thinning is the Reduction of memory requirement for storing the essential structural information presented in a pattern. Moreover, it simplifies the data structure required in pattern analysis. Thinning of shape has a wide range of application in image processing, machine vision, and pattern recognition (Li & Basu, 1991; Sinha 1987). The ability to extract distinctive features from the pattern plays a key role in enhancing the effectiveness and efficiency in the task of recognizing patterns. In the domain of character images, strokes are the commonly extracted digital patterns used for recognition. Strokes can be easily extracted from a thinned character image. However, the process of thinning generally produces spurious strokes and shape deformations

DOI: 10.4018/978-1-5225-1025-3.ch006

which later cause problems in the recognition task. Different thinning algorithms produce different degrees of distortion in character shape.

In the past several decades, many thinning algorithms have been developed (Lam et al., 1992; Vincze & K˝ov´ari, 2009). They are broadly classified into two groups: raster-scan based and medial-axis based. Raster-scan based methods are classified into two other categories:

- Sequential, and
- Parallel (Nagendraprasad et al., 1993).

Sequential algorithms consider one pixel at a time and visit all the pixels in the character by raster scanning or contour following (Arcelli, 1981; Arcelli & Baja, 1989; Beun, 1973; Rosenfeld, 1970; Wang & Zhang, 1989). Parallel thinning algorithms are based on iterative processing and they consider a pixel for removal based on the results of previous iteration. In 1966, Rutovitz (1966) proposed the first parallel thinning algorithm. Since then, many 2D parallel thinning algorithms have been proposed (Arcelli et al., 1975; Datta & Parui, 1994; Huang et al., 2003; Leung et al., 2000; Lu et al., 2010; Manzanera & Bernard, 2003; Pavlidis, 1982; Wang & Zhang, 1989; Zhang & Suen, 1984; Zhang & Wang, 1994, 1995; Zhu & Zhang, 2008).

A desirable property of thinning algorithms is to preserve the true shape of the given structure. But many of the raster-scan based character thinning methods cannot preserve the local properties or features of the character images properly (Couprie, 2005). As a result, they produce skeletons with some shape distortions. Medial-axis based methods generate a central or median line of the pattern directly in one pass without examining all the individual pixels (Martinez-Perez et al., 1987). However, the results suffer from local distortions especially near the junction points. Next, authors begin a brief description about few prominent thinning algorithms.

BACKGROUND

Thinning methods are classified into two types:

- Raster-scan based, and
- Medial-axis based.

Now, we will broadly discuss different types of thinning methods exist in literature.

Existing Thinning Methods Involving Raster Scan

The general procedure of this type of thinning methods is to define a set of masks and use them to repetitively remove the pixels of the edge until only the central pixels constituting the medial axis of the structure are left.

Datta and Parui (1994) have developed a parallel thinning algorithm to preserve the pixel connectivity and produces skeleton of one pixel thick. Each iteration of the algorithm is divided into four sub-iterations. These sub-iterations use two 1×3 and two 3×1 templates for removing boundary pixels: east, west, north, and south respectively. The method uses one 3×3 window to avoid the removal of critical point (which alters the connectivity) and endpoint (which shortens a leg of the skeleton). Han et al. (1997) have proposed a fully parallel, albeit a computationally expensive thinning algorithm which makes use of a 5×5 mask.

Leung et al. (2000) have proposed a contour following thinning method having no sub-iteration. They have used a look-up table to avoid the use of multiple templates for removing boundary pixel. The number of entries depends on the different possibilities of 8-connectivity of a pixel.

Huang et al. (2003) have also used the template based elimination rules for deleting boundary pixels. A new set of templates were formulated and pixel deletion is done based on the number of black pixels in the 8-neighbor connectivity. Deleting too many pixels from the image leads to information loss, i.e., the thinned image is an inadequate representation of the true character image. The method detects information loss if the ratio of the number of skeleton to contour pixels is less than a predefined threshold. It then compensates for it by replacing the thinned image by the contour image.

To preserve the true shape of skeleton, Lam and Suen (1991) have proposed a thinning method which takes into consideration the shape of the contours and placement of pixels in order to obtain skeletons that are more representative of the shapes of the original patterns. Zhu and Zhang (2008) have proposed a method which is based on the substitution of pixels of strokes or curves which are most valuable part for character recognition. The method attempts to preserve the shape of character images such as Chinese characters, English alphabets, and numerals after thinning.

Wu et al. (2015) have proposed to explore ring radius transform (RRT) to generate a radius map from Canny edges of Each input image to obtain its medial axis. This propose method is efficient to preserve the visual topology of characters in video is challenging in the field of document analysis and video text analysis due to low resolution and complex background. The method finally preserves the true shape of the character through radius values of medial axis pixels for the purpose of recognition with the Google Open source OCR (Tesseract). Experimental results comparing with the other existing methods show that the proposed method is generic and outperforms the existing methods in terms of obtaining skeleton, preserving visual topology and recognition rate. The method is also robust to handle characters of arbitrary orientations.

All the above discussed algorithms perform boundary pixel deletion using various thinning templates or values from look-up tables. Next, we discuss medial-axis based thinning algorithms.

Existing Thinning Methods Involving Medial Axis

Martinez-Perez et al. (1987) have introduced a medial-axis based method for thinning regular shapes. The method does not use templates or look-up tables. It generates medial points in between two parallel contour segments. Applying the procedure for all parallel contour segments yields a set of skeleton segments. During post-processing these segments are linked to get the final skeleton image. However, the method cannot be applied for thinning character images because it does not handle curved segments. Hilaire and Tombre (2002) have proposed an improved skeletonization method for handling images which are a combination of straight lines and curves. The method corrects topological errors at the junction points and improves the accuracy of junction point detection. The method has been applied on fragments of architectural drawings and synthetic data but not on character images.

Telea et al. (2004) have introduced a distance transform based method to simplify the skeleton structure of an image by removing few skeleton branches. The simplification is done by analyzing the quasi-stable points of the Bayesian energy function, parameterized by boundary of contour and internal structure of the skeleton. The experimental results show that it gives a multi-scale skeleton at various abstract levels.

To overcome a skeleton's instability of boundary deformation, Bai et al. (2007) have presented a new skeleton pruning method based on contour partitioning. The partitioning is done using discrete curve evolution (DCE) technique. The obtained skeletons are in accord with human visual perception and stable, even in the presence of significant noise and shape variations, and have the same topology as the original skeletons.

Bag & Harit (2011) have proposed a medial-axis based skeletonization method for printed and handwritten images. The inputs are takes as an isolated character images for different Indian languages, like Bangla, Hindi, Kannada, Tamil, etc. The proposed method is divided into three sub-parts:

- Contour extraction,
- Intermediate medial axis generation, and
- Extrapolation.

The overall performance of the proposed method is better than few other existing methods. Later they (2011) have proposed another contour following thinning strategy for handling isolated character images for different Indian languages. This approach helps to preserve the local features and true shape of the character. Additionally, it produces a set of vectorized strokes with the thin skeleton as by-product.

Some new techniques have been also used for image thinning for example:

- Graphical method,
- Euclidean distance map method,
- Group-wise approach,
- Wavelet,
- Neural networks,
- Critical kernels, and
- Pulse coupled neural networks.

Melhi et al. (2001) have proposed a method for thinning binary handwritten text images by generating graphical representations of words within the image. Choi et al. (2003) have used signed sequential Euclidean distance map to extract a well-connected Euclidean skeleton. Ward and Hamarneh (2010) have developed a group-wise skeletonization framework that yields a fuzzy significance measure for each branch. Tang and You (2003) have introduced a wavelet based scheme to extract skeleton of Ribbon-like shape. Krishnapuram and Chen (1993) have applied recurrent neural networks to image thinning. Altuwaijri and Bayoumi (1995) have used self-organizing neural networks to thin digital images. Gu et al. (2004) and Shang et al. (2007) have proposed a binary image thinning algorithm by using the auto-waves generated by pulse coupled neural network (PCNN). Bertrand and Couprie (2006) have introduced a new 2D parallel thinning algorithm using critical kernels. These new approaches have been primarily used for skeletonizing regular shapes, architectural drawings, English, and Chinese words.

It is also observed that writing with ink pens on poor quality paper when scanned with high precision scanner result in digitized character images with thick stroke and noisy contour. Such character images when thinned by any existing methods produce distortions as a name of spurious loops and spurious strokes. Junction point distortion and touching strokes are other properties of the thinned skeletons of degraded character images with thick stokes and unstructured handwriting. It is noted that there is no such methods exist in literature to tackle these types of distortions for very thick images. This inefficiency leads to make OCR unstable for feature extraction and classification. We have developed efficient and novel methods (Ghosh & Bag, 2013, 2014) for handling these type of distortions which will be a considerable contribution in modern day OCR systems.

MAIN FOCUS OF THE CHAPTER

Traditional thinning strategies lead to deformation of character shapes, especially where the strokes get branched out; i.e., the junction points in the skeleton. In this chapter we develop a medial-axis based thinning for performing thinning of character images. The algorithm is non-iterative and does not make use of any pre-defined templates or masks. It can preserve the local features and the true shape of the character. Additionally, it produces a set of vectorized strokes with the thin skeleton as by-product. The major contribution of our work is a thinning methodology which minimizes local distortions in the thinned skeleton by making use of shape characteristics of text and would offer better models for feature extraction and classification for optical character recognition. The major contribution of our work is a thinning methodology which minimizes local distortions in the thinned skeleton by making use of shape characteristics of text and would offer better models for feature extraction and classification for OCR.

Shortcomings of Existing Thinning Methods

Digital skeleton of character images, generated by thinning method, has a wide range of applications for shape analysis, feature extraction, and classification in the field of document image analysis in terms of optical character recognition (OCR). But thinning of character images is a big challenge. In the past several decades, many thinning algorithms have been developed which have attempted to solve various problems related to thinning. But the performance of the traditional thinning strategies falls short of expectation for characters of Indian scripts. Traditional thinning strategies lead to deformation of character shapes, especially where the strokes get branched out; i.e., the junction points in the skeleton. Apart from that removal of spurious strokes or loops in thinning is also a difficult problem. In this section, Authors examine the shortcomings of existing thinning methods and the motivation for our work. We have taken three representative thinning methods (Datta & Parui, 1994; Huang et al., 2003; Telea et al., 2004) for analyzing their shortcomings at the time of performing skeletonization of character images. Methods proposed by Datta & Parui and Huang et al. are iterative parallel methods and the one by Telea et al. is distance transform based method. We examine four different types of problems (see Figure 1) which affect the thinning results:

- Shape distortion at junction points,
- Spurious strokes at endpoints,
- Spurious strokes on high curvature region,
- Recession of stroke length.

Figure 1. Shortcomings of existing thinning methods: (a) input image; (b) thinned image (red circle indicates distortion at junction points and endpoints; red arrow indicates spurious strokes on high curvature region)

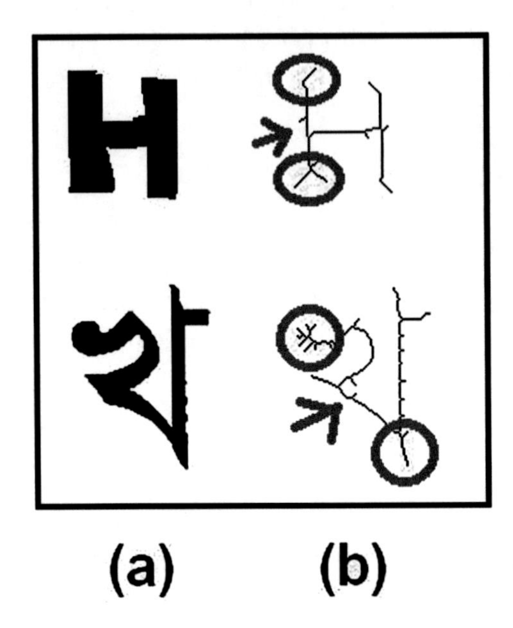

METHODOLOGY TO HANDLE SHAPE DISTORTIONS

To overcome these types of distortions, we have proposed a medial-axis based thinning strategy for character images. The advantages of the proposed technique are:

- It provides a vectorized output of the thin skeleton.
- The vectorized output is a collection of strokes. Hence further stroke segmentation is not required.
- Many of the spurious thinning branches which inevitably occur when applying other rasterized thinning algorithms do not occur in our proposed algorithm.
- Our approach is most suitable for thinning printed and handwritten isolated text alphabets. It gives correct output even in the presence of changing width of the strokes.

In the next section, we give in detail the proposed thinning methodology.

Thinning Methodology Based on Medial Axis

In this section we describe our proposed thinning strategy based on contour extraction and medial axis generation for text alphabets. Figure 2 shows the system architecture of the proposed method. The method involves 3 major steps:

1. Preprocessing and contour extraction,
2. Nearest opposite pixel detection to get medial axis,
3. Extrapolation of the intermediate skeleton segments.

Figure 2. System architecture of the proposed method

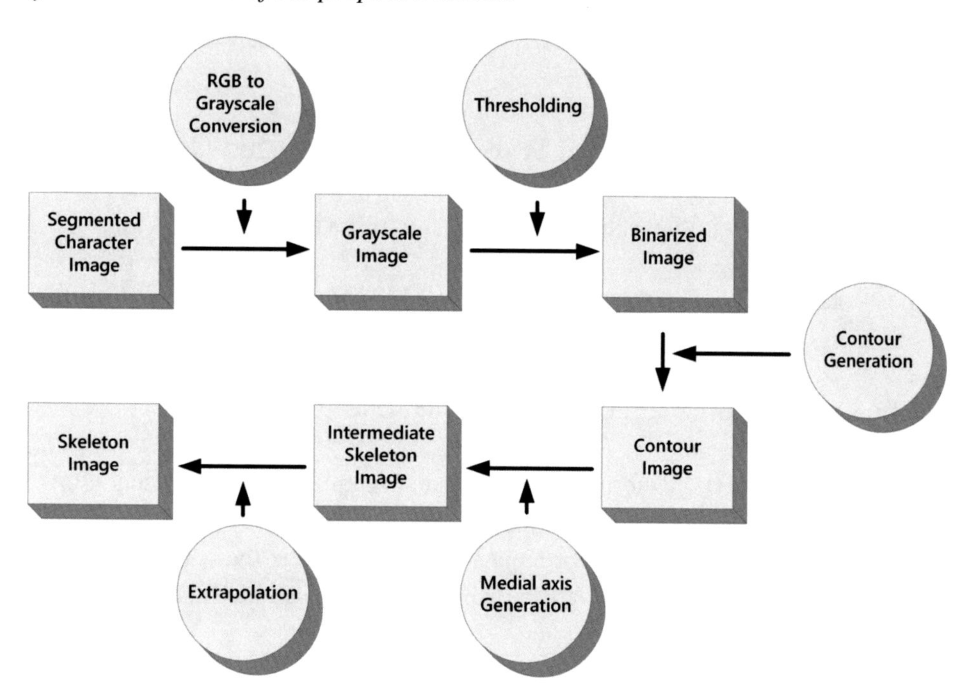

Preprocessing and Contour Extraction

Currently authors are working with isolated character images of English language and four different Indian languages (Bangla, Hindi, Kannada, and Tamil). The scanned character image is binarized by the method presented in (Bag et al., 2011). Given the binarized character image, we then extract its boundary contour using a 3×3 mask. The result of contour extraction for Figure 3(a) is shown in Figure 3(b).

Figure 3. The contour extracted from a binarized character image: (a) input image; (b) contour image; (c) medial axis segments (see in color); (d) skeleton image

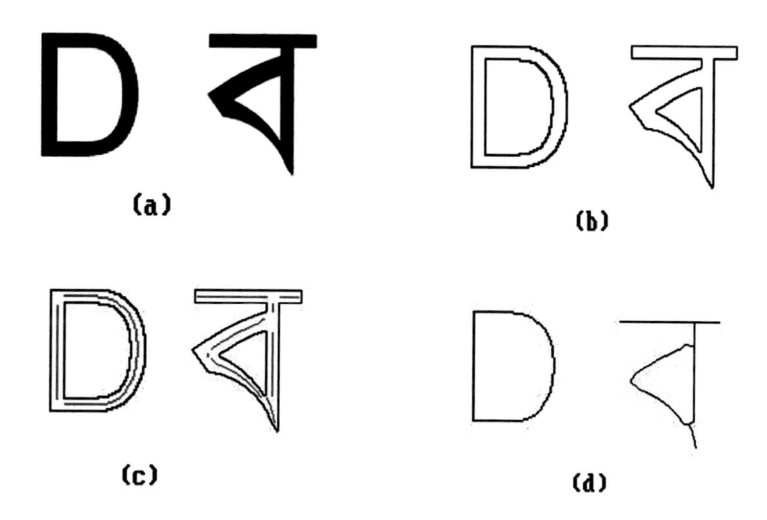

We then order the contour pixels so that we can traverse the contour in a particular direction. Our methodology for medial axis generation (described in next section) makes use of certain procedures which are given below.

Computing the Average Pen Width of Strokes in the Binarized Character Image

We estimate the average pen width of the strokes using the following formula:

Avg pen width = (total number of pixels in the character) / (0.5×number of contour pixels)

Computing the Local Orientation of a Thinned Stroke at Any Pixel

Let p_k be the pixel on the thinned skeleton at which we want to estimate the local orientation of the stroke. Let θ_{kj} be the orientations of the line segments joining the pixels p_k and p_j. The local orientation at p_k is computed as the average of the values θ_{kj} for values of j varying over the neighboring skeleton pixels p_k, i.e., j varies from $k-N$ to $k+N$ with $j \neq k$. If p_k happens to be the endpoint, the range for varying j is restricted from $k-N$ to $k-1$.

Nearest Opposite Pixel Detection to Get Medial Axis

In this section, authors describe the methodology for detecting the nearest opposite pixel pairs on the boundary contour to get the medial axis segments of the contour image.

Let $< x_1, x_2, \ldots, x_{i-1}, x_i, x_{i+1}, \ldots, x_n >$ be the ordered set of contour pixels of a contour image, where x_{i-1} and x_{i+1} are the predecessor and successor of the pixel x_i respectively. Each contour pixel x_i is represented as $[x_i^{Xcor}$ & $x_i^{Ycor}]$, where these values represent the width and height of the pixel respectively. Now, take a contour pixel x_i to find out the nearest opposite contour pixel by following the steps given below.

1. Obtain local orientation θ_{local}^{xi} at the pixel x_i by calculating the orientation of the line segment joining two pixels x_i and x_{i+1} in forward direction.
2. Compute the direction perpendicular to the local orientation θ_{local}^{xi}.

Define the straight line using the following equation given below

$$y - x_i^{Ycor} = M' \times (x - x_i^{Xcor})$$

where M' represents the gradient of the orientation perpendicular to the local orientation θ_{local}^{xi}. x_i is the starting pixel of the straight line.

4. Check all the contour pixels (excluding x_i and all other pixels those have already selected as a nearest opposite pixel pairs to get medial axis) to find out the contour pixel (y_i) satisfies the straight line defined above.
5. After getting the opposite contour pixel, we measure two parameters which help us to take a decision about the acceptability of the pixel pair $< x_i$ & $y_i >$ as a nearest opposite contour pixel pair.

a. Measure the City Block distance $dist_i$ in between two nearest opposite contour pixels x_i and y_i.
b. Obtain local orientation θ_{local}^{yi} at the pixel y_i.

6. Now, apply a set of rules based on the text-specific knowledge for selecting the pixel pair < x_i & y_i > as a nearest opposite contour pixels for generating the skeleton pixels. These rules are as follows
   a. **R1:** All the pixels in-between the pixel pair < x_i & y_i > should be black i.e., belong to the character region itself and not the background.
   b. **R2:** $dist_i$ must be ≤ *THRESHOLD*. This value depends on the pen width. It is changeable for different character font.
   c. **R3:** θ_{local}^{xi} must be equal to θ_{local}^{yi}.

We can conclude that if the pixel pair < x_i & y_i > satisfies all the above rules then mark the midpoint of the line joining x_i and y_i as a pixel on the medial axis. If the above conditions are not satisfied then the medial pixel is not marked since this is likely to be the junction point of two or more pen strokes, hence the ambiguity has to be resolved later.

7. Flag off the pixels x_i and y_i marked as processed and move to the pixel x_{i+1} and repeat the above steps.

The above steps are repeated for all the contour pixels, and for each pixel we consider only the pixels which have not flagged as already processed.

Figure 3(c) shows the intermediate skeleton image of the character as a set of segments. These set of segments now need to be extrapolated to get the final skeleton image. But before going to that step, we need little bit of noise removal by applying very simple mask based convolution technique.

Post-Processing

The intermediate skeleton image is not fully noise free. It contains few pixels those are not essential for generating skeleton image. These pixels are removed by applying a post-processing scheme. Authors observed that two types of noises are present in test images. One is isolated pixel and another one is 8-connected neighbor pixel of the contour of the test image. To maintain the connectivity of the skeleton image, we need to remove first type of noise and to maintain the skeleton along the mid zone of the image, we have to remove second type of noise. For both the cases, we have used a 3×3 mask centred on every black pixel of the intermediate skeleton image. If there is no background (black) pixel within the mask, then marked that pixel as an isolated pixel, and if there is at least one contour pixel in the mask, then marked that pixel as a contour touching pixel. Now, we remove all these marked pixels for making the image noise free.

Extrapolation of the Intermediate Skeleton Image

Before doing the extrapolation, at first we need to detect medial axis segments generated by the previously stated steps. We detect the medial axis segments by applying the concept of 8-connected component analysis. Here we have used 3×3 masks for checking the pixel connectivity and form the set $M = \{(M^s_1, M^e_1), (M^s_2, M^e_2), ..., (M^s_m, M^e_m), \}$, where M^s_i and M^e_i denote the start and end points of the ith medial axis segment. The start/end points now need to be extended so that the neighboring medial axis segments can be joined together. The steps to identify the neighboring medial axis segments are as follows:

1. For each start/end point of a medial axis segment find the close start/end points belonging to other medial axis segments. For this purpose, we consider that the distance between the two start/end points should be less than a threshold and that the line joining the two start/end points should be within the character region.
2. If a given start/end point has two or more start/end points of other segments close enough then we make a proximity set, e.g. $P = \{M^s_a, M^s_b, M^e_c\}$ would indicate start/end point of three medial axis segments *a*, *b*, and *c*.

Each proximity set would correspond to start/end points which belong to the junction region belonging to multiple pen strokes. Identifying the medial axis points in the junction region of pen strokes give skeleton points which tend to distort the true shape of the character. Our approach avoids identifying the medial axis points in such ambiguous regions and instead tries to extrapolate the start/end points of the medial axis segments in a proximity group such that the result would be very close to the true shape of the character. This is the major contribution of this work. The steps for extrapolation are as follows:

1. In a given proximity set *P* if any two medial axis segments have the same orientation then any one of the two close start/end points is extended to meet the other one.
2. If there is a medial axis segment whose orientation does not match with any other segment in the same proximity set, then its start/end point is extended till it meets the medial axis (extrapolated from other segments) or a junction (formed by extrapolating from other segments in *P*) on the medial axis or to the start/end point of the other medial axis segment. Of the three cases mentioned which one applies depends on the number of start/end points which are part of a proximity set.

The result of extrapolation is shown in Figure 3(d). Note that extrapolation result is almost the true shape of the character and does not produce spurious segments. Our approach gives correct thinning results even for very thick strokes.

EXPERIMENTAL RESULTS

We have tested our proposed algorithm on printed English and Bengali character images. We have collected characters from several printed documents as well as handwritings of different persons

written by black ink pen. Our main consideration is on only isolated character images. The overall dataset size is 10,550. Here we have taken two types of images: cleaned and noisy. All the programs are written in C++ using Opencv in the UNIX platform. The pen width has to be provided so as to be proportional to the average height of the characters.

Test Results of Cleaned Images

There are two outputs of the algorithm. First one is the set of vector sized medial axis segments and last one is the final skeleton image formed by performing the extrapolation of the medial axis segments. In Figure 4, there are three type of images:

- Contour of the input image (see Figure 4(a)),
- Medial axis segments (Figure 4(b)), and
- Skeleton image (Figure 4(c)).

Figure 4. Test results of cleaned images: (b) contour image; (c) medial axis segments (see in color); (d) skeleton image

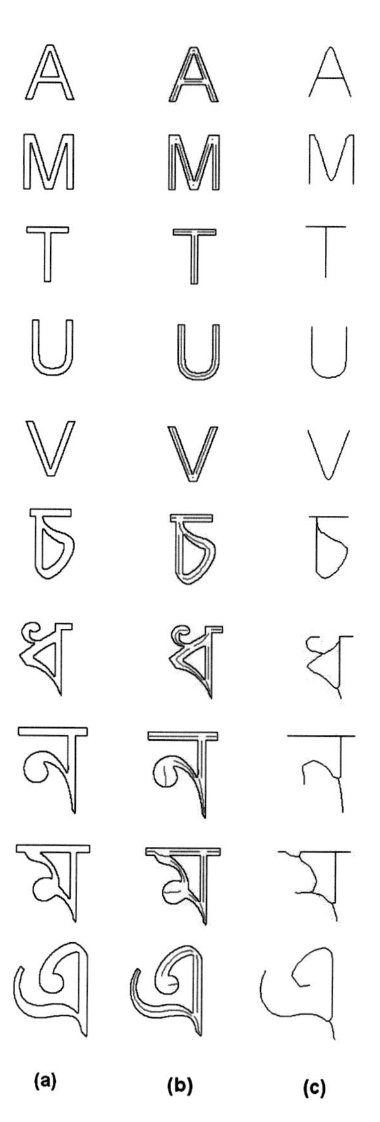

Test Results of Noisy Images

We have tested our algorithm on few noisy English character images. The noisy means the contour boundary of the images are not smooth. Figure 5 shows the rest results. By observing the outputs, we can conclude that the result is very promising comparing with other thinning method.

Comparison of Test Results with Other Thinning Methods

We have applied Datta's, Huang's, and Telea's methods on printed English and Bengali character images (Figure 6(a)) and compare the results with our proposed algorithm. We have seen that Datta's result (Figure 6(b)) produces few unwanted strokes which are not acceptable for stroke segmentation and character recognition. Huang's result (Figure 6(c)) suffers from the distortion at the junction points of the thinned image. Telea's result (Figure 6(d)) suffers from shortage of skeleton edges. Finally, after comparing the test results of our algorithm (Figure 6(e)) with all other test results, we have seen that our method gives much better result for thinning character images and maintains all the basic thinning properties as well. The major improvement is that the deformation of the skeletal structure at the junction point of the skeletal branches is not there with our results.

Figure 5. Test results of noisy images: (b) contour image; (c) medial axis segments (see in color); (d) skeleton image

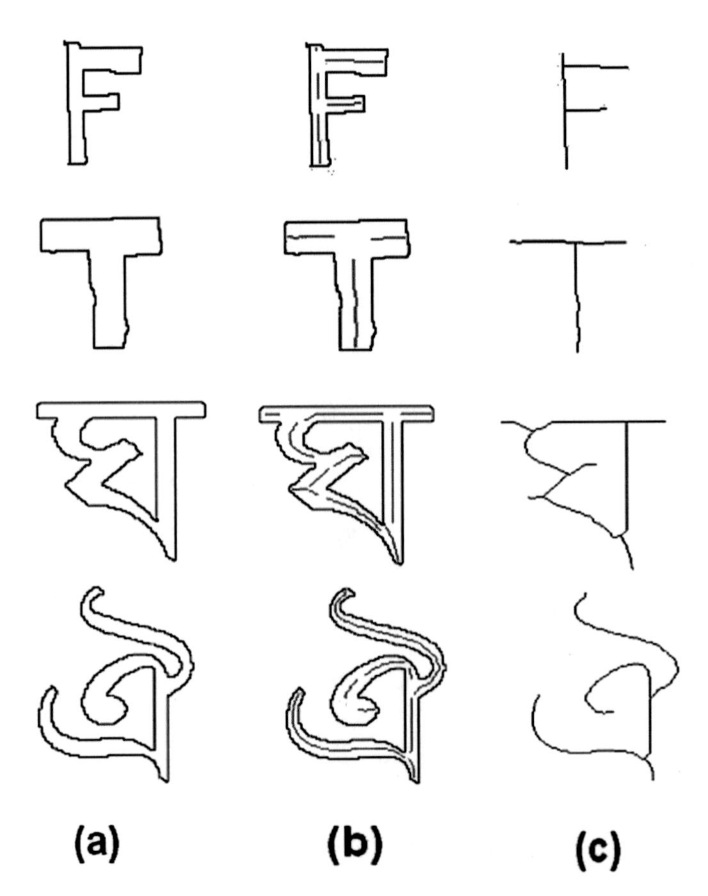

Figure 6. Comparison of results with other methods: (a) input image (b) Datta's method; (c) Huang's method; (d) Telea's method; (e) proposed method

CONCLUSION

This paper improves the performance of existent character thinning algorithms. The main challenge of thinning character images is to preserve the shape of characters after thinning. In spite of slight speed disadvantage, our algorithm gives quite promising result. The proposed algorithm avoids shape distortion by detecting the ambiguous regions and avoiding the identification of medial axis points within these regions. The resultant skeleton maintains the pixel connectivity and is very close to the medial axis. Additionally, our algorithm provides strokes of character images as an intermediate result during the thinning process. We have compared our test results with three other thinning algorithms and have observed better performance compared to all other algorithms. The proposed thinning approach has a good potential for applications to improve the performance of OCR in Indian script.

Future Plan

The use of document image processing techniques to automate methods for solving various real-world problems is increasing every day. Researchers of this domain aim at devising new methods that can enhance the performance of existing systems. At the same time a proper analysis and use of existing schemes to their full potential for solving application specific problems is also important. Thinning strategy in an integral part of character or object recognition methods because they help to improve the overall recognition accuracy. In our proposed work, we have proposed a medial-axis based thinning strategy to improve the performance of existing OCR system for Bangla language. We have a plan to apply this preprocessing strategy for such type of Indian languages having structural similarities with Bangla language (e.g., Hindi, Marathi, Gurumukhi, etc.).

REFERENCES

Altuwaijri, M., & Bayoumi, M. (1995). A new thinning algorithm for Arabic characters using self-organizing neural network. In *IEEE International Symposium on Circuits and Systems* (pp. 1824–1827). doi:10.1109/ISCAS.1995.523769

Arcelli, C. (1981). Pattern thinning by contour tracing. *Computer Graphics and Image Processing*, *17*(2), 130–144. doi:10.1016/0146-664X(81)90021-6

Arcelli, C., & Baja, G. S. D. (1989). A one-pass two-operation process to detect the skeletal pixels on the 4-distance transform. *IEEE Transactions on Pattern Analysis and Machine Intelligence*, *11*(4), 411–414. doi:10.1109/34.19037

Arcelli, C., Cordella, L., & Levialdi, S. (1975). Parallel thinning of binary pictures. *Electronics Letters*, *11*(7), 148–149. doi:10.1049/el:19750113

Bag, S., & Harit, G. (2011). An improved contour-based thinning method for character images Skeletonizing. *Pattern Recognition Letters*, *32*(14), 1836–1842. doi:10.1016/j.patrec.2011.07.001

Bag, S., & Harit, G. (2011). Skeletonizing character images using a modified medial axis-based strategy. *International Journal of Pattern Recognition and Artificial Intelligence*, *25*(07), 1035–1054. doi:10.1142/S0218001411009020

Bai, X., Latecki, L. J., & Liu, W. Y. (2007). Skeleton pruning by contour partitioning with discrete curve evolution. *IEEE Transactions on Pattern Analysis and Machine Intelligence*, *29*(3), 449–462. doi:10.1109/TPAMI.2007.59 PMID:17224615

Bertrand, G., & Couprie, M. (2006). New 2D parallel thinning algorithms based on critical kernels. In *International Workshop on Combinatorial Image Analysis* (pp. 45–59). doi:10.1007/11774938_5

Beun, M. (1973). A flexible method for automatic reading of handwritten numerals. *Philips Technical Review*, *31*, 89–101.

Choi, W. P., Lam, K. M., & Siu, W. C. (2003). Extraction of the Euclidean skeleton based on a connectivity criterion. *Pattern Recognition*, *36*(3), 721–729. doi:10.1016/S0031-3203(02)00098-5

Couprie, M. (2005). *Note on fifteen 2D parallel thinning algorithms.* Internal Report, Universit´e de Marne-la-Vall´ee, IGM2006-01.

Datta, A., & Parui, S. K. (1994). A robust parallel thinning algorithm for binary images. *Pattern Recognition, 27*(9), 1181–1192. doi:10.1016/0031-3203(94)90004-3

Ghosh, S., & Bag, S. (2013). A modified thinning strategy to handle junction point distortion for Bangla. In *IEEE Students'Technology Symposium* (pp. 52–56).

Ghosh, S., & Bag, S. (2013). An improvement on thinning to handle characters with noisy contour. In *National Conference on Computer Vision, Pattern Recognition, Image Processing and Graphics* (pp. 1–4). doi:10.1109/NCVPRIPG.2013.6776178

Gu, X., Yu, D., & Zhang, L. (2004). Image thinning using pulse coupled neural network. *Pattern Recognition Letters, 25*(9), 1075–1084. doi:10.1016/j.patrec.2004.03.005

Han, N. H., La, C. W., & Rhee, P. K. (1997). An efficient fully parallel thinning algorithm. In *International Conference on document Analysis and Recognition* (pp. 137–141).

Hilaire, X., & Tombre, K. (2002). Improving the accuracy of skeleton-based vectorization. In *International Workshop on Graphics Recognition* (pp. 273–288).

Huang, L., Wan, G., & Liu, C. (2003). An improved parallel thinning algorithm.*In International Conference on document Analysis and Recognition* (pp. 780–783).

Krishnapuram, R., & Chen, L. F. (1993). Implementation of parallel thinning algorithms using recurrent neural networks. *IEEE Transactions on Neural Networks, 4*(1), 142–147. doi:10.1109/72.182705 PMID:18267712

Lam, L., Lee, S. W., & Suen, C. Y. (1992). Thinning methodologies—A comprehensive survey. *IEEE Transactions on Pattern Analysis and Machine Intelligence, 14*(9), 869–885. doi:10.1109/34.161346

Lam, L., & Suen, C. Y. (1991). A dynamic shape preserving thinning algorithm. *Signal Processing, 22*(2), 199–208. doi:10.1016/0165-1684(91)90050-S

Leung, W., Ng, C. M., & Yu, P. C. (2000). Contour following parallel thinning for simple binary images. In *International Conference on Systems, Man, and Cybernetics* (pp. 1650–1655).

Li, X. B., & Basu, A. (1991). Variable-resolution character thinning. *Pattern Recognition Letters, 12*(4), 241–248. doi:10.1016/0167-8655(91)90038-N

Lu, S., Su, B., & Tan, C. L. (2010). Document image binarization using background estimation and stroke edges. *International Journal on Document Analysis and Recognition, 13*(4), 303–314. doi:10.1007/s10032-010-0130-8

Manzanera, A., & Bernard, T. M. (2003). Metrical properties of a collection of 2D parallel thinning algorithms. In *International Workshop on Combinatorial Image Analysis* (pp. 255–266). doi:10.1016/S1571-0653(04)00491-3

Martinez-Perez, M. P., Jimenez, J., & Navalon, J. L. (1987). A thinning algorithm based on contours. *Computer Vision Graphics and Image Processing, 39*(2), 186–201. doi:10.1016/S0734-189X(87)80165-2

Melhi, M., Ipson, S. S., & Booth, W. (2001). A novel triangulation procedure for thinning hand-written text. *Pattern Recognition Letters*, *22*(10), 1059–1071. doi:10.1016/S0167-8655(01)00038-1

Nagendraprasad, M. V., Wang, P. S. P., & Gupta, A. (1993). Algorithms for thinning and rethickening binary digital patterns. *Digital Signal Processing*, *3*(2), 97–102. doi:10.1006/dspr.1993.1014

Pavlidis, T. (1982). An asynchronous thinning algorithm. *Computer Graphics and Image Processing*, *20*(2), 133–157. doi:10.1016/0146-664X(82)90041-7

Rosenfeld, A. (1970). Connectivity in digital pictures. *Journal of Alternative and Complementary Medicine (New York, N.Y.)*, *17*, 146–160.

Rutovitz, D. (1966). Pattern recognition. *Journal of the Royal Statistical Society. Series A (General)*, *129*(4), 504–530. doi:10.2307/2982255

Shang, L., Yi, Z., & Ji, L. (2007). Binary image thinning using autowaves generated by PCNN. *Neural Processing Letters*, *25*(1), 49–62. doi:10.1007/s11063-006-9030-9

Sinha, R. M. K. (1987). A width-independent algorithm for character skeleton estimation. *Computer Vision Graphics and Image Processing*, *40*(3), 388–397. doi:10.1016/S0734-189X(87)80148-2

Tang, Y. Y., & You, X. G. (2003). Skeletonization of ribbon-like shapes based on a new wavelet function. *IEEE Transactions on Pattern Analysis and Machine Intelligence*, *25*(9), 1118–1133. doi:10.1109/TPAMI.2003.1227987

Telea, A., Sminchisescu, C., & Dickinson, S. (2004). Optimal inference for hierarchical skeleton abstraction. In *International Conference on Pattern Recognition* (pp. 19–22).

Vincze, M., & K˝ov'ari, B. (2009). Comparative survey of thinning algorithms. In *International Symposium of Hungarian Researchers on Computational Intelligence and Informatics* (pp. 173–184).

Wang, P. S. P., & Zhang, Y. Y. (1989). A fast and flexible thinning algorithm. *IEEE Transactions on Computers*, *38*(5), 741–745. doi:10.1109/12.24276

Ward, A. D., & Hamarneh, G. (2010). The groupwise medial axis transform for fuzzy skeletonization and pruning. *IEEE Transactions on Pattern Analysis and Machine Intelligence*, *32*(6), 1084–1096. doi:10.1109/TPAMI.2009.81 PMID:20431133

Wu, Y., Shivakumara, P., Wei, W., Lu, T., & Pal, U. (2015). A new ring radius transform-based thinning method for multi-oriented video characters. *International Journal on Document Recognition and Analysis*, *18*(2), 137–151. doi:10.1007/s10032-015-0238-y

Zhang, T. Y., & Suen, C. Y. (1984). A fast parallel algorithm for thinning digital patterns. *Communications of the ACM*, *27*(3), 236–239. doi:10.1145/357994.358023

Zhang, Y. Y., & Wang, P. S. P. (1994). A new parallel thinning methodology. *International Journal of Pattern Recognition and Artificial Intelligence*, *8*(05), 999–1011. doi:10.1142/S0218001494000504

Zhang, Y. Y., & Wang, P. S. P. (1995). Analysis and design of parallel thinning algorithms—A generic approach. *International Journal of Pattern Recognition and Artificial Intelligence*, *9*(05), 735–752. doi:10.1142/S0218001495000298

Zhu, X., & Zhang, S. (2008). A shape-adaptive thinning method for binary images. In *International Conference on Cyberworlds* (pp. 721–724). doi:10.1109/CW.2008.133

ADDITIONAL READINGS

Abu-Ain, W., Sheikh Abdullah, S. N. H., Abu-Ain, B. B. T., & Omar, K. (2013). Skeletonization algorithm for binary images. In *International Conference on Electrical Engineering and Informatics* (pp. 704–709).

Bag, S., Bhowmick, P., Harit, G., & Biswas, A. (2013). Character segmentation of handwritten Bangla text by vertex characterization of isothetic covers. In *National Conference on Computer Vision, Pattern Recognition, Image Processing and Graphics* (pp. 21–24).

Bag, S., & Harit, G. (2013). A survey on optical character recognition for Bangla and Devanagari scripts. *Sadhana, 38*(1), 133–168. doi:10.1007/s12046-013-0121-9

Bag, S., Harit, G., & Bhowmick, P. (2014). Recognition of Bangla compound characters using structural decomposition. *Pattern Recognition, 47*(3), 1187–1201. doi:10.1016/j.patcog.2013.08.026

Das, N., Acharya, K., Sarkar, R., Basu, S., Kundu, M., & Nasipuri, M. (2012). A novel GA-SVM based multistage approach for recognition of handwritten Bangla compound characters. In *International Conference on Information Systems Design and Intelligent Applications* (pp. 145–152). doi:10.1007/978-3-642-27443-5_17

Das, N., Basu, S., Sarkar, R., Kundu, M., Nasipuri, M., & Basu, D. K. (2009). Handwritten Bangla compound character recognition: Potential challenges and probable solution. In *Indian International Conference on Artificial Intelligence* (pp. 1901–1913).

Das, N., Das, B., Sarkar, R., Basu, S., Kundu, M., & Nasipuri, M. (2010). Handwritten Bangla basic and compound character recognition using MLP and SVM classifier. *Journal of Computing, 2.*

Garain, U., & Chaudhuri, B. B. (1998). Compound character recognition by run number based metric distance. In *International Symposium on Electronic Imaging: Science Technology* (pp. 90–97).

Gonzalez, R. C., & Woods, R. E. (2008). *Digital Image Processing. 3rd Volume.* USA: Prentice Hall.

Pal, U., & Chaudhuri, B. B. (2004). Indian script character recognition: A survey. *Pattern Recognition, 37*(9), 1887–1899. doi:10.1016/j.patcog.2004.02.003

Pal, U., Wakabayashi, T., & Kimura, F. (2007). Handwritten Bangla compound character recognition using gradient feature. In *International Conference on Information Technology* (pp. 208–213). doi:10.1109/ICIT.2007.62

Sarkar, R., Das, N., Basu, S., Kundu, M., Nasipuri, M., & Basu, D. K. (2008). A two-stage approach for segmentation of handwritten Bangla word images. In *International Conference on Frontiers in Handwriting Recognitions* (pp. 403–408).

Zhu, S., Yang, J., & Zhu, X. F. (2013). A thinning model for handwriting-like image skeleton. In Computer Engineering and Networking (pp. 667–693).

KEY TERMS AND DEFINITIONS

DCE: Discrete curve evolution technique is used for generating single pixel thin image.

OCR: Optical character recognition is a process to convert scanned images to their electronic form. It has wide application in the field of documents image processing.

PCNN: Pulse coupled neural network is used for image thinning.

RRT: Ring radius transform is used for image skeletonization.

Thinning: A technique to convert a thick pixel image into a single pixel thin image. During this process, it must preserve the true shape of the original image.

Section 3

Motion Detection in Video Applications and Miscellaneous Related Topics

Real time dynamic scene has been employed in a number of applications, which can be useful for three-dimensional (3D) animation videos in electronic games, 3D television, motion analysis, and gesture recognition. This process is proposed and discussed in this section. In addition, the developed artificial vision system allows allocating the position of objects located on the robot's Cartesian work plane. The frame by frame acquisition of images allows positioning in real time the links of a manipulator robot. However, cameras speed of frame acquisition restricts such operation, which is discussed in this section. Finally, the section includes other related topics.

Chapter 7
Multi-View RGB-D Synchronized Video Acquisition and Temporally Coherent 3D Animation Reconstruction Using Multiple Kinects

Naveed Ahmed
University of Sharjah, UAE

ABSTRACT

This chapter introduces a system for acquiring synchronized multi-view color and depth (RGB-D) video data using multiple off-the-shelf Microsoft Kinect and methods for reconstructing temporally coherent 3D animation from the multi-view RGB-D video data. The acquisition system is very cost-effective and provides a complete software-based synchronization of the camera system. It is shown that the data acquired by this framework can be registered in a global coordinate system and then can be used to reconstruct the 360-degree 3D animation of a dynamic scene. In addition, a number of algorithms to reconstruct a temporally-coherent representation of a 3D animation without using any template model or a-prior assumption about the underlying surface are also presented. It is shown that despite some limitations imposed by the hardware for the synchronous acquisition of the data, a reasonably accurate reconstruction of the animated 3D geometry can be obtained that can be used in a number of applications.

INTRODUCTION

Temporally coherent time-varying dynamic scene geometry has been employed in a number of applications. It can be used for 3D animation in digital entertainment productions, electronic games, 3D television, motion analysis, gesture recognition etc. First step in obtaining temporally coherent 3D video is to capture the shape, appearance and motion of a dynamic real-world object. One or more video cameras are employed for this acquisition, but unfortunately, data obtained by these video cameras has no temporal

DOI: 10.4018/978-1-5225-1025-3.ch007

consistency, as there is no relationship between the consecutive frames of a video stream. In addition, for a multi-view video, all the cameras have to be synchronized to extract temporal correspondences at each frame of the video. This synchronization is typically achieved by means of a hardware-based camera trigger, which acts as an external synchronizer. From the acquired synchronized data, in order to reconstruct a temporally coherent 3D animation, a spatial structure between cameras has to be established along with the temporal matching over the complete video data.

In this chapter, a system for acquiring synchronized dynamic 3D data using multiple RGB-D cameras along with three new method for capturing spatio-temporal coherence between RGB-D images captured from multiple RGB-D video cameras are presented. Synchronized multi-view video (MVV) data is used in a number of applications, e.g. motion capture, dynamic scene reconstruction, free-viewpoint video etc. Traditionally, the MVV recordings are acquired using synchronized color (RGB) cameras, which are later processed for use in a number of applications (Aguiar et al., 2008; Carranza et al., 2003; Starck et al., 2007; Theobalt et al., 2007; Vlasic et al., 2008). The acquisition setups used for these earlier works comprised of a dedicated system for capturing synchronous high quality RGB MVV recordings, which were then used to reconstruct dynamic 3D scene representation.

One of the earlier works in this area was presented by Carranza et al. (2003), who used eight multi-view recordings to reconstruct the motion and shape of a moving subject and applied it in the area of free-viewpoint video reconstruction. Theobalt et al. (2007) extended this work so that in addition to capturing the shape and motion they also captured surface reflectance properties of a dynamic object. Starck et al. (2007) presented a high quality surface reconstruction method that could capture detailed moving geometry from multi-view video recordings. Later de Aguiar et al. (2008) and Vlasic et al. (2008) presented new method for reconstructing really high quality of dynamic scene using multi-view video recordings. Both of their methods first obtained the shape of the real world object using a laser scanner and then deformed the shape to reconstruct the 3D animation. Ahmed et al. (2008) presented a method of dynamic scene reconstruction with time coherent information without the use of any template geometry, but unlike one of the presented method they did not explicitly include multiple matching criteria for extracting time coherence in their method.

With the arrival of depth sensors, especially low cost Microsoft Kinect (Microsoft, 2010), there has been a wave of interest in incorporating them in a number of research areas including the reconstruction of dynamic scene geometry. For example, one or more depth sensors are employed in 3D shape scanning and dense 3D reconstruction of static objects; pose, motion and 3D shape estimation (Baak et al., 2011; Berger et al., 2011; Weiss et al, 2011). These works show that despite the limitation of depth sensors, i.e. low resolution and high noise, it is possible to employ them to get high quality results. Recently, two methods have been presented to capture data using multiple depth sensors. Kim et al. (2008) presented a system for the fusion and calibration of RGB and depth sensors. Their system uses a dedicated hardware setup for the synchronization of the color and depth cameras. Recently, Berger et al. (2011) presented a method to capture motion using four Kinects but without any active synchronization between the sensors. These methods do not try to extract any spatial or temporal coherence information from the acquired dynamic data. More recently, Ye et al. (2014), and Zollhöfer et al. (2014) presented probabilistic methods for shape and pose estimations of dynamic objects using a single RGB-D camera. Unlike one of the presented methods, both of these methods deform a template mesh to achieve 3D animation reconstruction. On the other hand, the presented methods do not rely on any a-priori model and is more general because the 3D geometry is directly acquired from the RGB-D video data.

The goal of this chapter is to present a unified system comprising of multiple Kinects to perform synchronous capture using a software-only acquisition setup, and to reconstruct temporally coherent dynamic 3D scene geometry from dynamic RGB-D data. The acquisition system is highly scalable and can be extended to any number of cameras. It is shown that the data from this acquisition setup can be merged to reconstruct the dynamic 3D scene to a very good approximation of its real world counterpart. The system is very low cost, and only uses easily available software solutions to acquire, register, and process the data. To the knowledge of the authors this is the first system, which shows that acquisition, and time-coherent 3D animation reconstruction is possible using multiple Kinects. This work is an extension of the acquisition system presented by Ahmed et al. (2012). In principle, any type and any combination of RGB and depth hybrid color (RGB) and depth camera system which provides both the color and depth information at 30 frames per second can be used. This acquisition system can acquire synchronous streams of RGB-D data from multiple Microsoft Kinects.

The acquired multi-view RGB and depth data is not temporally coherent as each frame is independent of the other. The second part of this chapter presents multiple solutions, which can use both depth and color information and extract time coherence information from the dynamic three-dimensional content. The dynamic three-dimensional content is assumed to be in the form of a three-dimensional point cloud with color information associated with every point at every frame. It will be show that this information can be obtained very easily from the presented acquisition setup that provides not only the depth information of real world scene but also its color information. The methods are not limited to the data obtained by the Microsoft Kinect cameras but are equally suitable for the three-dimensional content obtained using a traditional acquisition setup of multi-view color cameras. Main benefit of using Microsoft Kinect cameras is that unlike the requirement of employing eight or more color cameras, only one Microsoft Kinect camera can be employed to get meaningful depth and color information of a real world dynamic scene. The methods are validated quantitatively using three error metrics that use silhouette, bounding and a deformation-based error. The main objectives of this chapter are to present:

1. Acquisition of data using one or more Microsoft Kinect camera and organize it in a form of a three-dimensional dynamic point cloud with the color information.
2. New methods of finding time coherent information from three-dimensional dynamic point cloud data using both color and depth information or only depth information and their comparative analysis. This data can either be acquired from Microsoft Kinect cameras as described in step 1 or a traditional setup of multi-view video acquisition using color cameras.

BACKGROUND

This work capitalizes on previous research in a number of areas, but primarily it derives from the areas of multi-view video (MVV) acquisition, 3D and free-viewpoint video, static and dynamic surface reconstruction from color and depth cameras.

MVV data has been used in a number of applications. In one of pioneering works on free-viewpoint video, Carranza et al. (2003) used eight color cameras to capture the shape, appearance and motion of a real-world actor. Starck et al. (2007) also used a high-definition acquisition system comprising of eight cameras to capture the moving actor. They were also able to capture the high-level cloth deformations on the actor. MVV acquisition is not limited to low number of cameras, rather Debevec et al. (2000),

Hawkins et al. (2005), and Einarsson et at. (2006) reconstructed a number of iterations of the so-called "light-stage" which used a large number of cameras to capture the static and dynamic objects under static and varying lighting conditions. The work on free-viewpoint video by Carranza et al. (2003) was extended by Theobalt et al. (2007) where, in addition to eight high resolution color cameras, they used to calibrated spot lights to not only acquire the shape, motion and appearance but also the surface reflectance properties of a moving person. The estimation of dynamic surface reflectance allowed rendering the reconstructed 3D animation in a virtual environment having starkly different lighting condition compared to the recording environment.

A number of methods have been proposed to reconstruct temporally consistent 3D animation from MVV data. De Aguiar et al. (2008) presented a method to reconstruct high quality temporal reconstruction of dynamic objects by means of a deformation based method. They first obtained a high quality template scan of the real-world person that was deformed over the course of the animation by means of an optimization method that ensured that the deformed model is consistent with the input MVV data. Similar approach was adopted by Vlasic et al. (2008) where the skeleton-based deformation was employed to track the high quality template mesh over the animation. On the contrary, Ahmed et al. (2008) first reconstructed temporally incoherent visual hulls from MVV data for each frame of MVV data. They tracked the first visual hull over the whole sequence by means of a dense correspondence finding method that maps one visual hull to the next. None of these methods employed depth cameras for the acquisition, and unlike those method, this work does not rely on any template data or 3D surface representation for reconstructing temporally coherent 3D animation.

With the advent of low cost depth sensors, especially Microsoft Kinect (Microsoft, 2007), there has been a wave of interest in incorporating depth sensors for the acquiring 3D static and dynamic content. One of the main benefits of using Kinect is that it provides both color and depth data simultaneously at 30 frames per second. Earlier works relied only on the color data where correspondences between cameras had to be used to reconstruct the depth information, now directly provided by Kinect. Ahmed et al. (2008) reconstructed time-varying visual hulls by similar means. It is not necessary to use Kinect for acquiring the depth information as it can also be obtained from other types of sensors, e.g. Time of Flight (ToF) sensors (Kim et al., 2008).

One or more depth sensors have been employed in a number of applications to reconstruct a three-dimensional representation or static and dynamic objects. Kim et al. (2009) presented a multi-view image and depth sensor fusion system to reconstruct 3D scene geometry. Castaneda et al. (2011) used two depth sensors for stereo-ToF acquisition of a static scene. Microsoft Kinect camera was employed by Weiss et al. (2011) for human shape reconstruction. Their method combines low-resolution image silhouettes with coarse range data to estimate a parametric model of the body. Similarly, Baak et al. (2011) employed a single depth camera in their pose estimation framework for tracking full-body motions. Pose estimation from a single depth sensor has been a hallmark of Kinect as an input device, and one of the seminal works in this area was presented by Girshick et al. (2011).

The low cost of Microsoft Kinect, coupled with the benefits of acquiring depth information directly from the sensor, has led to the use of multiple depth sensors in an acquisition system. In one of the pioneering works, Kim et al. (2008) presented the design and calibration of a system that enables simultaneous recording of dynamic scenes with multiple high-resolution video and low-resolution ToF depth cameras. Unlike the system presented in this chapter, their system relied on hardware trigger for the explicit synchronization of color and depth cameras. Berger et al. (2011) employed four Kinects for marker-less motion capture. Since their area of application was silhouette-based motion capture, they did not explore the use of multiple Kinects in generating dynamic scene geometry. They also assume that Kinects are synchronous and did not actively try to create a setup for the synchronous capture. For motion capture, it can be assumed that synchronization is not a primary requirement as shown by Hasler et al. (2009). Nevertheless, for a dynamic scene reconstruction setup, which merges the data from multiple cameras, a higher degree of synchronization is required to produce a correct 3D animation. Both of the methods (Berger et al., 2011; Kim et al., 2008) do not try to extract any time coherence information from the captured depth and color data.

This chapter presents a software-synchronized multi-view acquisition system using multiple Microsoft Kinects. It will be shown that using just of-the-shelf equipment, it is possible to create a highly scalable low-cost multi-view acquisition system. In addition, three new methods that can reconstruct temporally coherent 3D video from a dynamic 3D representation will be presented. The dynamic 3D representation can be acquired by the acquisition system presented in this chapter or any comparable MVV acquisition system.

MAIN FOCUS OF THE CHAPTER

This chapter focusses on two main topics. First, the synchronized acquisition of multi-view RGB-D data using Kinects is detailed. In the second part, a number of methods to reconstruct temporally-coherent 3D animation from RGB-D data are discussed.

Data Acquisition

The multi-view recording setup is comprised of multiple Kinect cameras. The acquisition system is tested using two, four and six cameras. For the acquisition with six cameras, three cameras are placed on each side of the room (Figure 1a). The four corner cameras are placed with the angle of 90 degrees between them. In between on each side, two additional cameras are placed that make the angle of 45 degrees with their two adjacent cameras (shown in red and yellow in Figure 1a). In principle, all Kinects emit the infrared laser at the same frequency, which is a potential source of problem when using multiple Kinects for simultaneous acquisition. Ideal angle between two Kinects would be 180 degrees for simultaneous acquisition without any interference. For this work, the interference issue was deliberately ignored because the aim was to test 360 acquisitions and observe how much problem is caused by the interference. The intuition that missing information from one camera will be filled by one of the other cameras turned out to be correct as shown by the results. The placement of the cameras allows to capture a dynamic object within an area of around 2m x 3m.

Each Kinect is connected to a dedicated machine comprising Intel Core i5 2.4 GHz with 4 GB of RAM running Windows 7 64 bit (Figure 2). In principal, this is not a big limitation as all comparable acquisition systems use a similar setup. Microsoft Kinect SDK is used for data acquisition (Microsoft, 2007). It provides a wrapper to query for the depth and RGB data using a synchronous interface. The synchronous interface manages a buffer where it holds the data and provides the depth or RGB data on query with their respective time stamp. This procedure introduces some gaps in the data, but for a multi-view synchronous capture these gaps are desirable if all Kinects can query the data at the same time.

Figure 1. The system pipeline: (a) six Kinects are used to acquire the RGB and depth images (only one frame from one camera is shown); (b) shows the 3D point cloud from one camera with the mapped RGB image; (c) shows the top down view of six merged 3D point clouds. The alignment of the cameras after the global registration is shown in (d) using the color-coded points. The final segmented and filtered point cloud is shown in (e).

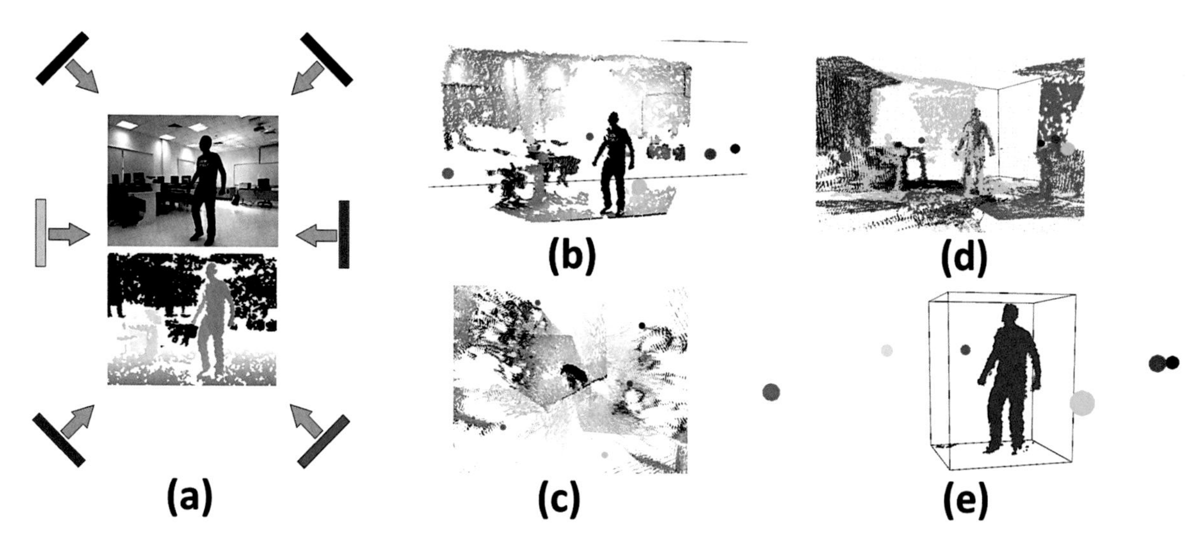

Figure 2. One Microsoft Kinect (circled) as used in our acquisition system.

The Kinect camera provides 640x480 pixels of RGB and depth data at the frame rate of 30 fps. Unfortunately this data is not hardware synchronized and there is a lag of 16 ms between depth and RGB data. If the data is acquired from each Kinect without any consideration to the acquisition from other cameras and assumed to be approximately synchronous, the video streams will start to drift temporally very quickly. This issue is circumvented by employing the following steps: First, the hardware setup is identical for each Kinect that assures that the data transfer and processing will take place at the same speed. Before the acquisition step, each machine is independently synchronized to a web-server (http://www.time.is) multiple times so that their internal clocks are synchronized. This web-server also indicates if the computer clock is running faster or slower than the web-server. The machines are synchronized till they all report the exact time. Thereafter all machines are programmed to start recording at the exact same time. This is done using the developed software interface, which accepts hour, minute and second as the starting time. Additionally, the number of frames to be recorded are provided. Since the frame rate of a Kinect, total number of frames to record, and starting time of the recording are known, and given that both RGB and depth frames are to be queried, exact clock ticks are calculated for querying each frame in advance. Each machine then uses the synchronous interface at the pre-calculated time to query for depth and RGB data alternatively. To minimize the I/O overhead, both RGB and depth data are stored in a buffer and once the recording is finished, data is written to the disk. Currently the time stamp information is not used for any post-recording synchronization but after comparing the timestamps for the 12 images (6 RGB and 6 depth) obtained for each frame, no noticeable differences were observed. Since all machines are querying the data at the exact time, the acquisition drift is kept under check.

Calibration and Segmentation

A multi-view acquisition system requires both local and global calibration. Local calibration provides camera specific parameters, or intrinsic parameters. On the other hand, the global calibration or extrinsic parameters provide the spatial mapping between the cameras.

For a Microsoft Kinect, which has two sensors, there is an additional level of local calibration. In the first step, both the depth and color sensors have to be calibrated to estimate their intrinsic parameters. Secondly, a mapping should be established between the depth and color sensors so that color data can be projected on the depth data. Finally, depth values are mapped to real-world distances in order to get 3D positions in a global coordinate system.

The intrinsic parameters are obtained using Microsoft Kinect SDK. Similarly the SDK is also used to map depth data to three-space real-world coordinate system. Using the internal calibration a 3D point cloud for each camera along with its mapping to the color data is obtained. An example of the 3D point cloud with depth to color mapping can be seen in Figure 4.

The final step for getting a dynamic 3D point cloud is to merge all the cameras together in a global unified coordinate system. This global registration is an important step because without it each point cloud would be in its own coordinate frame. To achieve this global registration, first a correspondences between the different cameras is established. This is achieved by recording the checkerboard pattern at different locations for each pair of adjacent camera as shown in Figure 3. The corners of the checkerboard provide the correspondences between two cameras are obtained using OpenCV, and additional correspondences are found using SIFT (Lowe, 1999). The correspondences are estimated in RGB space and from depth to RGB mapping the correspondences between the 3D point clouds is established. Once the correspondences between all adjacent cameras are found, one camera is selected as a reference camera

Figure 3. Extrinsic calibration for global registration - checkerboard as recorded from three cameras; corners from the checkerboard are used as some of the initial correspondences for the global registration.

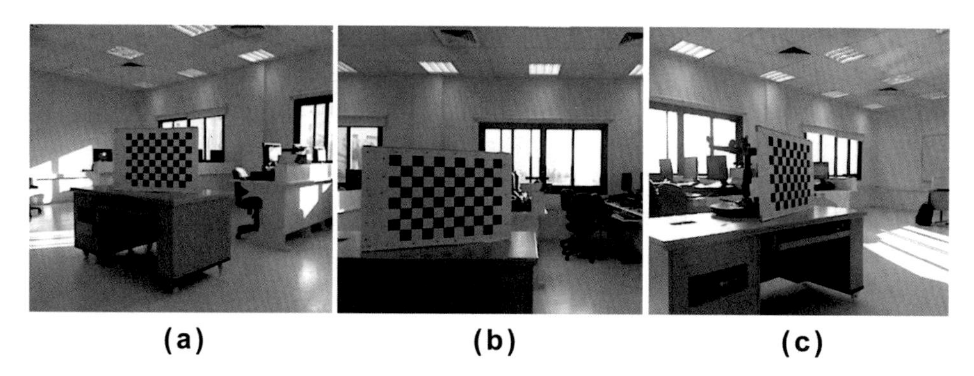

Figure 4. One frame of the dynamic 3D point cloud with RGB mapping can be seen in (a). (b) shows the merged point clouds from all cameras after global registration and segmentation.

and the correspondences are used as the starting point for the iterative closest point algorithm to find the rotation and translation transformations that maps one point cloud to the other. This transformation is found for each of two adjacent pairs and all cameras are mapped to a unified global coordinate system, which is coordinate frame of the reference camera. The global registration is a standard process and any relevant method can be applied for this step. In this work the Point Cloud Library (PCL) (Rusu & Cousins, 2011) is used to handle the point cloud data, because of its flexible data format (PCD) for storing point clouds.

Global registration gives a unique pair of rotation and translation transformations for each camera and it is applied to the corresponding depth data. The final result of the global registration is a 3D point cloud for each frame of the animation. Additionally, using the mapping between color and depth cameras, a color value with each point is associated. Thus, a dynamic 3D representation of a real world scene with RGB mapping is obtained. This representation is not time coherent because each frame is independent of the other. Examples of a 3D point cloud from one of the cameras can be seen in Figure 1b and 4a.

Final step before getting a dynamic 3D point cloud is to segment the scene so that the real-world actor can be separated from the background. The background subtraction is done using the depth data. First the acquisition room is recorded without the human actor and later the depth information of the background is used to subtract the real-world actor from the background. The result of Global registration and segmentation can be seen in Figure 1 and Figure 4.

In addition to the acquired data from the MVV setup, the data from Ahmed et al. (2008) is also used, which have a 3D visual hull representation at every time step and the corresponding color information. Dynamic point cloud data is extracted from the visual hull representation along with the RGB mapping. In the following sections, three different approaches of extracting the time-coherent representation of dynamic 3D content from the dynamic 3D point cloud data are explained. A flowchart depicting the algorithmic layout of the three approaches with respect to the acquired data can be seen in Figure 5.

Temporally Coherent 3D Animation Using Non-Linear Optimization

As explained in the previous section, the dynamic three-dimensional content obtained through either one or more Microsoft Kinects or a traditional multi-view video acquisition system completely lacks any temporal coherence. There is no connectivity from one point cloud to the next for each consecutive frame of the video. Thus the data is not very useful in extracting any meaningful information about the scene other than simple visualization. Even it is not visually pleasing, as the position of the points change so quickly from frame to frame that it distracts the viewer from the actual animation. In this section, the first of three methods to extract temporal coherence information from this dynamic 3D point clouds us-

Figure 5. Flowchart of the proposed methods starting from RGB-D data acquisition to temporally coherent 3D animation reconstruction

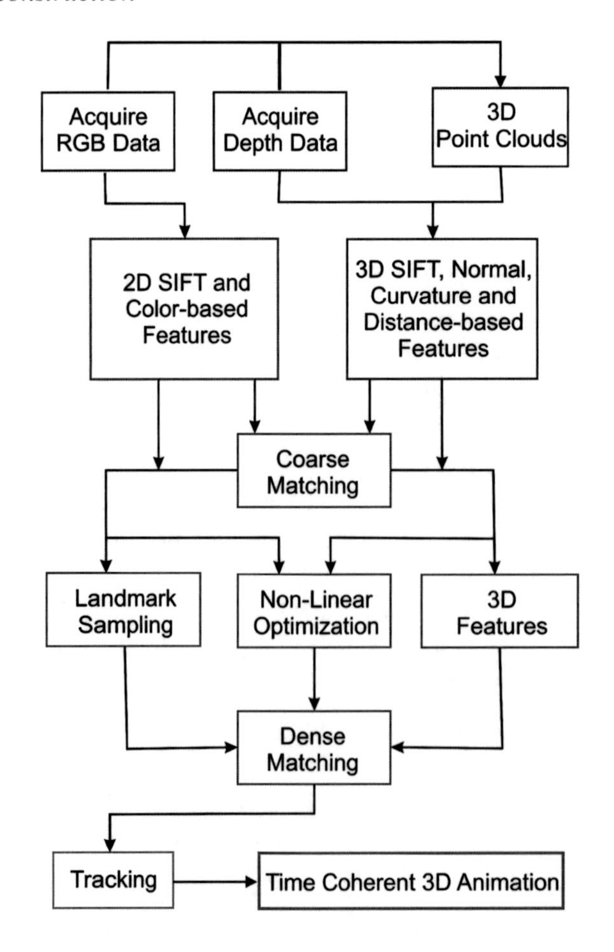

ing both geometric and color information through non-linear optimization is proposed. This coherence info will be found between two consecutive frames over the course of the animation. The aim of this method is to track a 3D point cloud throughout the entire animation using this coherence information.

The first step in this method is to extract the orientation or the normal at every point of all 3D point clouds. To approximate the normal, the normal to the plane that is tangent to the surface at that point is found. Due to the nature of the data being a point cloud, there is no actual surface, rather 10 nearest points are chosen to fit a plane and then find normal to that plane to find the orientation for each point. Normal at each point is treated as one of the feature of that point. Henceforth, for a 3D point x at frame i, the normal of that point will be referred as $\mathbf{N}(x_i)$. One way to match one point cloud to the next would be to just match the two points that have similar normals. This will mostly hold true as the animation is not very fast and at each frame there is no strong motion. But this relies on the assumption that each point has a completely unique orientation and therefore mapping from one frame to the next is trivial. In practice this is a false assumption because for every three dimensional object there are planar regions where the orientation of all the points are the same and only using the orientation information to match two 3D point clouds can never work because of the ambiguity in one to many mapping from one frame to the next.

To circumvent this ambiguity, another set of features using the color information are proposed. Since the mapping of the depth data to the color data is known, the color info at each point in the 3D point cloud is extracted. Thus, when matching two point clouds, in addition to the orientation, the color information is also matched to ensure that there is no obvious incorrect mapping of the point clouds. Henceforth, for a 3D point x at frame i, the color of that point will be referred as $\mathbf{C}(x_i)$.

The color and orientation information can give a partial matching but it is still ambiguous as there can be some regions without any meaning color-based features. Therefore two more criteria are introduced to make sure that a point is not mapped to another point of the same color and the same orientation but at farther distance. The data has two notions of distance: one obvious notion is the three-space distance between two three-dimensional points. This can be trivially found by finding the 3D Euclidean distance between two points. For a 3D point x at frame i the Euclidean distance of that point with the other 3D point at frame $i + 1$ will be referred as $\mathbf{D}(x_i)$. The second novel notion of the distance which is one of the major contributions of the algorithm is to use SIFT (Lowe, 1999) to find out the feature points in the 3D point cloud. SIFT is one of most well knows feature descriptors in the image space which is invariant under affine transformation and varying lighting conditions. Using SIFT feature points in the color image are found and consequently using the mapping between depth and RGB images, the feature 3D points for each frame are found. The SIFT matching is then used to find out the correspondences between the feature points at each frame. Once the feature point matches are identified, a distance to each point with respect to its nearest feature point is established. For a 3D point x at frame i the distance to its nearest feature point will be referred as $\mathbf{F}(x_i)$.

Assuming a match between two consecutive frames, e.g. i and $i + 1$, is being established. The match for every 3D point x at the frame i with the 3D points at the frame $i + 1$ is found using the following matching fun

$$M(x_i) = \alpha\left(1.0 - N(x_{i+1})\right) + \beta\left\|C\left(x_i\right) - C\left(x_{i+1}\right)\right\| + \gamma\left\|F\left(x_i\right) - F\left(x_{i+1}\right)\right\| + \delta D(x_i) \quad (1)$$

where x_i+1 is the 3D point at the frame $i+1$, which is used to evaluate the Equation 1. $\mathbf{M}(x_i)$ is the matching distance, 1.0-$\mathbf{N}(x_i)$-$\mathbf{N}(x_{i+1})$ is the angular difference in orientation, with the similar orientation resulting in a smaller value. $\|\mathbf{C}(x_i)-\mathbf{C}(x_{i+1})\|$ is the absolute difference of color components between (R, G, B) components of two 3D points. $\|\mathbf{F}(x_i)-\mathbf{F}(x_{i+1})\|$ is the absolute difference in the distance to the nearest feature point and $\mathbf{D}(x_i)$ is the 3D Euclidean distance between x_i and x_{i+1}. The four parameters $\alpha, \beta, \gamma, \delta$ are weighting parameters resulting in a convex combination of four terms, i.e. their sum is equal to 1 and their value is between 0 and 1. For this method, the values are set to $\alpha = 0.25, \beta = 0.2, \gamma = 0.5, \delta = 0.05$.

These values are chosen based on the reliability criteria for each of the term. The most weight is given to the difference to the nearest feature point because it is directly derived from SIFT and has a higher degree of accuracy. Least weight is chosen for $\mathbf{D}(x_i)$ because in principal the difference in 3D Euclidean position cannot be penalized because the change in position is a fundamental property of an animation. This term is only used to preserve the drift and avoid the local minima in case multiple points at frame $i + 1$ match the feature distance, orientation and the color. The matching point x_{i+1} is chosen as the one that minimizes of the convex combination. If two points result in the same value of $\mathbf{M}(x_i)$, then the point with smaller $\|\mathbf{F}(x_i)-\mathbf{F}(x_{i+1})\|$ is chosen as the matching point. In the unlikely case of same values for $\mathbf{M}(x_i)$ and $\|\mathbf{F}(x_i)-\mathbf{F}(x_{i+1})\|$, 1.0-$\mathbf{N}(x_i)$-$\mathbf{N}(x_{i+1})$ is used to find the matching point, followed by $\|\mathbf{C}(x_i)-\mathbf{C}(x_{i+1})\|$ and $\mathbf{D}(x_i)$.

The mapping starts from frame 0 to 1, and then trivially extended till the end of the sequence to find a single 3D point cloud that is tracked over the whole sequence. The results, validation and limitations of this method will be discussed in the Solutions and Recommendations section.

Temporally Coherent 3D Animation Using Landmark Sampling

In the previous section, a method to reconstruct a temporally coherent 3D animation from RGB-D data using the non-linear optimization was explained. The method is very accurate but it has one drawback of having of very high computation time, as the Equation 1 is needed to be solved for each 3D point in the point cloud. Thus in order to reconstruct a temporally coherent 3D animation efficiently with a little bit compromise on the accuracy, propose a new method to reconstruct the temporally coherent 3D animation using the landmark sampling is proposed.

This temporal coherence extraction scheme is based on matching landmarks over two consecutive frames. The process comprises of two steps:

1. Establishing reliable landmarks on each frame, and
2. Matching landmarks accurately.

These two steps are not discrete; rather an iterative process is proposed that first establishes a rough correspondence between two frames and then refines it to get an accurate match.

In the first step, SIFT features are extracted from each color image for the frames t_0 and t_1. Where t_0 is the first frame of the animation and t_1 is the second frame of the animation. Matching of the feature points gives a reliable matching in the RGB data for each of the six cameras. Since depth to RGB mapping is known, the mapping of a 3D point at t_0 to the corresponding 3D point at t_1 is implicitly established. Unfortunately the 3D correspondences are not accurate because depth to RGB mapping is many-to-one. Thus multiple 3D points match to a single pixel in the RGB image. Given a number of mappings from

3D points at t_0 to t_1, the approach proposed by Tevs et al. (2011) is used to randomly choose one of the mapping as the landmark. This gives the first rough map between the two point clouds.

In the second step, an iterative process is started that randomly picks one of the matching M_0 to M_1 found in the first step. Here the assumption is that M_0 is not just a single 3D point but a set of all 3D points that can potentially match to corresponding 3D points M_1 at frame #1. It is to be noted that in the first step one of the matching is randomly chosen as the coarse matching to facilitate the iterative process. Given the coarse matching from M_0 to M_1, three non-collinear nearest landmarks in the two point clouds are found with respect to the Euclidean distance. In practice three nearest collinear landmarks are never found but in case the three landmarks are collinear, the one at the farthest is to be discarded and the next closest one is to be selected. The non-collinear matches are required because once found the three 3D positions are used on each frame to construct a plane with normal pointing outwards to the point cloud. Assuming the nearest landmarks at frame #0 are L_{00}, L_{01} and L_{02} and on frame #1 are L_{10}, L_{11} and L_{12}. Two planes at each frame P_0 and P_1 are constructed with their normal being n_0 and n_1 respectively. Given the two planes, their normal and the root points, it is trivial to parameterize the matching points M_0 and M_1 with respect to P_0 and P_1:

$$M_0 = L_{00} + u\left(L_{01} - L_{00}\right) + v\left(L_{02} - L_{00}\right) \tag{2}$$

$$M_1 = L_{10} + u\left(L_{11} - L_{10}\right) + v\left(L_{12} - L_{10}\right) \tag{3}$$

where u and v are the two parameters that define the projection of each 3D position in M_0 and M_1 on P_0 and P_1. It is to be noted that the root points L_{00} *and* L_{10} *are* chosen randomly. This assumption is important because this step is repeated multiple times and the random selection reduces the bias in the estimation. Given the parameterization in Equations 2 and 3, for all 3D positions within the landmark matches M_0 to M_1 that are obtained in the first step, the new match is defined that has the minimum distance within the parameterized space, i.e. its u and v coordinates at t_0 and t_1. The second step is repeated multiple times, with the starting point chosen at random, and the root points also chosen at random. As shown by Tevs et al. (2011) that the random sampling with an iterative process is sufficient to correctly establish an unbiased mapping, thus a correct matching of two frames using a geometric based mapping algorithm is obtained, which uses color based matching as the starting point. The iterative process stops when the matching points are stabilized over the sequence of 5 iterations.

Once the mapping between t_0 and t_1 is established, it is propagated to the mapping between t_1 and t_2, ideally till the end of the sequence or unless it degenerates. The results, validation and limitations of this method will be discussed in the Solutions and Recommendations section.

Temporally Coherent 3D Animation Using 3D Features

In the previous two sections, two methods to reconstruct a temporally coherent 3D animation from RGB-D data using the non-linear optimization or landmark sampling were explained. Both of these methods rely on the color and depth data to extract temporal coherence. The RGB feature points are extensively used in the previous two methods to establish coarse correspondences between two frames of dynamic point clouds. In many situations the color camera is not available or the color data is not usable, e.g., actor is wearing plain clothes that cannot be used for feature or landmark extraction. In these cases where the color data is not available, or not usable, a solution to reconstruct temporally coherent 3D animation directly using the depth data and without making any use of the color data is proposed. Even though the method is not as accurate as the previous two methods, it still provides a solution where the previous two methods do not work.

To reconstruct time coherent animation the starting point is established by estimating a mapping between two consecutive frames of the dynamic scene sequence. It starts by extracting 3D features from the first two frames t_0 and t_1. These features are then matched to find a sparse mapping between the two frames. This sparse matching is used to estimate the motion between the two frames. If the object undergoes a simple motion, e.g., translation, then only one match between the two frames is sufficient to track the point cloud from one frame to the next. Three or more matches can estimate a rigid body transform. On the other hand, if the motion is non-linear, which is true in real-world recordings then the motion of every point in the point cloud is required. The motion of all the points in the point cloud by using the sparse matching is estimated as the starting point. In the subsequent steps, t_0 is tracked over the whole sequence, resulting in a time coherent animation. Thus, the time coherent animation reconstruction algorithm takes the following form:

1. Find 3D feature points at each frame.
2. Match two consecutive frames starting from t_0 & t_1.
3. Estimate motion of each point on t_0.
4. Using the estimated motion at t0, track it to t_1.
5. Loop from step 2 and track t_0 over the sequence.

In the first step, a number of 3D feature points for each frame of the 3D point cloud are found. The Point Cloud Library (Rusu & Cousins, 2011) is used to estimate the following 3D features:

1. Estimate 3D SIFT (Scovanner et al., 2007) over the depth image. The depth image is treated as an intensity image, and every feature point has a unique three-space location.
2. For every point on the point cloud, estimate its underlying curvature and normal.
3. Using the normal information from step 2, estimate Clustered Viewpoint Feature Histogram (CVFH) descriptor (Aldoma et al., 2011).

These 3D features are then used to find a sparse correspondence between t_0 and t_1. 3D SIFT features are matched over the two depth images. It provides with a one to one mapping for a sparse number of 3D positions. While matching 3D SIFT descriptors, curvature and normal are used to ensure that the matching is not an outlier. On the other hand, CVFH provides the matching clusters. Sparse matching approach is incorporated in earlier works, e.g., (Ahmed et al., 2007; Carranza et al., 2003). This method is significantly different from those works, because it is incorporating 3D features. Once the 3D features are established, the similar algorithm as of the previous method of landmark sampling is used to estimate the matching of every 3D point to the next frame. Similarly, once the mapping between t_0 and t_1 is established, it is trivially extended till the end of the sequence. The results, validation and limitations of this method will be discussed in the next section.

SOLUTIONS AND RECOMMENDATIONS

To test the software-synchronized multi-view RGB acquisition system, a number of sequences using different number of cameras are recorded. Each sequence is between 100 – 200 frames long. The sequences range from a simple walking motion to the fast boxing motion. Results from the acquisition system can be seen in Figure 1, 6(right). It can be seen that the dynamic depth maps are well aligned and the RGB image are also mapped accurately to the point cloud. As shown in the figures, the method is able to reconstruct a full 360 degree 3D animation of even the faster motion to a great degree of accuracy. No drift between the cameras was observed even though the acquisition setup is not hardware synchronized. The synchronous capture approach was validated by deliberately recording a sequence in which one camera was set to record with the standard asynchronous acquisition interface. This resulted in a temporal misalignment as can be seen in Figure 6(left). The results and validation show that the acquisition system is capable of synchronous capture of RGB-D data from multiple Kinects, which can be used for 3D animation reconstruction.

The acquisition system currently requires a dedicated machine for each Kinect. The acquisition uses machines with Intel Core i5 2.4 GHz with 4 GB of RAM running Windows 7 64 bit. In principal, this is not a significant limitation of the acquisition system because most of the comparable acquisition systems use similar arrangement. Using Microsoft's current SDK for Kinect it is possible to connect multiple Kinects to a single machine albeit using different USB hubs, but this setup is not yet tested. If multiple Kinects are attached to a single machine then a new study would be required to study the impact of processing cost and other input / output overheads. This acquisition system provides a good solution in terms of robustness, efficiency and scalability, as a new camera can easily be added to the system without compromising the acquisition as long as it is connected to a dedicated machine.

To test the temporally coherent 3D animation reconstruction methods two types of data sets are used. The first data set is recorded through the presented software-synchronized acquisition system. In addi-

Figure 6. The image on the left shows the acquisition drift (circled) if the software synchronization is not employed. The next two images show the result of synchronous acquisition with the alignment of the dynamic point clouds. The last image shows the fusion of both RGB and depth data.

tion, data from Ahmed et al. (2008), which is captured using eight color cameras with an acquisition system synchronized by the dedicated hardware. Figure 7a, qualitatively shows two consecutive frames without time coherence, whereas Figure 7b shows the same two frames with time coherence. As can be seen in Figure 7a there is no connectivity between the two frames, e.g. feet of the actor have different shape. Using time coherence one can visualize the animation with a single 3D point cloud tracked over the sequence, which can be seen in Figure 7b. All of the methods can reliably reconstruct the temporally coherent 3D animation over the whole sequence. It can be observed in the results that one point cloud can be reliable tracked from one frame to the next and consequently over the course of the animation. This results in generating a 3D animation that is temporally smooth.

In order to quantitatively validated the proposed method, 3D animation reconstruction is performed while measuring the silhouette-based overlap and also by comparing the bounding box between the temporally consistent animation and the non-coherent animation. For the silhouette based overlap, the temporally consistent point cloud is rendered from the viewpoint of one of the input cameras. Once a 3D point cloud is projected onto a 2D image plane, calculating the silhouette of projected 2D points is trivially limited to finding their convex hull. Both, the original non-coherent 3D point cloud and the spatio-temporally coherent 3D point cloud are rendered from the same camera view and their silhouettes are extracted. For the bounding box based comparison, the error measure computes the bounding box

Figure 7. (a) shows two consecutive frames from a dynamic 3D point cloud without any time coherence. (b) shows same two frames tracked using the time coherence. For example, at the feet, the point cloud changes dramatically from one frame to the next without the time coherence, whereas in (b) the point cloud remains consistent.

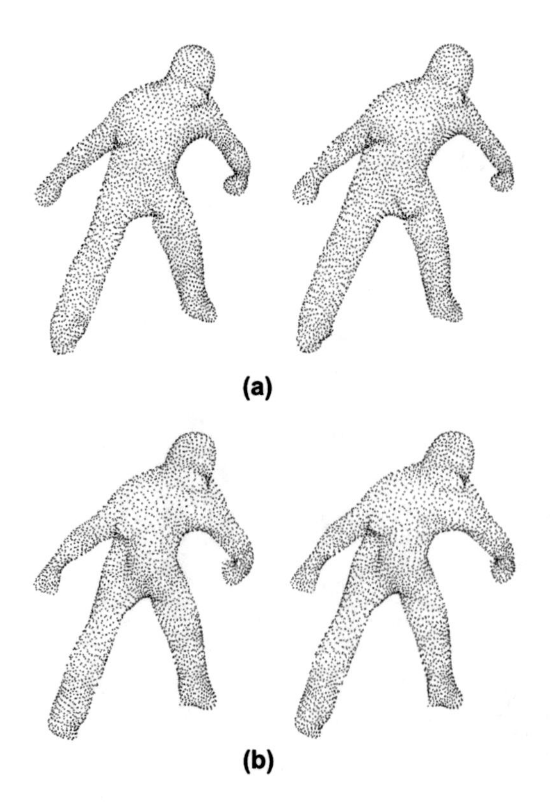

for the non-coherent 3D point cloud and count the 3D points in the temporally coherent point cloud that are not inside this bounding box. This measure also provides a good quantitative analysis in analyzing the goodness of the tracking algorithm and its temporal consistency. The two silhouettes are overlapped and the number of pixels that do not overlap for each frame are counted. In addition, a deformation based error measure to test how the distance between the feature points is changing over the whole animation is computed. Ideally, the distance should remain consistent over the sequence under the assumption that the point cloud only goes through isometric deformation. These three error measures:

- Silhouette,
- Bounding box, and
- Deformation

are used to validate the accuracy of the methods. In addition, the time complexity of each method is compared by simply running the algorithm over three sequences of 200 hundred frames each. The average time for all three sequences is normalized to obtain on average per frame computation time.

For the silhouette based overlap, on average the temporally coherent 3D animation reconstruction using non-linear optimization performs best with the average error of 2.8% in terms of silhouette overlap comparison. The error for landmark sampling is 3.2%, whereas expectedly the error for 3D features based temporally coherent 3D animation reconstruction increases to 3.5%. The graph of silhouette based overlap error can be seen in Figure 8. It can be seen that even though the error increases with time, which is expected for every tracking based algorithm it remains really low and does not deter the quality of the 3D animation. Similar pattern emerge for the other two errors. Table 1 lists the average error for all three methods and clearly the accuracy of the non-linear method is higher compared to the other two methods. This is due the fact that it uses a number of terms for the matching that results in the better quality of time-coherent 3D animation. On the other hand, as can be seen in the further discussion in this section that this accuracy comes at the cost of much higher time complexity.

The time complexity results can be seen in Table 2. Landmark sampling is the fastest method with the processing speed of 20 frames per second. 3D features based method is slightly slower at 18 frames per minute. The main reason for the 3D features based method's higher time complexity is that geometric based features calculation is much more computationally expensive. The non-linear optimization based method is the slowest at 12 frames per minute. Since, for each 3D point the method has to minimize the Equation 1 that has a number of terms, thus, the computational cost of these calculations is very high.

Based on these results the effectiveness of each method can be analyzed and the methods according to their strength and weaknesses can be recommended. First, the 3D features based method can only be recommended if the color data is not available or is not usable. It has a benefit of using only depth data for the temporally coherent 3D animation reconstruction but accuracy of the reconstruction is worst among the three methods. Between the non-linear optimization based method and the landmark sampling based method the choice depends on the preference of higher accuracy over the time complexity. The landmark sampling based method is significantly faster than the non-linear optimization based method but it is also less accurate. The non-linear optimization based method should only be chosen when the accuracy is paramount, otherwise the landmark sampling based method would suffice.

Figure 8. The silhouettes overlap error measure for each time step of the spatiotemporal coherent animation

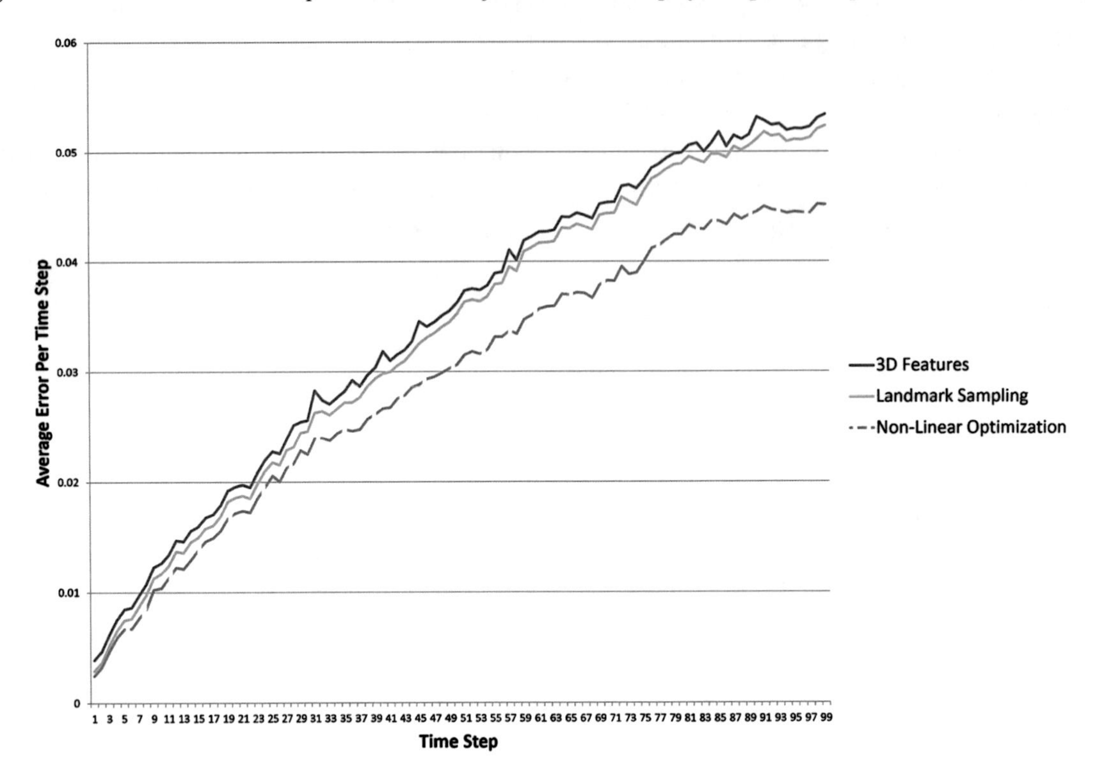

Table 1. Accuracy error measure

	Average Silhouette Error	Average Bounding Box Error	Average Deformation Error
Non-linear Optimization	2.8%	1.8%	2.45%
Landmark Sampling	3.2%	2.1%	2.8%
3D Features	3.5%	2.35%	3.1%

Table 2. Time complexity

	Frames per Minute
Non-linear Optimization	12
Landmark Sampling	20
3D Features	18

All three methods for reconstructing temporally coherent 3D animation are subject to some limitations. For the non-linear optimization based method, most notably, the method only employs one feature point in the matching function (Equation 1). This feature point is the nearest in terms of 3D Euclidean distance under whereas ideally one should look for more than one nearest feature points relative to the geodesic distance, as shown by Ahmed et al. (2008). The method circumvents this issue by only using only one nearest feature point in terms of 3D Euclidean distance, which given the high number of feature points does not pose any issues for this method. Other approach would be to find the body segments and use multiple feature points from the nearest segments. Additionally the method relies on empirical justification for the steps that require heuristics, e.g. the choice of the values of coefficients in the objective function (Equation 1). Employing machine learning techniques to estimate proper values of parameters may well circumvent these heuristics. Similarly landmark sampling and 3D features based methods are limited in their scope because there is no underlying surface representation of the 3D point cloud. These two methods use more than one landmark or feature and thus are more prone to error if the features are closer to each other in terms of Euclidean distance but far apart in terms of actual surface distance. Since the methods are dealing with the point clouds with a very high random noise, therefore the surface reconstruction is not an option. A true dynamic surface reconstruction from the depth data acquired by Kinect is a complete research problem in itself. Limiting the number of features to smaller number, i.e. 5 allows the methods to circumvent this problem. All the methods depend heavily on the relationship of the motion with the acquisition frame-rate. If the motion is very fast then 30 frames per second data from Kinect will not be able very sharp and will result in the blurred data. This type of data will result in unreliable or very low number of feature points. This is not a principal limitation of the methods as higher frame rate cameras can be used to rectify this problem.

Despite the limitations, it is shown that using multiple Microsoft Kinect cameras it is possible to capture synchronized dynamic 3D point clouds with the color information. It is also shown that given a dynamic three-dimensional content representation in the form of dynamic 3D point clouds with or without color information, it is possible to reconstruct temporally coherent 3D animation of a real-world object.

FUTURE RESEARCH DIRECTIONS

As discussed in the previous section, both the acquisition system and time coherent 3D animation reconstruction methods have some limitations that point to the future work of this project. In addition, this work opens up new areas of applications where time coherent 3D animation can be employed.

The acquisition system presented in this work is comprised of Kinect v1, released by Microsoft in 2010. It captures RGB data at the native resolution of 640x480. The resolution of RGB data from today's standards is extremely low. It is four times less than an acceptable minimum level of high definition video data. The native resolution of the depth data is 320x240, which is upscaled to 640x480. In addition, the sampling density of the depth data is very low, resulting in a sparse representation of the 3D surface with high level of discontinuities. The frame rate for both RGB and depth data is 30 frames per second. The quality of the RGB and depth data results in a number of limitations that are circumvented by the algorithms using techniques that limit their effectiveness to some degree.

One basic extension of the acquisition system would be to use the version 2 of Microsoft Kinect sensor. The new version captures RGB data in full high definition 1920x1080. The availability of the high definition RGB data will facilitate greatly in all the algorithms that rely on the color data. It opens additional avenues to manipulate RGB data to capture completely new features that are possible to be detected by the low resolution data. Additionally, the native resolution of the depth data is increased to 512x424. While this does not look like a bigger increase compared to the RGB data, the underlying depth capturing mechanism is completely changed. Now, instead of projection based depth capture, a time of flight sensor with much higher density is employed. The new sensor returns very high quality of depth data that has a higher sampling density with less discontinuities. One obvious benefit of having high quality depth data would be for the algorithms that rely on 3D features for finding the correspondences. Any method that calculates the normal or the curvature of a 3D point will benefit greatly from the new data. In addition, surface reconstruction of dynamic data should be performed to find out the quality of dynamic 3D surface. If a plausible dynamic 3D surface could be reconstructed, it will completely change the algorithms that at the moment rely on Euclidean distance between the feature points. Using the Euclidean distance is a stopgap measure to approximate the surface distance between the feature points, but it is not very reliable if the points are distributed all over the surface. Currently, all the algorithms use maximum five nearest feature points in their calculations, because Euclidean distance cannot guarantee the nearest feature points on the surface itself. Having the surface available will result in replacing Euclidean distance with the geodesic distance and any number of feature points could be used. This would result in much higher accuracy of reconstructed 3D animation.

The reconstructed time coherent 3D animation should also be validated by employing it in a number of applications that will test its goodness. For example, a temporally coherent 3D animation of a person can be used in motion analysis to determine the type of motion the person is going through. This information would not be possible to extract from the non-coherent 3D animation. Similarly, time coherent 3D animation can be easily parameterized to generate a compact animation compared to a non-coherent animation that has a very high storage cost. It opens new ways in which 3D animations can be used in a multitude of applications that require a compact representation in the areas of CGI movies or motion editing. In addition, temporally coherent 3D animation can be employed in the areas of free-viewpoint video or 3D video. The free-viewpoint video allows the user to watch a traditional movie from any angle that allows for more creative content delivery via traditional mediums of television, cinema or the new paradigms that are based on the virtual or augmented reality.

CONCLUSION

This chapter presented:

1. A system for software-based synchronized 3D video acquisition, and
2. Three methods for temporally coherent 3D animation reconstruction of a real-world scene from the dynamic RGB-D data.

It was shown that such a representation could be reconstructed using one or more Microsoft Kinect cameras. Microsoft Kinect provides both color and depth information of a scene. The acquisition system combines multiple Kinect cameras and captures a complete three dimensional dynamic scene. The acquisition system is scalable and can be used with data obtained from any number of cameras. In addition, the chapter presented three methods reconstruct temporally coherent 3D animation from the acquired dynamic RGB-D data. The temporally coherent reconstruction methods can be applied to any three-dimensional representation of the data, as long it is comprised of 3D point clouds with or without color information.

- The first method makes use of both color and RGB-D data and presents a non-linear optimization solution for extracting temporal coherence.
- The second method uses landmark sampling to achieve the same results with decreased time complexity but at the cost of lesser accuracy.
- Finally, the third method only relies on the depth data to reconstruct time coherent 3D animation in cases where the color data is not available or does not contain enough features. This method, though less accurate than the other two methods, can still reconstruct temporally coherent 3D animation for any type of RGB-D data.

The methods were validated using a number of error measures, and their capability was demonstrated by applying them on the data obtained using multiple acquisition setups, and in future they could be extended to increase the robustness of the tracking algorithm and enhanced by exploring the possibilities in the area of scene analysis and dynamic surface reconstruction.

REFERENCES

Aguiar, E. D., Stoll, C., Theobalt, C., Ahmed, N., Seidel, H. P., & Thrun, S. (2008). Performance capture from sparse multi-view video. Association for Computing Machinery Transactions on Graphics, 27(3), 98:1-98:10.

Ahmed, N. (2012). A system for 360-degree acquisition and 3D animation reconstruction using multiple RGB-D cameras. In *Proceedings of the 25th International Conference on Computer Animation and Social Agents* (vol. 1, pp. 9-12). Wiley.

Ahmed, N., Theobalt, C., Rossl, C., Thrun, S., & Seidel, H. P. (2008). Dense correspondence finding for parametrization-free animation reconstruction from video. *Proceedings of Computer Vision and Pattern Recognition, 1*, 1–8.

Aldoma, A., Blodow, N., Gossow, D., Gedikli, S., Rusu, R. B., Vincze, M., & Bradski, G. (2011). CAD-Model Recognition and 6 DOF Pose Estimation. In *Proceedings of International Conference on Computer Vision 3D Representation and Recognition workshop* (vol. 1, pp. 585-592). IEEE.

Baak, A., Muller, M., Bharaj, G., Seidel, H. P., & Theobalt, C. (2011). A data-driven approach for real-time full body pose reconstruction from a depth camera. In *Proceedings of International Conference on Computer Vision* (vol. 1, pp. 1092-1099). IEEE. doi:10.1109/ICCV.2011.6126356

Berger, K., Ruhl, K., Schroeder, Y., Bruemmer, C., Scholz, A., & Magnor, M. A. (2011). Markerless motion capture using multiple color-depth sensors. *Proceedings of Vision Modelling and Visualization, 1*, 317–324.

Carranza, J., Theobalt, C., Magnor, M. A., & Seidel, H. P. (2003). Free-viewpoint video of human actors. *ACM Transactions on Graphics*, *22*(3), 569–577. doi:10.1145/882262.882309

Castaneda, V., Mateus, D., & Navab, N. (2011). Stereo time-of-flight. In *Proceedings of International Conference on Computer Vision* (vol. 1, pp. 650-657). IEEE.

Debevec, P. E., Hawkins, T., Tchou, C., Duiker, H. P., Sarokin, W., & Sagar, M. (2000). Acquiring the reflectance field of a human face. *Proceedings of SIGGRAPH, 1*, 145–156.

Einarsson, P., Chabert, C. F., Jones, A., Ma, W. C., Lamond, B., Hawkins, T., & Debevec, P. E. et al. (2006). Relighting human locomotion with flowed reflectance fields. In *Proceedings of Eurographics Symposium on Rendering* (vol. 1, pp. 183-194). Eurographics Association.

Girshick, R., Shotton, J., Kohli, P., Criminisi, A., & Fitzgibbon, A. (2011). Efficient regression of general-activity human poses from depth images. In *Proceedings of International Conference on Computer Vision* (vol. 1, pp. 856-863). IEEE. doi:10.1109/ICCV.2011.6126270

Hasler, N., Rosenhahn, B., Thormahlen, T., Wand, M., Gall, J., & Seidel, H. P. (2009). Markerless motion capture with unsynchronized moving cameras. *Proceedings of Computer Vision and Pattern Recognition, 1*, 65–73.

Hawkins, T., Einarsson, P., & Debevec, P. E. (2005). A dual light stage. In *Proceedings of Eurographics Symposium on Rendering* (vol. 1, pp. 91-98). Eurographics Association.

Kim, Y. M., Chan, D., Theobalt, C., & Thrun, S. (2008). Design and calibration of a multi-view tof sensor fusion system. In *Proceedings of IEEE CVPR Workshop on Time-of-flight Computer Vision* (vol. 1, pp. 1-7). IEEE.

Kim, Y. M., Theobalt, C., Diebel, J., Kosecka, J., Micusik, B., & Thrun, S. (2009). Multi-view image and ToF sensor fusion for dense 3d reconstruction. In *Proceedings of IEEE Workshop on 3-D Digital Imaging and Modeling* (vol. 1, pp. 1542-1549). IEEE. doi:10.1109/ICCVW.2009.5457430

Lowe, D. G. (1999). Object recognition from local scale-invariant features. In *Proceedings of International Conference on Computer Vision* (vol. 1, pp. 1150-1157). IEEE.

Microsoft. (2010). *Kinect for Microsoft windows and Xbox 360.* Retrieved January 14, 2016, from http://www.kinectforwindows.org/

Rusu, R. B., & Cousins, S. (2011). 3D is here: Point Cloud Library. In *Proceedings of International Conference on Robotics and Automation* (vol. 1., pp. 1-8). IEEE.

Scovanner, P., Ali, S., & Shah, M. (2007). A 3-dimensional SIFT descriptor and its application to action recognition. In *Proceedings of the 15th international conference on Multimedia* (*vol. 1*, pp 357–360). ACM Press New York. doi:10.1145/1291233.1291311

Starck, J., & Hilton, A. (2007). Surface capture for performance-based animation. *IEEE Computer Graphics and Applications, 27*(3), 21–31. doi:10.1109/MCG.2007.68 PMID:17523359

Tevs, A., Berner, A., Wand, M., Ihrke, I., & Seidel, H. P. (2011). Intrinsic Shape Matching by Planned Landmark Sampling. *Computer Graphics Forum, 30*(2), 543–552. doi:10.1111/j.1467-8659.2011.01879.x

Theobalt, C., Ahmed, N., Ziegler, G., & Seidel, H. P. (2007). High-quality reconstruction of virtual actors from multi-view video streams. *IEEE Signal Processing Magazine, 24*(6), 45–57. doi:10.1109/MSP.2007.905701

Vlasic, D., Baran, I., Matusik, W., & Popovic, J. (2007). Articulated mesh animation from multi-view silhouettes. Association for Computing Machinery Transactions on Graphics, 27(3), 97:1-97:9.

Weiss, A., Hirshberg, D., & Black, M. J. (2011). Home 3D body scans from noisy image and range data. In *Proceedings of International Conference on Computer Vision* (vol. 1, pp. 1951-1958). IEEE. doi:10.1109/ICCV.2011.6126465

Ye, M., & Yang, R. (2014). Real-time simultaneous pose and shape estimation for articulated objects using a single depth camera. In *Proceedings of IEEE Conference on Computer Vision and Pattern Recognition* (vol. 1, pp. 2345-2352). IEEE. doi:10.1109/CVPR.2014.301

Zollhöfer, M., Nießner, M., Izadi, S., Rehmann, C., Zach, C., Fisher, M., Wu, C., Fitzgibbon, A., Loop, C., Theobalt, C., & Stamminger, M. (2014). Real-time non-rigid reconstruction using an RGB-D camera. *Association for Computing Machinery Transactions on Graphics, 33*(4), 156:1-156:12.

KEY TERMS AND DEFINITIONS

3D Animation: Movement of three-dimensional representation of data done using the computer. It can be created by an artist or automatically through an algorithm.

3D Reconstruction: An algorithmic process to extract 3D information from 2D data.

Background Segmentation: An algorithmic process that separates the foreground in an image from its background.

Camera Calibration: An algorithmic process that determines all the parameters of a camera that are used to capture an image using it.

Free-Viewpoint Video: Also termed as a 3D video is a special type of video that can be watched from any viewpoint and the user has control over the view.

Kinect: A motion sensing input device by Microsoft, originally created for Xbox 360. I houses two cameras, one depth and one color.

Multi-View Video: Video that is capture from multiple cameras capturing the same scene from different viewpoints.

RGB-D Data: Data comprising of color and depth information. Color information is encoded in the form of three red, green and blue color values. Depth information is encoded in a single depth value.

Time Coherence: Property of the animated data where each frame of animation can be linked to previous or next frame through some parameters rather than existing completely independent of each other.

Chapter 8
On the Use of Motion Vectors for 2D and 3D Error Concealment in H.264/AVC Video

Hugo R. Marins
Universidade Federal Fluminense, Brazil

Vania V. Estrela
Universidade Federal Fluminense, Brazil

ABSTRACT

The fundamental principles of the coding/decoding H.264/AVC standard are introduced emphasizing the role of motion estimation and motion compensation (MC) in error concealment using intra- and inter-frame motion estimates, along with other features such as the integer transform, quantization options, entropy coding possibilities, deblocking filter, among other provisions. Efficient MC is one of the certain reasons for H.264/AVC superior performance compared to its antecedents. The H.264/AVC has selective intra-prediction and optimized inter-prediction methods to reduce temporal and spatial redundancy more efficiently. Motion compensation/prediction using variable block sizes and directional intra-prediction to choose the adequate modes help decide the best coding. Unfortunately, motion treatment is a computationally-demanding component of a video codec. The H.264/AVC standard has solved problems its predecessors faced when it comes to image quality and coding efficiency, but many of its advantages require an increase in computing complexity.

INTRODUCTION

Because 3D video has gotten crescent importance in multimedia and there are lots of legacy 2D videos, both 2D+depth and the H.264/AVC standard still need attention. Most multimedia data streams contain audio, video, and some metadata, but for these streams to be useful in stored or transmitted form, they must be encapsulated together in a container format.

DOI: 10.4018/978-1-5225-1025-3.ch008

To offer better video quality and more flexibility compared to previous standards, the H.264/AVC (Wiegand et al., 2003) video coding standard has been devised by the Joint Video Team (JVT) to deliver significant efficiency, simple syntax specifications, and seamless integration of video coding into all current protocols and multiplex architectures. H.264/AVC supports several video applications like broadcasting, streaming, and conferencing over fixed and wireless networks with different transport protocols.

Error Concealment (EC) involves recreating lost video data using already received information. Due to redundancy in both the spatial and temporal domains, the lost data can be estimated from existing information via Motion Estimation (ME), also known as Motion Prediction. The main types of EC include the following techniques: spatial (intra-frame), temporal (inter-frame) and hybrid. The last technique is a combination of the spatial and temporal strategies (Fleury et al., 2013).

Efficient Motion Compensation (MC) is one of the key reasons for the H.264/AVC superior performance compared to its predecessors. Unfortunately, motion estimation/compensation (MEMC) is the most computationally-intensive part of a video encoder.

A video is organized as a sequence of frames, where each frame is an image consisting of pixels. The H.264/AVC divides every frame into several Macroblocks (MBs). A MB is a Processing Unit (PU) in video compression formats relying on linear block transforms, such as the Discrete Cosine Transform (DCT). The H.264/AVC main profile supports a 4×4 transform block size while its high profile allows for a transform block size of either 4×4 or 8×8, tailored on a per-MB basis (ITU-T, 2013). The MBs are grouped into partitions called slices.

There are three different types of frames:

- Intra-picture frames (I-frames),
- Unidirectional predicted frames (P-frames), and
- Bidirectional predicted frames (B-frames).

These frames form a sequence called a Group of Pictures (GOP) as shown in Figure 1.

In preliminary standards such as H.261, MPEG-1 Part 2, and H.262/MPEG-2 Part 2, MC is performed with one Motion Vector (MV) per MB. Diversely from the division into transform blocks, an MB can be split into prediction blocks with multiple variable-sized prediction blocks (partitions) in the H.264/AVC.

Figure 1. Typical GOP

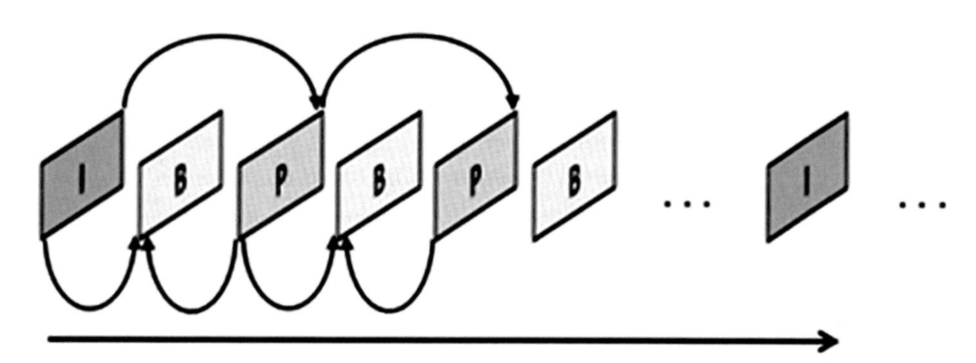

The H.264/AVC videos coding layer design is based on the concepts of Motion Compensated Prediction (MCP) and transform coding. Within this framework, some important functionalities deserve mention such as:

- Intra-prediction in spatial domain,
- Hierarchical transform with (4×4, 8×8) integer DCT transforms,
- Multiple reference pictures in inter-prediction,
- Generalized bidirectional prediction,
- Weighted prediction,
- Deblocking filter,
- Context-Based Adaptive Variable Length Coding (CAVLC)
- Context Adaptive Binary Arithmetic Coding (CABAC),
- Entropy coding, and
- Parameter setting.

Flexible MB Ordering (FMO) allows to partition a picture into regions (slices) where each region can be decoded independently, flexible slice size that enhance coding efficiency by reducing the header data, arbitrary slice ordering allows each slice to be decoded independently and they can be sent out of order, thus reducing end-to-end delay in some networks, redundant slices, Switched P (SP)/Switched (SI) slices for error resilience.

Multimedia data transmitted over communication channels are affected by various types of distortions that cause deviations of the received packets compared with the original packets sent by the transmitter side. The compressed video has to be segmented into packets and multiplexed with other types of data. Packets may be missing or degraded, due to either traffic congestion or link failure (Coelho et al. 2012b; Coelho et al. 2013).

Packet Loss and Its Effect on Video

The H.264/AVC video codec is an excellent coding format that includes all forms of video compression ranging from low bit-rate Internet streaming purposes to HDTV broadcast and Digital Cinema applications. The H.264/AVC standard yields the same quality of images with a bitrate savings of 50% over that produced by previous standards. This codec distinguishes conceptually between a Video Coding Layer (VCL) and a Network Abstraction Layer (NAL).

The VCL is the signal processing section of the codec namely, mechanisms such as transform, quantization, and motion compensated prediction; and a loop filter. It follows the general concept of most of today's video codecs, MB-based coders that uses inter-picture MC prediction and transform coding of the residual signal (Cui et al., 2009). The VCL encoder outputs slices:

- A bit string that has the MB information about a number of MBs, and
- The information of the slice header (containing the spatial address of the first MB in the slice, the initial quantization parameter, and similar data).

The NAL encoder encapsulates the slice received from the VCL encoder to form NAL Units (NALUs) suitable for transmission over packet networks or application in packet-oriented multiplex environments. A NALU has a one-byte header and the payload byte string. The header indicates the NALU type, if it contains bit errors or syntax infringements in the NALU payload, and data on the relative importance of the NALU for the decoding process.

IP packets contain lower hardware-level protocols for the delivery over various networks, and they encapsulate higher transport- and application-level protocols from streaming and other us s. In the case of streaming H.264/AVC over IP networks, numerous protocols, such as Real-Time Protocol (RTP) and User Datagram Protocol (UDP), are sent in the IP payload, where each has its header and payload that recursively transmits another protocol packet. For instance, H.264/AVC data is carried by an RTP packet which in turn is carried by a UDP packet and, then, is transmitted as an IP packet. The UDP sends the media stream as a series of small packets. Although the protocol does not guarantee delivery, it is simple and efficient. The receiving application has to detect loss or corruption and recover data using error correction techniques. The flow may suffer a dropout, if data is lost. The RTP is a native internet protocol intended for transmission of real time data over IP networks and it does not provide any multiplexing capability. Instead, each stream is sent as a detached RTP stream and relies on underlying encapsulation (normally UDP) to provide multiplexing over an IP network without the need of an explicit de-multiplexer on the client. Each RTP stream must carry timing information for the client side to synchronize streams when required.

Such errors are likely to damage a Group of Blocks (GOB) in the decoder. An error also propagates due to the high correlation between neighboring frames and degrades the quality of consecutive frames. The RTP over UDP is a recommended and commonly used mechanism employed with the H.264/AVC streaming. Various approaches have achieved error resilience to deal with the above problem. One of the ways to overcome this problem is using EC at the decoder.

Packetized multimedia transmission is prone to hindrances, caused by packet losses. H.264/AVC data is coded as packets called Network Abstraction Layer (NAL) Units. NALUs provide unique features to adjust well to an extensive variety of communication networks. For this reason, each NALU or group of NALUs can be directly packetized without having to subdivide the stream into network packets, as is the case with transport stream (TS) packets (Chong et al., 2007).

Unlike TS packetization, the RTP Payload Format for H.264/AVC video is specially designed for packetizing NALUs. This scheme is more robust because more important NALUs such as those containing slice headers are usually small enough to fit within individual RTP packets and are thus not fragmented, as is the case with TS packets. Therefore, when packet loss occurs, a loss of a single packet that contains only part of a slice header, as in TS packetization, prevents the slice header from being decoded correctly. As a result, the entire slice cannot be decoded.

The majority of NALUs have visual slice data. HD videos encoded in H.264, characteristically have 8 slices per frame. Each of these slices is as wide as a frame and consists of 8 or 9 rows of MBs.

Lost packets straightforwardly correspond to missing NALUs, which produce lost slice information. Therefore, packet losses that arise during wireless streaming of H.264/AVC HD video appear as lost whole slices or parts of slices. Loss of an entire slice may happen due to the following reasons:

1. Packet loss may be severe enough to affect an entire slice as well as neighboring slices or even a whole frame, and

2. Certain NALUs contain slice header data. If a packet containing such information is gone, the rest of the slice cannot be decoded.

EC is required to recreate missing video data due to network errors; otherwise visual quality will be further degraded. The easiest form of EC is frame copy, which involves just replicating video data from the preceding frame in place of the missing information in the present frame. Although this is better than just ignoring the lost slices, it is not visually satisfactory. More notably, lost slices lead to error propagation since spoiled frames are often reference pictures for next frames. Support for multiple reference frames provides considerable compression when there is periodic motion, translating motion, occlusions as well as alternating camera angles that switch back and forth between two different scenes

So, it is essential to employ more sophisticated methods other than pure EC techniques to ameliorate the concealment of corrupted frames and to limit error propagation. The video content moves in a somewhat deterministic way through time producing MVs that are correlated spatially or temporally. Moreover, what happens in one part of a picture, in general, determines or is determined by another part of the frame. For example, consider an animal moving across a screen.

The next sections will review fundamental aspects of the H.264/AVC standard (refer to Figure 2) with emphasis on the characteristics that improve EC performance by means of a more rational use of MEMC. Because the H.264 format is not flexible at the encoder stage and the decoder is non-normative, most of the discussion in this work revolves around the decoder stage.

OVERVIEW OF THE H.264/AVC CODEC

A codec is a system capable of encoding/compressing/encrypting or decoding/decompressing/decrypting a digital data signal.

Many codecs are lossy, meaning that they reduce quality by some amount to achieve compression. Often, depending on the codec and the settings used, this compression type is virtually indistinguishable from the original uncompressed data. Lower data rates also reduce cost and improve performance when sending data.

There are also many lossless codecs typically store data in a compressed form while retaining all of the information from the original stream. If preserving the original quality of the data stream is greater than the data sizes, lossless codecs are preferred. If the data have to undergo further processing by several lossy codecs, then there will be quality degradation that results in useless data.

Figure 2. Scope of video coding standardization: only the syntax and semantics of the bitstream and its decoding are defined.

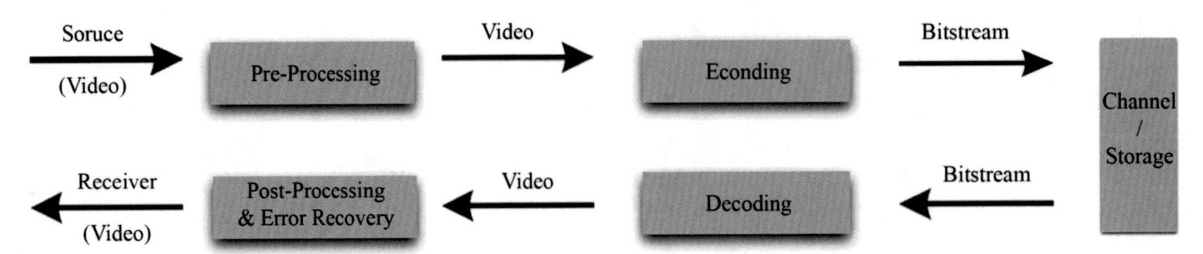

H.264/AVC Encoder

The block diagram for the H.264/AVC encoder is shown in Figure 3. The encoder may select between intra- and inter-coding for block-shaped regions of each frame. To enclose various applications, profiles are defined. The main Profile includes support for interlaced video inter-coding using B slices, inter-coding using weighted prediction and CABAC (Richardson, 2010).

MVs are a form of storing the compressed image changes from one frame to the next. These vectors may be used creatively to detect and track motion and to find an alternative to conventional video decoding using phase shifting.

The initial compression step involves changing from the RGB to the YCrCb color space. YCrCb uses three components: luma, red chrominance (chroma), and blue chroma. This alteration mimics the human eye that is less sensitive to chroma than it is to luma. Subsequently, chroma data is sampled at a quarter of the rate used for luma data.

Spatial redundancy attenuation comes from separating the images into 16×16-pixel MBs. Each MB contains 16×16 luma pixels and 8×8 red/blue chroma pixels. The luma block is then split into four 8×8 blocks. After that, DCT is performed on six 8×8 blocks and the resulting coefficients are quantized, filtered, and then stored.

The next stage in compression is intended to reduce temporal redundancy. A series of frames is divided into a GOP containing I, P, or B frames. The usual method breaks a video into 15-frame GOPs where (i) an I-frame is always the first picture, (ii) it is frequent to have two B-frames after the I-frame, followed by a P-frame, two B-frames, and so on: ...IBBPBB...

Figure 3. The H.264/AVC encoder

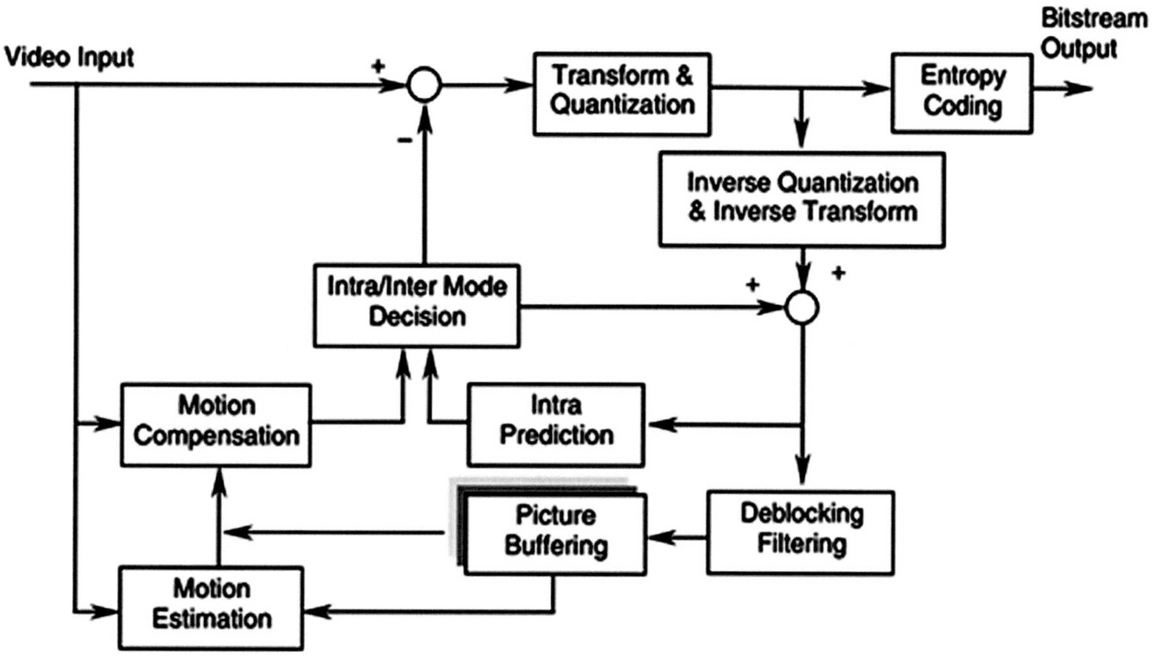

Figure 4. Possible prediction directions for INTRA_4×4 mode

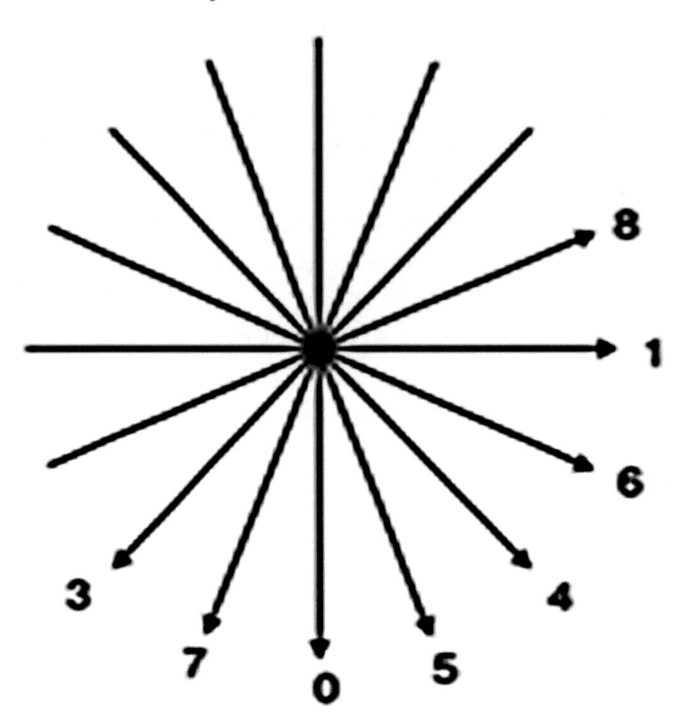

Intra-coding uses various spatial prediction modes (Figure 3) to reduce spatial redundancy in the source signal for a single frame. Intra-prediction uses the data estimated from nearby blocks for sending information about the current MB instead of the actual pixel data. Finally, the MVs or intra-prediction modes are combined with the quantized transform coefficient information and encoded using entropy code such as CAVLC or CABAC.

Some advantages of spatial domain estimations are:

1. The MB prediction from the neighbouring pixels (top/left) in the raster scan would be much efficient as compared to the transform domain estimates;
2. Facilitates compression for intra blocks in a inter frame; and
3. Since compression improves, a flexible bit-rate control can be achieved.

Inter-coding (predictive or bi-predictive) uses MVs from some previously decoded pictures for block-based inter-prediction to reduce temporal redundancy among different pictures. Prediction is obtained from the deblocking filtered signal of previous reconstructed pictures. The de-blocking filter is to reduce the blocking artifacts at the block boundaries. The H.264/MPEG-4 AVC format defines a deblocking filter that manages 16×16 MBs and 4×4 block boundaries. In the case of MBs, the filter is intended to remove artifacts that may result from adjacent MBs having different estimation types (e.g. motion vs. intra estimation), and/or different quantizer scale. In the case of blocks, the filter takes away artifacts resulting from transforms, quantization, and MV differences between adjacent blocks.

Classifying a frame as I, P, or B determines the way temporal redundancies are encoded. An I-frame is encoded without relying on other frames. However, a P-frame encoding necessitates breaking the image into MBs, and subsequently using a matched filter, or a comparable system, to match each MB to a 16×16-pixel region of the last I-frame. Once the best match is found, the MV (the vector pointing from the center of the MB to the center of the area that is the closest equivalent) is assigned to that MB, and the error between the DCT coefficients of the current MB and the region it is being compared to the encoded I-frame. A B-frame differs from a P-frame only in that the above step is performed twice: relating an MB in a B-picture to a point to the last I- or P-frame, and relating the same MB to a point in the next I- or P-frame. The DCT coefficients of the two sections are averaged, and the consequential error (residual) is encoded. Finally, the video stream compression involves using a variable length Huffman coding algorithm, for instance, to lessen redundancies.

H.264/AVC Decoder

Unlike the encoder part, the H.254/AVC has a non-normative decoder. The H.264/AVC decoder (Figure 5) has an EC section, where different MV EC procedures can be found. It defines the syntax and semantics of the bitstream as well as what the decoder needs to do to decode the video bitstream.

The format does not define how encoding or other video pre-processing is done. At the same time, the standardization preserves the interoperability requirement for any communications framework.

Coding efficiency and easy integration of the coded stream into all types of protocols and network architectures promote efficient transmission. Still, only a good combination of network adaptation and video coding can bring the best achievable performance for a video communication system. Therefore H.264/AVC consists of two conceptual layers. The VCL defines the efficient representation of the video, and the NAL converts the VCL representation into a format suitable for specific transport layers or storage media.

The EC algorithms for H.264/AVC non-normative decoder rely on two basic assumptions:

Figure 5. H.264 decoder

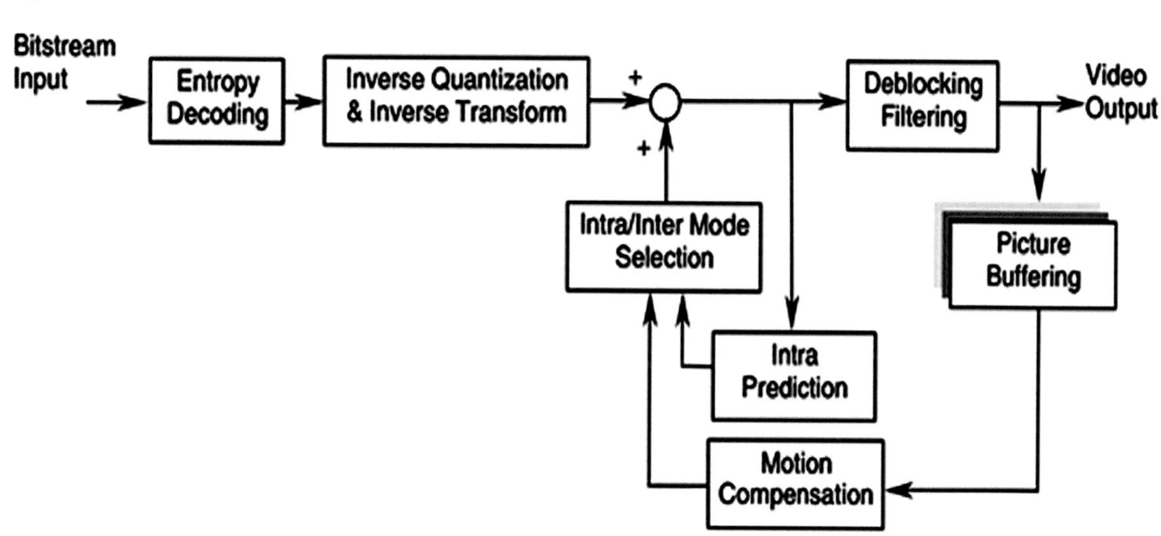

1. Erroneous or incomplete slices are discarded; and
2. Received video is stored in the correct decoding order.

When all received picture slices have been decoded, the skipped slices are concealed according to the EC algorithms. In practice, record is kept in an MB based status map of the frame.

H.264 prediction enables the delivering of high quality video at low compression rates, since only the most significant portions of a frame are coded and the rest can be reconstructed by the decoder via prediction. There are two types of prediction: intra- and inter-prediction as illustrated in Figure 6.

Inter-prediction uses samples from past and/or future frames which results in P-MBs and B-MBs. A P-MB calls for unidirectional MB prediction from samples of a previously decoded frame.

A B-MB is a consequence of bidirectional prediction from samples coming from two previously decoded frames (typically, a past P-frame and a future P-frame). Hence, each B-frame requires the decoding of two P-frames to provide the references for the bidirectional prediction. The frame decoding order is different from the display order. For example, given two P-frames and one B-frame, the initial P-frame is displayed first, followed by the B-frame, and finally by the second P-frame., in other words if the decoding order is IPBPBPBPB..., the equivalent display order is IBPBPBPBP.

Figure 6. Different types of MBs and the relationship between them in a GOP

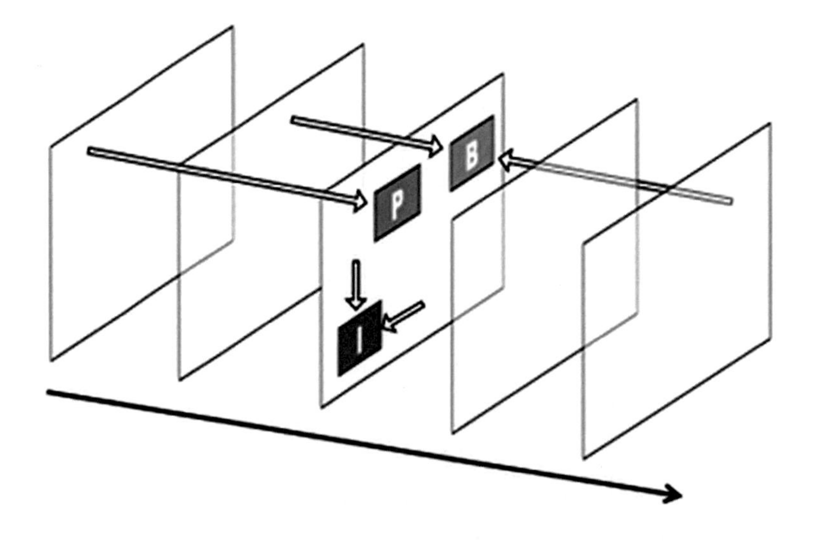

Figure 7. Illustration of the relationships between I-frames and P-frames and their MBs

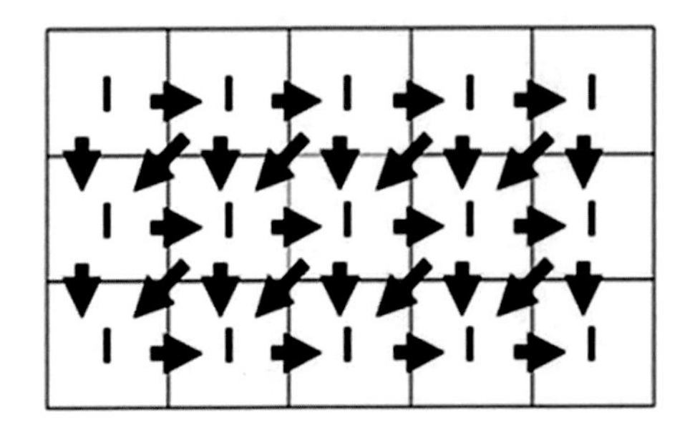

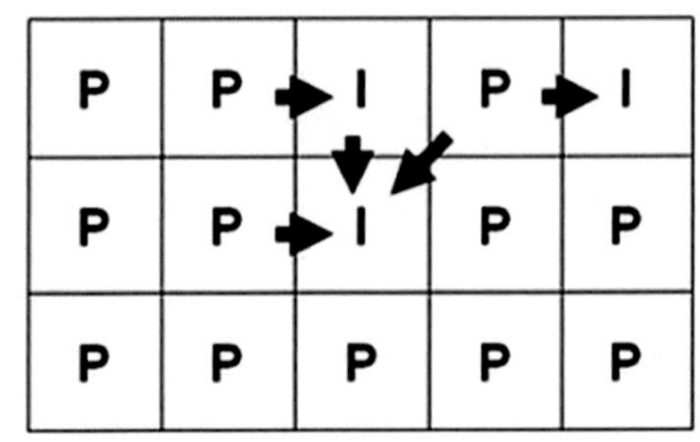

Inter-prediction is performed by using luma and chroma samples from previously coded frames to forecast samples in a current frame. For every block in the current frame that is to be inter-predicted, a best-match corresponding block is searched for in previous frame(s) ME. These algorithms are more localized and thus much less costly than doing a full-frame search for each block. An MV between the current block and the best-matching block in the corresponding reference frame is shown in Figure 6. Inter-prediction has an associated prediction error similar to the residual (Cui et al., 2009) obtained in intra-prediction that has to be encoded with the MV. Again, this improves coding efficiency compared to encoding a complete image block. During decoding, the MV is used to reference a block in a previously decoded frame via MC. The prediction error that results from inter-coding is usually much smaller than the residual that is due to intra-coding. However, in certain situations, the prediction error is too big for efficient coding. A prominent example of such a situation is an object moving in a complex manner, e.g. stretching, contortion, etc. Another significant example of a situation where the prediction error would be unfavorably large is a frame's contents at the edges, especially if they do not have suitable references in past frames. This is because as new video parts arrive, new features are most likely to appear in a scene. For instance, for a video that is panning to the left, new features would consistently appear at the left edge and thus have no suitable references in past frames. Situations such as these two examples result in intra-coding being preferred to inter-coding.

The I-MBs exploit spatial neighborhoods by referencing past decoded MBs in the same video and the P-MBs take advantage of temporal locality by referencing MBs from a previously decoded frame. Both types of MBs also include a variable quantity of residual data that cannot be inferred from prior MBs. I-MBs have strict data dependencies within the same frame. On the contrary, P-MBs do not depend on other MBs within the same frame. As an outcome, it is reasonable for both P-frames and B-frames to include I-MBs. For example, Figure 6 shows an I-frame with only I-MBs and a P-frame with both P-MBs and I-MBs (Chong et al., 2007).

Both inter-prediction and intra-prediction result in efficient compression in H.264. Rather than encoding the complete image data, only MVs and prediction error are encoded in the case of inter-coded blocks. Additionally, the residual and the prediction mode are encoded in the case of intra-coded blocks. During H.264/AVC decoding, MVs are decoded and used to perform MC. Prediction error is also decoded and added to the visual data that is produced by MC. Intra-coded data is also decoded during H.264/AVC decoding. A prediction mode is decoded and used to reconstruct each MB using its neighboring MBs. The residual is also decoded and then added to the result produced by the prediction mode.

When an MV is lost, its corresponding block cannot be decoded because there is no information for referencing a past reference block. Moreover, if this undercoded block can be an erroneous reference for a future inter-predicted block. This degradation progression continues until an intra-coded block is met causing error propagation. Consequently, to conceal inter-prediction errors within a particular frame and, by extension, within future inter-predicted frames, a lost block needs to be estimated by MV recovery, which is a type of EC that can be done by inferring from data within the current frame (spatial) or from other frames (temporal) or a combination of both (spatial-temporal). This results in an estimated MV that can be used for MC, which yields an estimated block.

MOTION AND ERROR CONCEALMENT

Intra-Frame Concealment

Spatial EC techniques involve utilizing the redundancy provided by surrounding MVs to cover a candidate MV. This may include replacing the lost MV with the median or mean of surrounding MVs, interpolation of missing MVs using surrounding ones, or some variation of these. The decoder can use mean and median MVs. However, it tends to default to frame copy or weighted pixel averaging (WPA), depending on decoder settings. Visually, WPA produces blurry distortion. WPA is done on a per-MB basis and is based on the assumption of spatial continuity. Each pixel in a lost MB is concealed by averaging the four orthogonally closest pixels from surrounding MBs. Each neighboring pixel is weighted according to its distance from the pixel to be concealed, with greater weight being attributed to spatially closer pixels.

Most of the existing work in MV recovery for video is limited to resolutions below HD. Also, a significant portion of the existing work involves MPEG video, which does not exhibit intra-coded and inter-coded MBs mixed together in the same frame in a manner similar to H.264. Furthermore, with the advent of H.264/AVC, it is possible to encode lower definition video at very low bitrates, thus making each frame small enough to fit within just a few network packets. Therefore, with sub-HD resolution video, errors due to network packet losses are often manifested as whole frame losses. HD video data is typically too large to have packet losses result in as many dropped frames as sub-HD video. Instead, portions of frames are often lost, making it possible to utilize incomplete frame information for EC. It is worth noting that a lot of the existing MV recovery work does not consider the impact of MVs neighboring the ones selected for EC. This consideration is important as the neighboring MVs expose spatial relationships.

Intra-prediction uses neighboring samples from the current frame. Consequently, intra-MBs (I-MBs) are created. A frame that contains only I-MBs is called an I-frame. Each MB is associated with 1 to 16 MVs ensuring backward compatibility with previous standards.

Both luma and chroma samples are weighted combinations of the components of the RGB color space. Luma is visually more significant to the Human Visual System (HVS) than chroma data. H.264/AVC allows for partitioning of MBs into sub-blocks. Due to the significance of the luma component to the HVS, it has a variety of block sizes: 16×16, 8×8 and 4×4, which makes intra-prediction more accurate.

The choice of the intra-prediction block size for the luma component is usually a tradeoff between prediction accuracy and the cost of coding the prediction mode. Different block sizes correspond to different prediction modes. Figure 8 shows prediction modes for 4×4 and 16×16 block sizes. The arrows indicate the directions of extrapolation and interpolation. Interpolation is performed when the current block is gradient-filled between two samples, e.g. modes 3 and 8 in Figure 8b. Extrapolation occurs when only one of the source samples is involved, e.g. vertical extrapolation in mode 0.

Once an intra-prediction block is generated, it is subtracted from the original block to form a residual. The more accurate the prediction, the smaller the residual size (Figure 9). Then, the residual and the prediction mode information are encoded. This improves coding efficiency compared to encoding a complete image block.

Figure 8. Intra-prediction modes

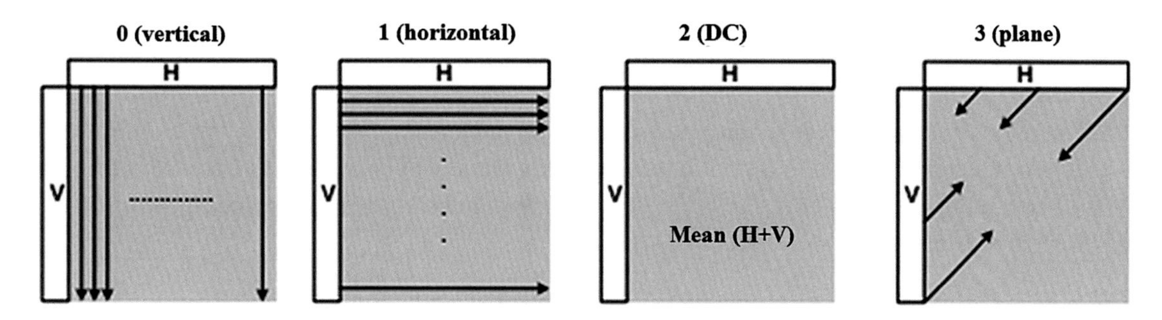

(a) 16x16 intra-prediction modes.

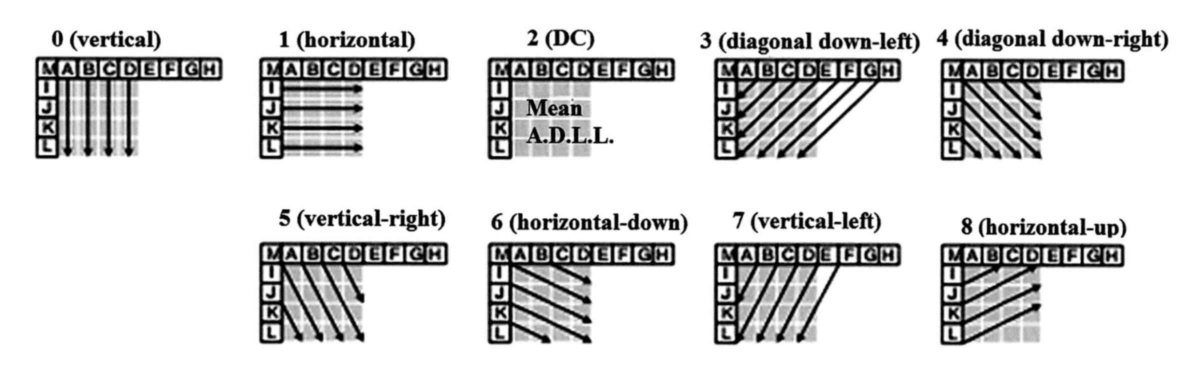

(b) 4x4 intra-prediction modes.

Figure 9. Residuals from intra-prediction

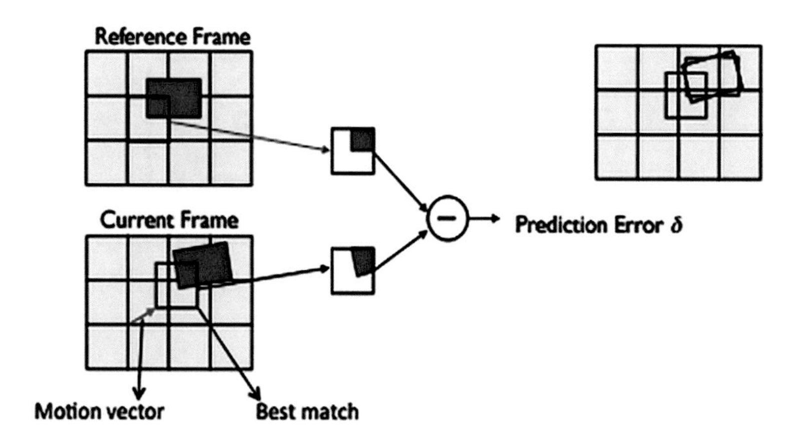

Intra-frame and inter-frame mode selection are features introduced in the H.264 standard. In the H.264/AVC, the MB decision mode in inter-frames is the most computationally expensive process due to the use of tools such as the variable block size MC (Figure 10), quarter-pixel MC, and intra-prediction, to name a few.

Intra-frame mode selection dramatically reduces spatial redundancy in I-frames, while inter-frame mode selection radically affects the output quality of P-/B frames by selecting an optimal block size with MV(s) or a mode for each MB. Unfortunately, this feature requires large encoding time when a full-search method is utilized.

Inter-Frame Concealment

For an inter-predicted MB in H.264/AVC, a separate MV is specified for each partition. Correspondingly, in an intra-predicted MB, where samples are estimated via extrapolation using information on the edges of nearby blocks. The prediction direction is specified on a per-partition basis.

Corrupted B-frames are handled differently. Since B-MBs are bi-directionally predicted, each lost MB in a B-frame requires two MV estimates. The two MVs are generated by temporally weighting the MVs of a future P-frame as shown in Figure 11. The first set of estimated MVs refers to a past refer-

Figure 10. Variable block size MC: MB partitioning and a sub-MB for motion compensated prediction

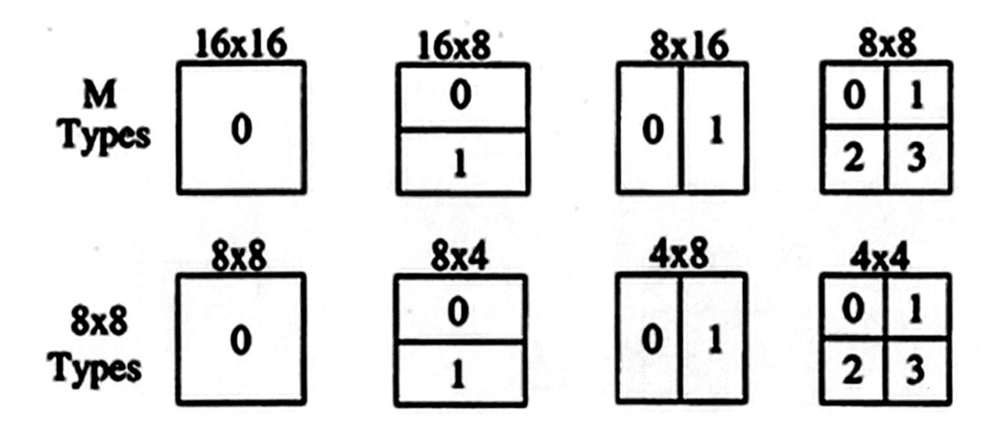

Figure 11. B-frame generation by means of temporal weighting of the next P-frame MVs

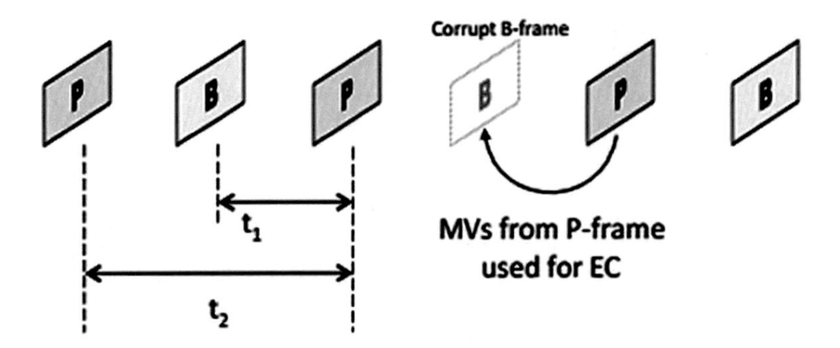

ence frame. This is generated by multiplying the MVs from the next P-frame by t_1/t_2 . The second set of estimated MVs, which points backwards, is generated by multiplying the same MVs from the next P-frame by $(t_1-t_2)/t_2$.

A lost MB is estimated using each of these techniques by estimating an MV for a 16×16 block of pixels and then performing MC using the estimated MVs. For each technique and corresponding estimated MV, there is a resultant estimated MB. Each estimated MB is then compared against neighboring MBs via a boundary matching algorithm that calculates the total absolute pixel value difference between the edge pixels of the estimated MB and the adjacent pixels of surrounding MBs as seen in Figure 12 (Shinfeng et al., 2011; Wang et al. 2010). The estimated MB with the best matching between its boundary pixels and adjacent pixels is picked as the final estimated MB for the sake of EC. This MB candidate system forms the framework for developing MV recovery methods adding a new estimated MB to the candidate system.

Hybrid Error Concealment

MV Recovery (MVR) uses mathematical models to recuperate incorrect motion fields. For instance, the Lagrange Interpolation (LAGI) is a fast MVR technique which increases the quality of the recovered video (Zheng et al., 2003). Among the existing MVR algorithms, the LAGI technique is frequently used to recover the lost MVs in H.264/AVC encoded video. Its low computation cost comes from the fact that LAGI is an MVR-based technique and, consequently, it is a good choice for real time video streaming applications (Zheng et al., 2003).

Putting that single frame in the context of neighboring frames adds a time dimension thus revealing whether an object is stationary or if it is indeed moving in a particular direction. Such conclusions can be drawn from correlations witnessed within a single frame or in the context of multiple frames. The underlying concept of these correlations is usually motion. By exploiting the spatial and temporal relationships that exist among MVs, it is possible to perform EC via MVR.

The decoder can recover an MV from an MB using available adjacent MVs. The decoder can apply frame copy, when it encounters a lost slice or part of a slice in a frame. In this case, all the lost MBs

Figure 12. Pixel comparison within the boundary matching algorithm

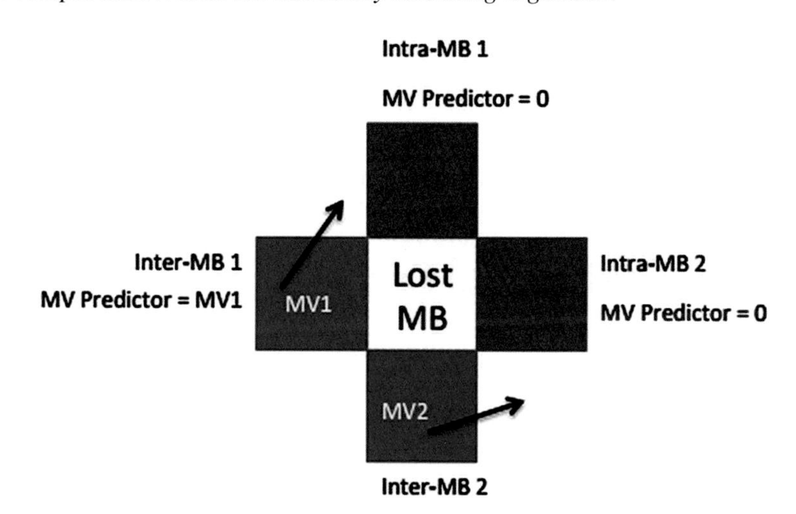

are initially set to have zero MVs by default. For each MB, MV predictors are then brought together. The MV predictors are based on the MVs of neighboring MBs, whether estimated or appropriately decoded. If a neighboring block is intra-coded, its corresponding MV predictor is zero. Since each MB is surrounded by a maximum of four neighbors (orthogonally), each lost MB can have up to four MV predictors as shown in Figure 13.

The order in which MBs within a missing region are concealed is performed in an inward spiral order within that region. The MBs on the outer edges are concealed first as they are initially the only ones with neighboring MVs. Once they are completely concealed, the decoder moves one MB row towards the middle of the lost region and the process is repeated until the whole slice is completely concealed. For each row of MBs, the even-placed MBs are concealed first, followed by the odd-placed MBs. Figure 13 illustrates that the inter-MBs can be either properly decoded or estimated via EC. The MV predictors are then used to generate estimates. These estimates, together with last MV and zero MV, form a set of MB candidates from which the best MB estimate is picked based on how well it matches its neighboring MBs.

Temporal EC techniques involve using frames other than the one containing losses. Multiple frames bring in a temporal dimension to the problem for the reason that pictures progress in time. The simplest temporal EC technique relies on frame and motion copy. Frame copy is appropriate only for slow-moving or still scenes. Motion copy involves repetition of MVs from previously decoded and/or error concealed reference frames in the decoder buffer into the lost frame. The MVs are then scaled depending on their frame distance from the current frame. This is suitable for slow-moving scenes or scenes with regular continuous motion. Motion copy can be improved by replication of MVs from the next reference inter-coded frame (typically a P-frame) too. A simple average is then calculated to come up with the final MVs for MC. Overlapped block MC (OBMC) is a type of deblocking filter used to reduce blocking artifacts with a window function.

Figure 13. Four MV predictors for a lost MB: MV1, MV2, zero, and zero

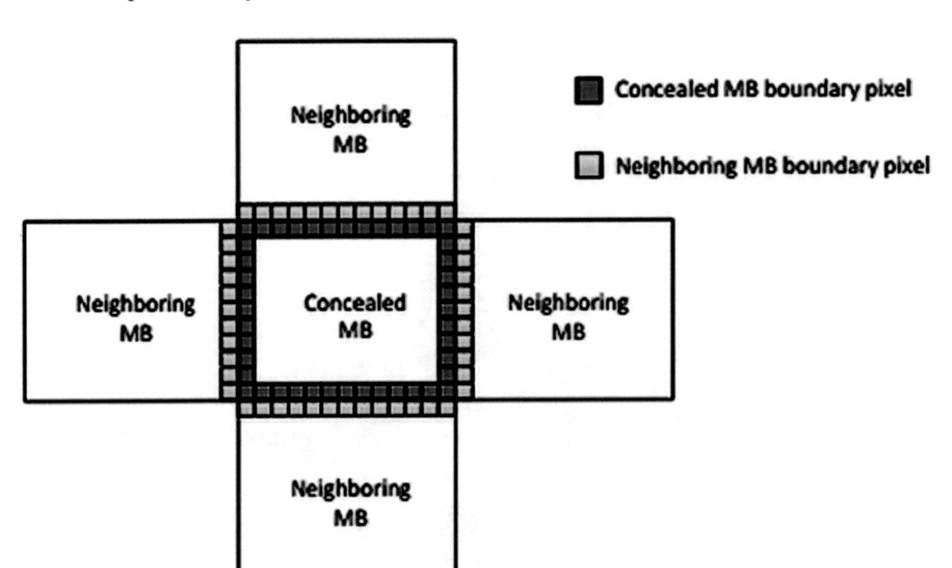

Motion copy outcomes can be refined by recursively calculating MV differences between adjacent frame pairs until the differences stop change. These differences are then added to motion replication results. This scheme adds spatial correlation by iteratively defining little by little a smaller locale from which to select MVs for difference computations and thus this method is unsuitable for real-time video decoding.

Using OBMC, Song et al. proposed MV Overlapped Block EC (MVOBEC) (Song et al., 2007). This method involves superposition of MVs using an overlapping window. An EC technique using MV Extrapolation (MVE) of MBs from the last two decoded pictures to account for lost MVs in the current frame is suggested by (Peng et al., 2002). The extrapolated MBs overlap the lost MB. Lost MVs are then estimated by considering the extrapolated MBs that possess the largest overlap with the lost MB. For more accuracy, each 8×8 block is subdivided into four 4×4 blocks. Each sub-block is then associated with the MV of the extrapolated MB that overlaps it most. If a sub-block is not overlapped by an extrapolated MB, then its recovered MV is simply the closest available recovered MV to its left. This method is only suitable for slow-moving or still scenes. However, frame copy alone is suitable for slow-moving or still scenes. Furthermore, MVE is inadequate for fast motion as blocking artifacts readily occur.

The accuracy of MVE has been improved by performing pixel-based MVE (PMVE) (Chen et al., 2007). Pixel-based MVs are generated by interpolating the original MVs, which are based on blocks. For each lost pixel in the current frame, an MV is estimated by averaging the MVs of all the overlapping forward-extrapolated MVs. If a pixel is not overlapped by an extrapolated MB, its estimated MV is simply a copy of the MV corresponding to the same spatial location in the previous frame. In addition to forward extrapolation, this method performs backward extrapolation to recover an additional set of MVs and do motion compensation for yet another estimation of the lost information. The forward and backward PMVE values are combined using a simple average. Backward estimation requires that future frames be added to the decoding frame buffer for calculation purposes, which can turn out to be an expensive process for HD video. Yan et al. improved PMVE by employing hybrid motion vector extrapolation (HMVE) (Yan et al., 2010). HMVE uses extrapolated MVs of blocks as in MVE as well pixels. Unlike PMVE, HMVE classifies pixels in accordance with coverage by extrapolated MBs prior to extrapolating pixel-based MVs. Using the two sets of extrapolations, a displacement-based threshold is used to discard wrongly extrapolated MVs. HMVE is incorporated into an object-based EC method that detects edge pixels and uses true motion estimation (TME) to detect objects and perform pixel-based recovery (Tai et al., 2010). Both HMVE and TME are pixel-based algorithms that would be computationally challenging in real-time HD video delivery.

Several optical-flow-based methods have been implemented for EC using MVs. The Multi-frame Motion Average (MMA) method involves generating an MV history (MVH) for each pixel in the last correctly decoded frame by averaging the MVs that point to that pixel in 2 to 5 levels, where each level is a backward step into the reference frame buffer (Belfiori et al., 2003). The resultant MVs are then projected onto the missing frame by estimating a Forward MV (FMV) for each pixel. Median filtering then spatially regularizes the FMV field and the missing frame is reconstructed at half-pel resolution. Missing pixels are then interpolated through median filtering and the final result is down-sampled to full-pel resolution. This method can be improved and made less complex by simply generating an FMV field using the last correctly decoded frame and performing temporal regularization, which is weighting each FMV according to the corresponding frame distance of the reference frame (Belfiori et al., 2005). An additional improvement to the MMA method comes from a block-based, rather than pixel-based, approach (Baccichet et al., 2004). Despite the reduction in recovery performance, this method is less complex than MMA. These optical-flow-inspired methods are complex, particularly when the operations are pixel based and/or half-pel-based, especially with regard to HD video. Furthermore, median filtering compounds the complexity.

SOLUTIONS AND RECOMMENDATIONS

Real-time transmission of video data in network environments, such as wireless and Internet, requires high compression efficiency and network friendly design. H.264/AVC aims at achieving improved compression performance and a network-friendly video representation for different types of applications, such as conversational, storage, and streaming. The robustness of coding schemes to channel losses is evaluated using video sequences containing various motion activity levels. Efficient implementation of compression algorithms is also crucial to lower the production cost of a codec.

There are some implementation challenges related to intra-prediction. First, the dependence of the blocks prediction on their neighbors, some of which may be part of present MB, may turn parallel processing of block data unfeasible. Second, each of the 16 blocks in a given MB can select any of the nine prediction modes, since each mode utilizes a totally different mathematical weighting function to obtain the predicted data from the samples.

Coarse quantization from any of the block-based integer image transforms generates distressing rough blocking artifacts at the block boundaries of the image. MC of the MB by data interpolation using previous reference frames might never deliver a perfect match, because edge discontinuities may come out for the copied blocks. When the later P- or B-frames reference these images having blocky edges, the blocking artifacts further propagates to the current blocks worsening the condition even more.

This can de remediated with a deblocking filter consisting of strength computation, decision making and the best smoothed boundary. If the deblocking filter is not forcibly employed, then frames may suffer from blocky artifacts due to the fact that past frames are used as reference.

The Boundary Strength (BS) is a kind of adaptive filtering for a specified edge based on certain conditions. H264/MPEG4 AVC has four types of BSs, where for the case of a chroma block, its BS comes from the BS of the corresponding luma MB.

The deblocking filter can be a hardware block based on specialized processors in the decoder loop. This stage has to be the last one of the decoder, to guarantee that all the top/left neighbors have been entirely reconstructed and are accessible as inputs for deblocking the present MB. Furthermore, block edges filtering of any slice can be halted by means of flag according to the situation. Vertical edges are filtered first (left to right) followed by horizontal edges (top to bottom). Multiresolution ME schemes and adaptive rate-distortion models can be combined with a fast inter-prediction mode decision and an adequate motion search algorithm.

Some fast ME strategies result from MV merging and splitting for variable size block ME, which explores the correlation of MVs of overlapping blocks for both the fast integer-pel and fractional-pel search. A low-complexity intra-prediction mode selection scheme without significant rate-distortion performance degradation is possible via spatial and transform domain features of the target block to filter out the majority of candidate modes.

Schemes to enhance MC and improve the coding efficiency can use the new DIRECT MB type to exploit the MV temporal correlation and an enhancement to the existing SKIP within Predictive P-slices. Fast mode selection algorithms for both intra-frames and inter-frames are required to fully explore the standard possibilities. In H.264/AVC, various MB types are used for improving the compression efficiency. However, it is not efficient for these MB types to be coded by separate syntax elements.

Flexible MB Ordering (FMO) is an important error resiliency scheme used in H.264/AVC because of its benefits in case of packet loss over IP networks.

There are still no standardized performance assessment metrics for EC methods. Common metrics to evaluate the quality of reconstruction are the Mean Square Error (MSE), the Peak Signal-to-Noise Ratio (PSNR), and the Structural Similarity Index Metric (SSIM) (Kung et al., 2006; Kolkeri et al., 2009) are used. To test performance evaluation methods, an H.264/AVC (Akramullah, 2014) video codec is used. H.264/AVC has succeeded for several years in broadcasting traditional 2D videos. This feature is very appealing for mobile networks due to the restricted bandwidth in these environments.

FUTURE RESEARCH DIRECTIONS

H.264 EC can be biased towards objects, meaning that based on the correctly decoded MBs in a current lossy frame in conjunction with previous frame(s). It is possible to recognize objects and perform EC on them before dealing with the rest of the frame to ensure a better concealment of objects as they tend to be the most visually significant portions of a frame and a video as a whole.

There needs to be a more systematic way of determining the speed of a video. An elementary beginning to this would measure average MV lengths throughout the video. With such a system in place, it would be possible to employ simpler EC techniques such as frame copy in low motion regions and more sophisticated methods such as spatial-temporal MV recovery in medium and high motion areas. Besides, systematically determining the speed of a video, would reveal how it varies with the complexity of spatial- temporal relationships among MVs.

Another area of improvement is to develop temporal techniques for MV recovery. The dominance of the last MV method as an EC technique is an indicator of a significant temporal relationship among MVs creating an opportunity for future work involving temporal MV recovery. A simple way to start investigating the temporal relationship of MVs is by modeling MVs prior to a lost one in previous frames and the same spatial location through time. Time series analysis would be a valuable tool in this case. In statistics and signal processing, a time series is a sequence of data points collected at discrete time intervals typically according to a particular sampling frequency. Each video sequence can be modeled as a time series, like for example a moving average that can be formalized and used for forecasting via a mathematically defined trend of previous MVs. Forecasted MVs will be used to estimate lost MVs. By calculating the moving average of an MV through time, it is possible to predict the value of a missing MV using previously available MVs.

The temporal characteristics of MVs can be thought of as considering a single block of video and observing how the MVs within the block change with time. Based on the content of the video as well as the surrounding MVs, there most likely exist equations that can be used to define the temporal trend of MVs. These equations would be the basis of EC by, perhaps, interpolating missing MVs based on properly decoded ones. The possibility of EC through temporal MV recovery introduces extra complexity into EC, where the primary task will be the decision-making processes that choose between either spatially or temporally generated MV estimates. On the other hand, a unified spatial-temporal realization can be used, e.g. a simple average or a dynamic weighted average of both spatial and temporal MV estimates.

Scalability

Scalability is an efficient layering technique that is not entirely supported in the current specification of H.264/AVC (Sullivan et al., 2012). The High-Efficiency Video Coding (HEVC) standard brings in lots of interesting developments to assist implementations.

Scene Change Detection

Scene Change Detection (SCD) schemes used with old formats like MPEG-1 or MPEG-2 are not straightforwardly applicable to the compressed bit stream of H.264/AVC.

SCD can be performed to decide whether the scene change occurs or not to increase EC performance. Supposing the motion activity is lower, zero MVs can be used. Otherwise, the MVs are then determined with an adequate MVR algorithm. Therefore, it is desirable that the EC incorporates SCD, motion activity detection and MV retrieval (Coelho et al.,2012a; Lin et al., 2010).

Except for the first intra-frame, the preceding decoded frame may also look like the present intra-frame, particularly when the scene change does not occur. And also, the current spatial interpolation algorithms often undergo blurring in the edge regions of the image.

The SCD based on motion activities helps to make coding/decoding decisions. Accordingly, a further major problem arises from the computing efficiency of the employed SCD algorithm since such procedure has to be done in the decoder.

Besides, conventional EC techniques have not reflected on the influences of scene change for inter-frames. Contrasting to intra-frame EC, inter-frame EC only uses temporal correlations in the H.264/AVC non-normative decoder. Since the temporal correlations cause a sudden interruption when there is a scene change, any MV recuperation based on MC that mentions the preceding frame would be improper. In this case, spatial EC would be better than temporal EC.

Conclusively, it is promising to guarantee adaptive selection and use of the spatial and temporal data for EC. SCD plays a key role in such selections. Conventional SCD methods are frequently based on color or shape examination in the spatial domain or DCT coefficients investigation in the frequency domain, and, hence, these methods are normally time-consuming. Since real-time processing should be unquestionably present in any realistic video decoder, these schemes are not appropriate for SCD in EC.

CONCLUSION

Important features from H.264/AVC:

- Prediction block size 4×4,4×8,8v4,8×8, 8×16,16×8,16×16,
- Intra-prediction support for spatial domain MV estimation,
- Entropy coding by means of CAVLC, CABAC,
- Provision for multiple reference frames,
- Weighted prediction,
- Deblocking filter,
- Integer Transforms (4×4 and 8×8 in high profile).

Knowledge on H.264/AVC is important in video applications. Due to the lossy nature of existing channels and the high propensity of the H.264/AVC video format to these losses, EC is a noteworthy process in multimedia broadcast and it is related to MV recovery. This is done by exploiting the spatial and temporal correlations between MVs, so that lost MVs are estimated based on the available MVs within the same space (frame) or using inter-frame relationships.

The H.264/AVC performs prediction of intra-blocks in the spatial domain rather than in frequency domain like other codecs. It uses the reconstructed, but unfiltered MB data from the neighboring MBs to estimate the current MB coefficients. There is an important block size tradeoff: large blocks produce fewer MVs, but contain big residuals; while small blocks have more MVs with small residuals.

REFERENCES

Akramullah, S. (2014). Digital video concepts, methods, and metrics: Quality, compression, performance, and power trade-off analysis. *Apress.*

Baccichet, P., & Chimienti, A. (2004). A low complexity concealment algorithm for the whole-frame loss in H.264/AVC. In *Proceedings of the 6th IEEE Workshop on Multimedia Signal Processing* (pp. 279– 282). IEEE. doi:10.1109/MMSP.2004.1436547

Belfiore, S., Grangetto, M., Magli, E., & Olmo, G., G. (2003). An error concealment algorithm for streaming video. In *Proceedings of the 2003 International Conference on Image Processing (ICIP 2003)*. IEEE. doi:10.1109/ICIP.2003.1247328

Belfiore, S., Grangetto, M., Magli, E., & Olmo, G. (2005). Concealment of whole-frame losses for wireless low bit-rate video based on multiframe optical flow estimation. *IEEE Transactions on Multimedia*, *7*(2), 316–329. doi:10.1109/TMM.2005.843347

Chen, Y., Keman, Y., & Jiang, L. (2004). An error concealment algorithm for entire frame loss in video transmission. In *Proceedings of the 2004 Picture Coding Symposium (2004 PCS*).

Chong, J., Satish, N., Catanzaro, B. C., Ravindran, K., & Keutzer, K. (2007). Efficient parallelization of H.264 decoding with macro block level scheduling. In *Proceedings of the 2007 IEEE International Conference on Multimedia and Expo (ICME 2007)* (pp. 1874-1877). doi:10.1109/ICME.2007.4285040

Coelho, A. M., & Estrela, V. V. (2012a). EM-based mixture models applied to video event detection. In P. Sanguansat (Ed.), *Principal component analysis – Engineering applications* (pp. 101–124). InTech. doi:10.5772/38129

Coelho, A. M., & Estrela, V. V. (2013). State-of-the art motion estimation in the context of 3D TV. In R. A. Farrugia & C. J. Debono (Eds.), *Multimedia networking and coding* (pp. 148–173). doi:10.4018/978-1-4666-2660-7.ch006

Coelho, A. M., Estrela, V. V., Fernandes, S. R., & do Carmo, F. P. (2012b). Error concealment by means of motion refinement and regularized Bregman divergence. In H. Yin, J. A. F. Costa & G. Barreto (Eds.), *Proceedings of the Intelligent Data Engineering and Automated Learning - IDEAL 2012* (Vol. 7435, pp. 650-657). doi:10.1007/978-3-642-32639-4_78

Cui, Y., Deng, Z., & Ren, W. (2009). Novel temporal error concealment algorithm based on residue restoration, In *Proceedings of the 2009 EEE International Conference on Wireless Communications, Networking and Mobile Computing* (pp. 1-4). doi:10.1109/WICOM.2009.5302239

Fleury, M., Altaf, M., Moiron, S., Qadri, N., & Ghanbari, M. (2013). Source coding methods for robust wireless video streaming. In R. A. Farrugia & C. J. Debono (Eds.), *Multimedia networking and coding*. Retreived from http://www.igi-global.com/chapter/source-coding-methods-robust-wireless/73139

ITU-T. (2013). *Information technology - Generic coding of moving pictures and associated audio information: Video*. Retrieved from http://www.itu.int/rec/T-REC-H.264

Kolkeri, V. S., Lee, J. H., & Rao, K. R. (2009). Error concealment techniques in H.264/AVC for wireless video transmission in mobile networks. *International Journal of Advances in Engineering Science, 2*(2).

Kung, W., Kim, C., & Kuo, C. (2006). Spatial and temporal error concealment techniques for video transmission over noisy channels. *IEEE Transactions on Circuits and Systems for Video Technology, 16*(7), 789–802. doi:10.1109/TCSVT.2006.877391

Lin, W., Sun, M. T., Li, H., & Ho, H. M. (2010). A new shot change detection method using information from motion estimation. In *PCM'10 Proceedings of the Advances in Multimedia Information Processing, and 11th Pacific Rim conference on Multimedia: Part II* (pp. 264-275). Springer-Verlag Berlin. doi:10.1007/978-3-642-15696-0_25

Peng, Q., Yang, T., & Zhu, C. (2002). Block-based temporal error concealment for video packet using motion vector extrapolation. In *Proceedings of the 2002 IEEE International Conference on Communications, Circuits and Systems and West Sino Expositions* (vol. 1, pp. 10–14). IEEE doi:10.1109/ICCCAS.2002.1180560

Richardson, I. E. G. (2010). *The H.264 advanced video compression standard*. John Wiley & Sons, Ltd. doi:10.1002/9780470989418

Shinfeng, D. L., Wang, C. C., Chuang, C. Y., & Fu, K. R. (2011). A hybrid error concealment technique for H.264/AVC based on boundary distortion estimation. In D. S. L. Javier (Ed.), Advances on video coding. InTech. Retrieved from http://www.intechopen.com/books/recent-advances-on-video-coding/a-hybrid-error-concealment-techniquefor-h-264-avc-based-on-boundary-distortion-estimation

Song, K., Chung, T., Kim, Y., Oh, Y., & Kim, C. (2007). Error concealment of H.264/AVC video frames for mobile video broadcasting. *IEEE Transactions on Consumer Electronics, 53*(2), 704–711. doi:10.1109/TCE.2007.381749

Sullivan, G. J., Ohm, J.-R., Han, W.-J., & Wiegand, T. (2012). Overview of the high efficiency video coding (HEVC) standard. *IEEE Transactions on Circuits and Systems for Video Technology, 22*(12), 1649–1668. doi:10.1109/TCSVT.2012.2221191

Tai, S., Hong, C. S., & Fu, C. (2010). An object-based full frame concealment strategy for H.264/AVC using true motion estimation. In *Proceedings of the 2010 Fourth Pacific-Rim Symposium on Image and Video Technology (PSIVT 2010)* (pp. 214–219). IEEE. doi:10.1109/PSIVT.2010.43

Wang, C.-C., Chuang, C.-Y., & Lin, S. D. (2010). An integrated spatial error concealment technique for H.264/AVC based-on boundary distortion estimation, In *Proceedings of the 2010 Fifth International Conference on Innovative Computing, Information and Control* (pp. 1-4).

Wiegand, T., Sullivan, G.-J., Bjontegaard, G., & Luthra, A. (2003). Overview of the H.264/AVC video coding standard. *IEEE Transactions on Circuits and Systems for Video Technology*, *13*(7), 560–576. doi:10.1109/TCSVT.2003.815165

Yan, B., & Gharavi, H. (2010). A hybrid frame concealment algorithm for H.264/AVC. *IEEE Transactions on Image Processing*, *19*(1), 98–107. doi:10.1109/TIP.2009.2032311 PMID:19758866

Zhan, X., & Zhu, X. (2009). Refined spatial error concealment with directional entropy, In *Proceedings of the 2009 IEEE International Conference on Wireless Communications, Networking and Mobile Computing* (pp. 1-4). doi:10.1109/WICOM.2009.5302608

Zheng, J. H., & Chau, L. P. (2003). Motion vector recovery algorithm for digital video using Lagrange Interpolation. *IEEE Transactions on Broadcasting*, *49*(4), 383–389. doi:10.1109/TBC.2003.819050

ADDITIONAL READING

Altaf, M., Fleury, M., & Ghanbari, M. (2011). Error resilient video stream switching for mobile wireless channels. *International Journal of Mobile Multimedia*, *7*(3), 216–235.

Ferré, P., Agrafiotis, D., & Bull, D. (2010). A video error resilience redundant slices algorithm and its performance relative to other fixed redundancy schemes. *Image Communication*, *25*(3), 163–178.

Hamzaoui, R., Stanković, V., & Xiong, Z. (2007). Forward error control for packet loss and corruption. In M. van der Schaar & P. A. Chou (Eds.), *Multimedia over IP and wireless networks* (pp. 271–292). Burlington, MA: Academic Press. doi:10.1016/B978-012088480-3/50010-2

Jia, J., Choi, H.-C., Kim, J-G., Kim, H.-K., & Chang, Y. (2007) Improved redundant picture coding using polyphase downsampling. *Electronics and telecommunications Research Institute Journal, 29*(1), pp. 18-26.

Kumar, S., Xu, L., Mandal, M., & Panchanathan, S. (2006). Error resiliency schemes in H. 264/AVC standard. *Journal of Visual Communication and Image Representation*, *17*(2), 425–450. doi:10.1016/j.jvcir.2005.04.006

Nguyen, K., Nguyen, T., & Cheung, S. (2010). Video streaming with network coding. *Journal of Signal Processing Systems for Signal, Image, and Video Technology*, *57*(3), 319–333. doi:10.1007/s11265-009-0342-7

Ostermann, J., Bormans, J., List, P., Marpe, D., Narroschke, M., Pereira, F., Stockhammer, T., & Wedi, T. (2004). Video coding with H. 264/AVC: Tools, performance and complexity. *IEEE Circuits and Systems Magazine,* 4, 7:28.

Salama, F., Shroff, N. B., & Delp, E. J. (1998). Error concealment in encoded video. In *Image Recovery Techniques for Image Compression Applications*. Norwell, MA: Kluwer.

Schwarz, H., Marpe, D., & Wiegand, T. (2007). Overview of the scalable video coding extension of the H.264/AVC standard. *IEEE Transactions on Circuits and Systems for Video Technology*, *17*(9), 1103–1120. doi:10.1109/TCSVT.2007.905532

Shi, Y. Q., & Sun, H. (2008). *Image and video compression for multimedia engineering: Fundamentals, algorithms, and standards* (2nd ed.). Boca Raton, Fl: CRC Press. doi:10.1201/9781420007268

Son, N., & Jeong, S. (2008). An effective error concealment for H.264/AVC. In *IEEE 8th International Conference on Computer and Information Technology Workshops* (pp. 385-390).

Stockhammer, T., & Zia, W. (2007). Error-resilient coding and decoding strategies for video communication. In M. van der Schaar & P. A. Chou (Eds.), *Multimedia over IP and wireless networks* (pp. 13–58). Burlington, MA: Academic Press. doi:10.1016/B978-012088480-3/50003-5

Sullivan, G., & Wiegand, T. (2005). Video compression — From concepts to the H. 264/AVC standard. *Proceedings of the IEEE*, *93*(18), 31.

KEY TERMS AND DEFINITIONS

Block Matching: Motion estimation technique that matches blocks between adjacent frames, aiming at minimizing a dissimilarity measure.

Macroblock: A processing unit in image and video compression formats based on linear block transforms, which can be divided into transform blocks that consist of prediction blocks.

Motion Compensation: A procedure used to predict a video picture, given its previous and/or future frames with the help of the motion existing in the video.

Optical Flow: The apparent motion of a brightness (or intensity) pattern. It captures the spatio-temporal variation of pixel intensities between neighboring frames.

Chapter 9
Vision-Based Protective Devices

M. Dolores Moreno-Rabel
Universidad de Extremadura, Spain

J. Álvaro Fernández-Muñoz
Universidad de Extremadura, Spain

ABSTRACT

Machine Safety is a growing technical discipline with a strong basis in the development of electrical and electronic devices, commonly known as Safety Protective Devices (SPDs). SPDs are designed to avoid or at least mitigate those risks associated with a particular human-machinery interaction. Ranging from conceptually simple electromechanical Emergency Stop Devices (ESDs) to the more complex Active Optoelectronic Protective Devices (AOPDs), a place for Real-Time Digital Video Processing has recently been open for research in Machine Safety. This chapter is intended to explore the standardized features of the so-called Vision-Based Protective Devices (VBPDs), their current technical development and principal applications in Machine Safety, with a stress on prominent vision-related implementation issues.

INTRODUCTION

In the manufacturing industry context, production consists of processing, assembling, and transporting materials. In the last decades, industrial machines have been increasingly used to reduce the burden from workers to assist in production. As a result, a wider range of machines are currently designed, produced, marketed and used for multiple industrial purposes on a global-scale, networked scenario. However, manufacturing is still not possible without the intended action of a trained worker who operates a machine. Since humans are prone to make mistakes, ensuring safety irrespective of worker operating experience on a machine is mandatory. The same applies to machines (and their hardware and software components), given that they also fail specially during machine maintenance and setup operations.

In the context of machinery, the purpose of safety is to protect persons from harm (i.e. physical injury or damage to health). A machine must be safe since its design. As stated in Machinery Directive 2006/42/EC (European Commission, 2006), it is responsibility of both builder and supplier to ensure that a machine is designed and constructed to be safe, so that it can be used in its intended manner, configured and maintained throughout all phases of its life, causing minimal risk to persons and the environment.

DOI: 10.4018/978-1-5225-1025-3.ch009

For this purpose, the machine designer must carry out a technical procedure to identify both hazards (i.e. potential source of harm) and risks (i.e. severity and probability of occurrence of a harm) associated with the machinery. In turn, these hazards and risks determine the so-called machine danger zone (i.e. any space within and/or around the machinery in which a person can be exposed to risk of injury or damage to health), in order to assess which measures are suitable and where they should be installed. Optionally, a warning zone (i.e. any space which surrounds the danger zone in which a person is close but not exposed to a hazard) may also be considered in the installation and utilization of some measures. From a Machine Safety viewpoint, this technical procedure —known as Risk Management— must only be finished when the machine is safe.

In general terms, a Safety Measure is a measure (i.e. a particular action which is intended to achieve an effect) that is taken to increase or ensure safety, or protection from danger. In the Machine Safety context, Safety Measures are those measures intended to achieve the necessary risk reduction for a particular machine, according to a standardized Risk Assessment procedure, which must be implemented by the machine designer and/or by the user. Plainly stated, Safety Measures are thus responsible to ensure that machines are mechanically and functionally safe. As for the machine designer, the Safety Measures under consideration include:

- Inherently Safe Design Measures,
- Safeguarding Devices,
- Complementary Safety Measures, and
- Information for Use.

On the other hand, Safety Measures applied by the user include organizational issues such as:

- Safe working procedures,
- Supervision and maintenance plans,
- Permit-to-work systems,
- The provision and use of additional Safeguarding Devices,
- Personal Protective Equipment (PPE), and
- Personnel training.

The present chapter is devoted to a specific group of Safety Measures known as Vision-Based Protective Devices (VBPDs). VBPDs belong to a wider group of Safety Measures called Electro-Sensitive Protective Equipment (ESPEs). An ESPE is an electronic Safety Protective Device (SPD) which is designed to provide the worker with a kind of protection not based on the physical separation of persons at risk from the risk itself, such as the protection given by a fixed metallic fence surrounding a machine. Instead, in ESPEs (and thus in VBPDs) worker protection is achieved through temporal separation (Ridley & Pearce, 2006). ESPE basic functionality follows. As long as there is somebody within a pre-defined area (i.e. the machine danger zone), no hazardous machine functions are initiated, and such functions are stopped if already started. A certain amount of time —the so-called stopping time or run-down time— is required to stop these machine functions. A machine stopping time fully depends on the machinery itself and its specific workplace environment (Goernemann & Stubenrauch, 2013). The ESPE must detect the approach of a person to the hazardous area in a timely manner and, depending on the application, the presence of the person in the hazardous area. VBPDs are a recently developed type of ESPE which is designed to perform this basic ESPE functionality through image and video processing.

BACKGROUND

In the following discussion, which has been split into a small number of parts for organizational convenience, a brief introduction of several relevant terms, definitions and topics related with the main topic of the chapter is given.

Safety Measures

Nowadays, Safety Measures are typically classified into four main groups following the hierarchical guidelines given in standard ISO 12100 (ISO 12100, 2010), as shown in Table 1. A brief description of these groups follows (Ridley & Pearce, 2006; Caputo et al., 2013):

Inherently Safe Design Measures

A group of Safety Measures which either eliminate hazards or reduce associated risks by changing the machine design, its operating characteristics or even its working process, any of which without the use of Safeguarding Devices. According to Table 1, this group of Safety Measures includes provisions for stability and maintainability and general technical knowledge of machine design, among others. In those cases where Inherently Safe Design Measures are not sufficient to obtain the intended risk reduction, Safeguarding Devices and Complementary Safety Measures must be used.

Table 1. Classification of safety measures

<table>
<tr><td colspan="2">Inherently Safe Design Measures</td><td>• Geometrical and physical factors.
• Provisions for stability and maintainability.
• Ergonomic principles.
• General technical knowledge of machine design.</td></tr>
<tr><td colspan="2">Complementary Safety Measures</td><td>• Emergency Stop Devices (ESDs).
• Measures for escape and rescue of trapped persons.
• Measures for isolation and energy dissipation.
• Provisions for easy and safe handling of machines.
• Measures for safe access to machinery.</td></tr>
<tr><td rowspan="2">Safeguarding Devices</td><td>Safety Guards (SGs)</td><td>• Fixed Guards.
• Movable Guards.
• Adjustable Guards.</td></tr>
<tr><td>Safety Protective Devices (SPDs)</td><td>• Safety Control Devices (SCDs).
• Access Detection Devices (ADDs).</td></tr>
<tr><td colspan="2">Supplementary Safety Measures</td><td>• Personal Protective Equipment (PPE).
• Warning and Mandatory Signs.
• Signal and Warning Devices.</td></tr>
</table>

Source: (ISO 12100, 2010).

Safeguarding Devices

As Table 1 shows, a Safeguarding Device is either a Safety Guard (SG) or a Safety Protective Device (SPD), which is designed to detect or to prevent inadvertent access to a hazard. The underlying safety principle is that any machine part, function, or process that may cause injury must be safeguarded. Thus, these Safety Measures provide a worker with protection from the hazards produced by the machinery and, as a result, they perform a safety function (i.e. a function of a component or equipment concerned with or contributing to a freedom from risk, whose failure can result in an immediate increase of its associated risks) (Ridley & Pearce, 2006). A SG can be defined as a part of a machine or a physical barrier between a worker and a dangerous part of the machine, which can be fixed, movable or adjustable, and is specifically designed to provide protection, whereas a SPD can be defined as an equipment, device or mean, other than a SG, which functions to prevent a worker from being harmed by a dangerous part of a machine. SPDs can act on their own (as control devices, SCDs) or in combination with SGs in order to reduce risks.

Complementary Safety Measures

This group encompasses those Safety Measures which are neither Inherently Safe Design Measures, nor Safeguarding Devices, nor Information for Use. Probably, Emergency Stop Devices (ESDs) are the most important type of Safety Measures included in this group. ESDs are designed to perform, after manual actuation, an emergency stop function (i.e. a function that is intended to avert arising, or to reduce existing hazards to persons and damage to machinery or to work in progress and to be started by a single human action). As Table 1 shows, other Safety Measures included (but not limited to) in this group are those designed for escape and rescue of trapped persons, isolation and dissipation of energy, safe access to machinery, and safe handling of machines and their heavy parts. However, as they do not perform a safety function by themselves, all of these Safety Measures must be always used as a complement of a previously chosen, particular set of Safeguarding Devices, and not a substitute for it (Ridley & Pearce, 2006).

Supplementary Safety Measures

This group contains any Safety Measure which must be provided to reduce residual risks, i.e. the remaining risks after all previously described Safety Measures have been considered and implemented. Any residual risk must be documented in the machine operating instructions, i.e. the information for use. As shown in Table 1, PPEs, warning and mandatory action signs, and signal and warning devices are also included in this group (Ridley & Pearce, 2006).

Machine Hazards

Workers are exposed to hazards during their workday due to the inevitable interaction between man and machinery in the workplace, since machines are used as working tools. This exposure puts workers at different kinds of risk, which in turn are directly linked to occupational accidents that may cause harm and even death to the worker. Hazards thus occur where machine work areas, as determined by its operating output, and human workspaces overlap. Industrial hazards include

- Mechanical (e.g. crush, shear, impalement, entanglement),
- Toxic (e.g. chemical, biological, radiation),
- Heat and flame,
- Cold,
- Electrical, and
- Optical (e.g. laser, welding, flash).

Machine hazards are usually classified into three groups: mechanical, non-mechanical, and access hazards (Ridley & Pearce, 2006):

Mechanical Hazards

Given that machines have moving parts, the action of these moving parts may have sufficient force in motion to cause injury to people. Typical machinery sources of mechanical hazards include machinery and equipment with moving parts that either can be reached by people (e.g. in-running nips); that can inadvertently reach people through extensible parts (e.g. booms, arms); that can eject objects (e.g. parts, components, products or waste items) that may strike a person with sufficient force to cause harm, and mobile machinery and equipment (e.g. forklifts, pallet jacks) which are operated in areas where people may gain access.

Non-Mechanical Hazards

Non-mechanical hazards associated with machinery and equipment can include harmful emissions, contained fluids or gas under pressure, chemicals and chemical by-products, electricity and noise, all of which can cause serious injury if not adequately controlled. Where people are at risk of injury due to harmful emissions from machinery and equipment, the emissions should be controlled at their source.

Access Hazards

Workers must be provided with safe means of access that is suitable for the work they perform in, on and around machinery and equipment. A stable work platform suited to the nature of the work that allows for good posture relative to the work performed, sure footing, safe environment and fall prevention (if a fall may occur) is thus a basic requirement for any industrial facility.

Preventing machinery hazards begins by eliminating those mechanisms that facilitate hazardous conditions. Two safety strategies are generally used to achieve this goal: isolation and temporal separation. The isolation strategy is based on ensuring a physical, spatial separation between human and machine workspaces; whereas the temporal separation strategy ensures that the amount of time needed for a worker to reach the hazardous machine area is always bigger than the time needed for the hazard to disappear, which is closely related to the machine stopping time in the majority of practical situations (Ridley & Pearce, 2006).

Access Detection Devices (ADDs)

In a controlled workplace, access needs can be predicted and access planning must occur in advance. People need access to machinery and equipment in the workplace (either continually or occasionally) for tasks such as:

- Operation,
- Maintenance,
- Repair,
- Installation,
- Service, or
- Cleaning.

These tasks are examples of access that can be predicted. However, even unpredicted access situations must be properly safeguarded. Safeguarding Devices are Safety Measures which are designed to detect or prevent inadvertent access to a hazard. According to the hierarchical Risk Reduction procedure given in the international standard ISO 12100: Safety of machinery – General principles for design – Risk assessment and risk reduction (ISO 12100, 2010), Safeguarding Devices must be used when Inherently Safe Design Measures cannot reasonably be applied or when, once implemented, there are still risks that need to be reduced or eliminated. Because of their lower cost, SGs must be selected in the first place as a mean of risk reduction. If they fail to achieve a sufficient reduction, or if a SG is not applicable, then SPDs must be used if possible.

According to the relevant literature on Safeguarding Devices, given that SGs impose a physical barrier between the workers and machinery (which provides workers with protection against projectiles, fluids, mist and other types of risk), they are often used when access to a hazard is infrequent, and to prevent access while the machine is active within its working cycle and it takes a long time to be stopped (Caputo et al., 2013). In contrast, if workers must frequently access to the machine danger zone, the use of one or more Access Detection Devices (ADDs) is usually a more suitable choice. This is because ADDs (which are a type of SPD, according to Table 1) impose a virtual, not physical barrier on which they perform the detection of persons. In other words, ADDs are used to sense the presence of persons, but not to restrict their access to the machine danger zone. For this reason, ADDs (also known as Presence Sensing Devices) allow for greater productivity and relatively are a more ergonomic, healthy solution when compared to SGs. In addition, most ADDs provide the worker with a greater freedom of movement and simplicity of use. Nevertheless, this type of SPD entirely relies on its ability to both sense and switch for the provision of safety (Ridley & Pearce, 2006; Caputo et al., 2013).

ADDs can be used by multiple workers and provide them with shared (and individual) protection. However, these Safeguarding Devices do not provide protection against projectiles, fluids, mist, or other types of hazards (Ridley & Pearce, 2006). The two main functions for which they are designed are (Caputo et al., 2013):

1. To switch off or disable power when a person is detected in its monitored zone, when e.g. entering the danger zone, and
2. To prevent power enabling or switching on when a person is inside the monitored zone.

At first thought, these two operational functions may seem to be one and the same thing, but although they are obviously linked, and quite often achieved by the same equipment, they actually are two separate functions. To achieve the first one (switch off or power off when entering the danger zone) some form of trip device is usually employed, i.e. a device which detects that a part of a person has gone beyond a certain point and provides a signal to trip the power off. If the worker is then able to continue past this tripping point and his presence is no longer detected, then the second function (switching on prevention while in the danger zone) may not be achieved. As long as somebody is within a pre-defined zone, no hazardous machine function is allowed to continue its execution or to be initiated (Ridley & Pearce, 2006).

A certain amount of time (the so-called stopping or run-down time) is always required to stop these functions (Caputo et al., 2013; Goernemann & Stubenrauch, 2013). ADDs are usually best suited to machinery with short stopping time. Given that a person can reach or even directly walk into the danger zone, it is obviously necessary that the time taken for the motion to stop should be less than the time required for the person to reach the hazard after tripping the device. In order to properly render both presence detection and machine stop functions, according to the international standards ISO 13856-1: Pressure-sensitive protective devices – Part 1: General principles for design and testing of pressure-sensitive mats and pressure-sensitive floors (ISO 13856-1, 2013), and IEC 61496-1: Safety of machinery – Electro-sensitive protective equipment – Part 1: General requirements and tests (IEC 61496-1, 2012), ADDs must comprise at least the three following elements:

- **Sensor or Sensing Device:** It is the input part of an ADD which uses electro-sensitive or pressure-sensitive means to determine the event or state that the ADD is intended to detect.
- **Output Signal Switching Device (OSSD):** It is the output circuit component of an ADD, which is connected to the Machine Control System (MCS) through the control unit. OSSD switches to the off state as soon as the ADD sensing device is activated or interrupted during its intended operation, causing the machine to stop.
- **Control Unit:** Configured as a linking I/O control device, the ADD control unit responds to the condition of the sensor(s) and controls and monitors the state of the OSSD. Typically this control unit is also able to monitor the externally connected switching elements, i.e. acting like a Monitoring Safety Relay (MSR), and to give a start readiness status output to the MCS. In some ADDs, the control unit may also contain the device power supply.

In order to determine whether the use of a particular ADD is feasible or not, both ADDs position and effective sensing zone (i.e. the area on which the application of an actuating force or detecting presence causes the signal sent from the sensor to the control unit to change its state) must be considered. Both parameters should meet the requirements of international standards ISO 13855: Safety of machinery – Positioning of safeguards with respect to the approach speeds of parts of the human body (ISO 13855, 2010), and ISO 13857: Safety of machinery – Safety distances to prevent hazard zones being reached by upper and lower limbs (ISO 13857, 2008), regarding both approach speed and control response time.

Currently available ADDs are able to perform presence detection by means of e.g.:

- A pressure sensor,
- A light beam,
- A doppler radar,
- An electrical contact,

- A set of video cameras, or
- An electromagnetic field (Ridley & Pearce, 2006).

As Table 2 shows, these Safeguarding Devices can be divided into two different families, according to their working principle. On the one hand, the family of Pressure-Sensitive Protective Devices (PSPDs) includes a variety of Safeguarding Devices such as:

- Pressure-Sensitive Safety (PSS) Mats,
- Floors,
- Edges,
- Bars,
- Bumpers,
- Plates,
- Wires, and
- Electric Cables.

And on the other hand, the family of ADDs not based on pressure sensing, known as Electro-Sensitive Protective Equipment (ESPEs), includes:

- Radio-Frequency Presence-Sensing Devices (RFPSDs),
- Safety Light Curtains (SLCs),
- Light Beam Safety Devices (LBSDs),
- Safety Laser Scanners (SLSs), and
- Vision-Based Protective Devices (VBPDs) (Ridley & Pearce, 2006; Goernemann & Stubenrauch, 2013).

Electro-Sensitive Protective Equipment (ESPEs)

As previously discussed, if a worker has to frequently or regularly access a machine and is therefore exposed to a hazard, choosing an ESPE instead of a (mechanical) SG is a more suitable solution from a productivity viewpoint. Since ESPEs do not impose any physical barrier to the worker as any SG does, machine access time will be reduced as SG opening and closing times are discarded (Ridley & Pearce, 2006). However, it must be also considered that their use is not possible on machines with lengthy run-

Table 2. Classification of Access Detection Devices (ADDs)

Pressure-Sensitive Protective Devices (PSPDs)	• Pressure-Sensitive Safety Mats & Floors. • Pressure-Sensitive Safety Edges & Bars. • Pressure-Sensitive Safety Bumpers & Plates. • Pressure-Sensitive Safety Wires & Electric Cables.
Electro-Sensitive Protective Equipment (ESPEs)	• Radio-Frequency Presence Sensing Devices (RFPSDs). • Safety Light Curtains (SLCs). • Light Beam Safety Devices (LBSDs). • Safety Laser Scanners (SLSs). • Vision-Based Protective Devices (VBPDs).

down times, as these machines typically require unrealizable minimum distances. In such cases, SGs must be used (Goernemann & Stubenrauch, 2013).

To make access detection possible, an ESPE comprises the three main elements of any ADD: sensor, OSSD and control unit, where the sensor uses electro-sensitive means to determine the event or state that this ADD type is intended to detect. For example, in an opto-electronic device the sensing function would detect an opaque object entering the detection zone. Regardless their technological basis and principle of operation, safety requirements for ESPEs are established in standard IEC 61496-1 (IEC 61496-1, 2012). In addition, standard ISO 13857 (ISO 13857, 2008) explicitly determines the minimum safety distances where ESPEs must be positioned in the industrial workplace. Given the close relationship existing between the underlying technology of VBPDs and other optical-based ESPEs (namely SLCs, LBSDs and SLSs), in the following paragraphs a brief discussion of their fundamental characteristics is addressed.

Safety Light Curtains (SLCs)

Also known as Active Opto-electronic Protective Devices (AOPDs), SLCs are most simply described as photoelectric presence sensing switches specifically designed to protect workers from injuries related to hazardous machine motion. This type of ESPE provides access detection to a machine danger zone through a parallel set of light-emitting diode (LED) light beams (normally infrared) arranged such as to create a virtual wall that must be penetrated by the worker to reach the danger zone, or direct access (e.g. of the hand) to the machine point of operation, i.e. the point where the machine is performing its intended task (e.g. cutting, shaping or stocking) on the material. Detection occurs when an opaque object fully interrupts one or more light beams (Ridley & Pearce, 2006).

According to the international standards IEC 61496-1 (IEC 61496-1, 2012) and IEC 61496-2: Safety of machinery – Electro-sensitive protective equipment – Part 2: Particular requirements for equipment using active opto-electronic protective devices (AOPDs) (IEC 61496-2, 2013), the detection capability of the sensing device should be taken into account in the design and use of SLCs. This detection capability is directly related to the ESPE minimum detectable object size within the defined detection zone (which is also sometimes referred to as protective field or protective zone). In general, ESPE detection capability is determined by the sum of beam separation and effective beam diameter. This definition ensures that an opaque object of such size will always interrupt at least one light beam, and will be therefore detected regardless of its position in the protective field (Goernemann & Stubenrauch, 2013). When this detection capability is large (i.e. when the SLC is configured to be only capable of detecting large-sized objects), SLCs are also referred to as Safety Light Grids (IEC 61496-2, 2013). For a SLC, the minimum detectable element is either a finger or a hand, so that the detection capability of SLCs must be chosen taking into account these considerations. Finally, the so-called effective aperture angle (EAA) —i.e. the maximum angle of deviation from the optical alignment of the emitting and receiving element(s) within which the AOPD normally operates— must be properly verified.

Modern SLC control microprocessor units incorporate specific beam activation control, which enables on/off beam switching in rapid succession, instead of the basic synchronized beam on/off switch. This improves SLC resistance to interference from other sources of light, thus increasing device reliability accordingly. In addition, given that state-of-the-art AOPDs incorporate sender/receiver automatic synchronization for each beam through digital optical links, improved reliability is achieved via beam coding. With this technique, a receiver can only respond to light beams from its code-paired sender. Further functionalities such as blanking can also be integrated in SLCs with beam coding. These include improved productivity features such as blanking, which allow certain object shapes to pass through the SLC protective zone without triggering the safety function (off state) (Goernemann & Stubenrauch, 2013).

Light Beam Safety Devices (LBSDs)

Also classified into the AOPD group, LBSDs are functionally similar to SLCs. They provide perimeter access detection through a set of properly configured Single Beam Safety Devices (SBSDs), or Multiple Beam Safety Devices (MBSDs). Thus the main difference between SBSDs and MBSDs is the number of light beams utilized for achieving the intended access detection. LBSDs light beams must be arranged so as to create a virtual wall around the machine danger zone, but these beams do not have designated a standardized detection zone. Thus LBSDs are actuated when one or more beams are interrupted (IEC 61496-2, 2013). SBSDs are also known as Single Beam Safety Light Barriers. These ESPEs supersede the concept of a curtain made of multiple light beams with the use of a single wider one (Ridley & Pearce, 2006). As a result, LBSDs have become a considerably more cost-effective solution than SLCs, mainly due to their simplicity of construction and the reduced number of sensors to be processed by the control unit (Goernemann & Stubenrauch, 2013). LBSDs are well suited for applications where the detection zone is not specified and thus the detection capability can be wider, typically covering a full human body. LBSD design and test specifications are established in standards IEC 61496-1 (IEC 61496-1, 2012) and IEC 61496-2 (IEC 61496-2, 2013).

Safety Laser Scanners (SLSs)

Technically known as Active Opto-electronic Protective Devices responsive to Diffuse Reflection (AOPDDRs), this type of ESPE is mainly used for stationary and mobile hazardous zone protection (Ridley & Pearce, 2006; Caputo et al., 2013). A SLS is an AOPD with a sensor function produced by opto-electronic infrared light sender and receiver elements, which detects the self-generated light diffuse reflection of an object in a pre-defined two-dimensional protective field. In other words, a SLS scans its surroundings with laser beams in two dimensions and monitors a detection zone that should contain the machine danger zone. SLS operating principle is time-of-flight (TOF) measurement. Very much like microwave radar, a SLS transmits very short light pulses while an electronic stopwatch runs simultaneously (Caputo et al, 2013). If a light pulse strikes an object, it is reflected and received by the scanner. The scanner calculates the distance to the object based on the TOF. A uniform speed rotating mirror in the scanner detects the light pulses so that a section of a circle around the scanner is covered. The scanner determines the exact position of the object from the measured distance and the angle of rotation of the mirror (Ridley & Pearce, 2006).

State-of-the-art AOPDDRs allow simultaneous monitoring of several zones and seamless zone switching during their operation (Ridley & Pearce, 2006; Caputo et al., 2013). For example, this type of ESPE can be used for the adjustment of the monitored zone to the speed of a conveyor, or to graduate response (warning/protective field) to prevent unnecessary interruptions in operation (Goernemann & Stubenrauch, 2013). Because of their method of operation, SLSs periodically cover the monitored zone with localization accuracy between 30 mm and 150 mm. SLS detection capability is independent of the distance to the object, the angle between the individual light beam pulses, and the shape and size of the transmitted beam. With their active scanning principle, SLSs need no external receivers or reflectors to perform their safety function. AOPDDRs must be able to reliably detect objects with extremely low reflectivity (e.g. black work clothing), a fact that limits the range of practical laser emitting and receiving elements used in their construction (Ridley & Pearce, 2006; Goernemann & Stubenrauch, 2013). AOPDDRs must be designed and tested according to the safety requirements of the international standard IEC 61496-3: Safety of machinery – Electro-sensitive protective equipment – Part 3: Particular requirements for Active Opto-electronic Protective Devices responsive to Diffuse Reflection (AOPDDR) (IEC 61496-3, 2008).

MAIN FOCUS OF THE CHAPTER

This chapter focuses on a family of ESPEs known as Vision-Based Protective Devices (VBPDs). The following paragraphs are devoted to technically address VBPD functionality, main types and parameters.

VBPD Functionality

VBPDs are the most recently developed and standardized ESPE family to date. VBPDs must be designed and tested according to the safety requirements of the international standard IEC 61496-4: Safety of machinery – Electro-sensitive protective equipment – Particular requirements for equipment using vision based protective devices (VBPD) (IEC 61496-4, 2007). In contrast to other closely related ESPEs, such as SLCs, LBSDs and SLSs, a VBPD utilizes one or more imaging sensors (i.e. cameras) and active illumination elements which operate in the visible near infrared part of the EM spectrum, in order to detect persons or objects in a defined field of view (FOV) through digital image and video processing techniques (Goernemann & Stubenrauch, 2013).

As shown in Figure 1, FOV is a parameter that relates the visible area covered at parallel planes located at different perpendicular distances *D* from a camera imaging sensor. The so-called image plane collects the reflected light from every object within the camera FOV, as projected through the lens (and the focal plane) on the imaging sensor (González & Woods, 2008). Camera FOV may also be expressed as an angle (FOV°) similarly to the EAA parameter of a SLC light beam. Thus, FOV° relates the not occluded (i.e. visible) area size and the distance *D* to the sensor. Camera FOV is usually controlled by one or more optical lenses mounted at a distance *f* from the imaging sensor. This distance *f* is commonly known as the focal length. The focal length of a lens is defined as the distance from the optical center of the lens to the so-called focal point, which is located on the imaging sensor if the viewed subject (at infinity) is in focus. Camera lens projects on the sensor the part of the viewed scene that falls into the area covered by the FOV at a magnification scale determined by *f*. Nevertheless, as the imaging sensor is usually implemented as a rectangular grid of multiple individual sensors —e.g. a Charge-Coupled Device (CCD)— a different FOV is obtained for each image dimension, horizontal (H) and vertical (V). These two FOV dimensions are also related by the so-called aspect ratio AR , which usually depends on the effective aspect ratio of the imaging sensor matrix. In this case, the angular FOV is determined for the sensor diagonal length $AR = FOV_H / FOV_V$. In a rectilinear distortionless lens, S_d (González & Woods, 2008).

Typically, the VBPD sensing device comprises one or more digital video cameras which are intended to monitor a detection zone, which is generally formed as the overlapped combination of each individual camera FOV. As Figure 1 shows, this detection zone should —at best— fully include the machine danger zone surrounded by a warning zone. In fact, even the hazardous machine itself is completely within

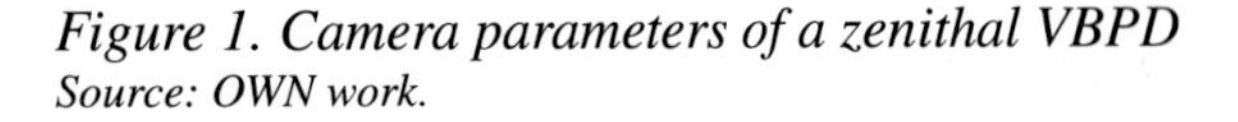

Figure 1. Camera parameters of a zenithal VBPD
Source: OWN work.

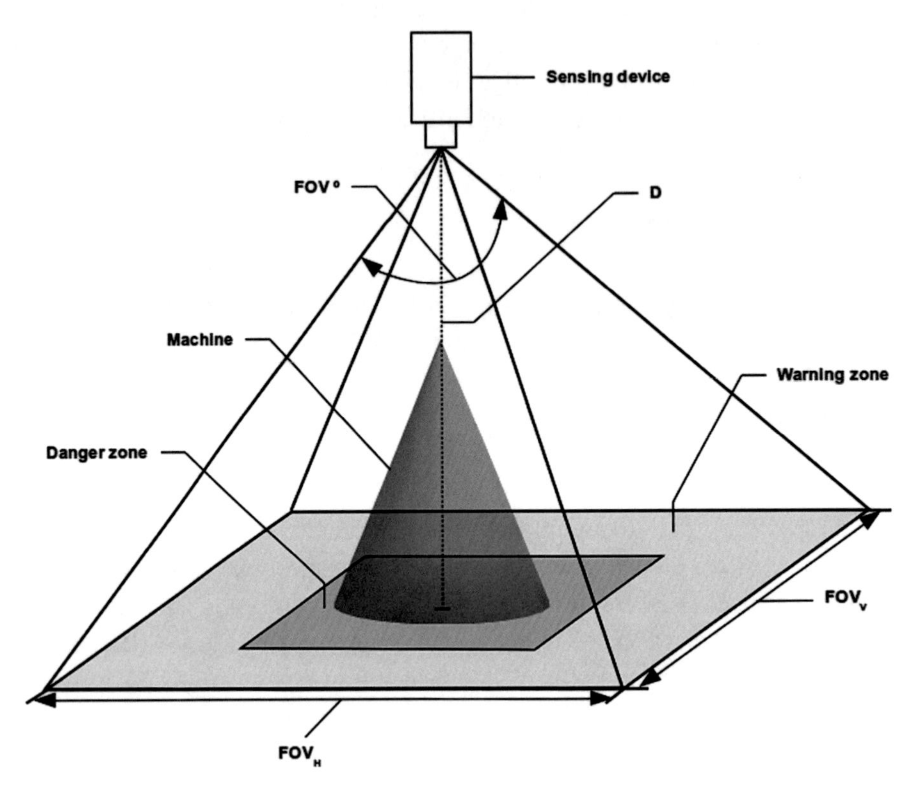

the FOV in this figure, where a VBPD is located in a zenithal (i.e. overhead) position relative to the monitored workplace. However, this is rarely the case given that VBPDs are typically used as ADDs, i.e. their safety functionality is similar to that of a SLC:

- To detect persons as they enter the danger zone, and
- To stop machine functions accordingly.

This means that the only dimensional requisite for VBPD FOVs is to fully contain any border between warning and danger zones that can be accessible to a person who trespasses it while approaching the hazardous machine. Any visible (i.e. opaque) object of a minimum given projected size on the image plane should modify the captured image scene with respect to a preferably static, free of risk scenario, thus interrupting the VBPD FOV in such a way that access detection is achieved (Goernemann & Stubenrauch, 2013). Danger and warning zone dimensions are determined according to international standards ISO 13855 (ISO 13855, 2010) and ISO 13857 (ISO 13857, 2008). Protective zone dimensions are set for those areas around the danger zone where, although presence of persons cannot cause an accident, access detection must be performed. Thus these dimensions largely depend on the access detection method to be used.

Currently available VBPDs are based on three different camera technologies:

- Standard 2-D,
- Stereo, and
- TOF (IEC 61496-4, 2007).

TOF cameras, also known as depth cameras, are able to resolve distance to objects inside the FOV by measuring the TOF of a light signal between the camera and these objects for each point of the captured depth image. TOF camera resolution is generally lower (from 1:5 to 1:10) than standard 2-D video cameras. However, typical TOF camera frame rates exceeds standard video frame rates, as they provide up to 160 depth image frames per second (fps) which must be compared with standard 30 fps of typical MJPEG cameras (Schuon et al., 2008).

As Figure 2 illustrates, in stereo vision VBPDs two cameras with a controlled spatial relationship between their respective image planes are used to obtain two slightly different views of a scene, in a physical arrangement similar to the human eyes. By comparing these two images, relative depth information can be obtained in the form of left point (PL) to right point (PR) disparities, which are inversely proportional to the differences in distance to the imaged objects. To compare the images, the two camera FOVs must be superimposed in a stereoscopic device which must compute image disparities to obtain the so-called depth map (Tan & Arai, 2011). Thus a stereo FOV can be defined for the volume depicted in shade in Figure 2, which accounts for the volume where disparities (and depth information) can be computed (González & Woods, 2008). Nowadays, several principles can be used to carry out access detection using VBPDs, including the following (Goernemann & Stubenrauch, 2013):

Figure 2. Standard geometry of a binocular stereo imaging system
Source: Own work.

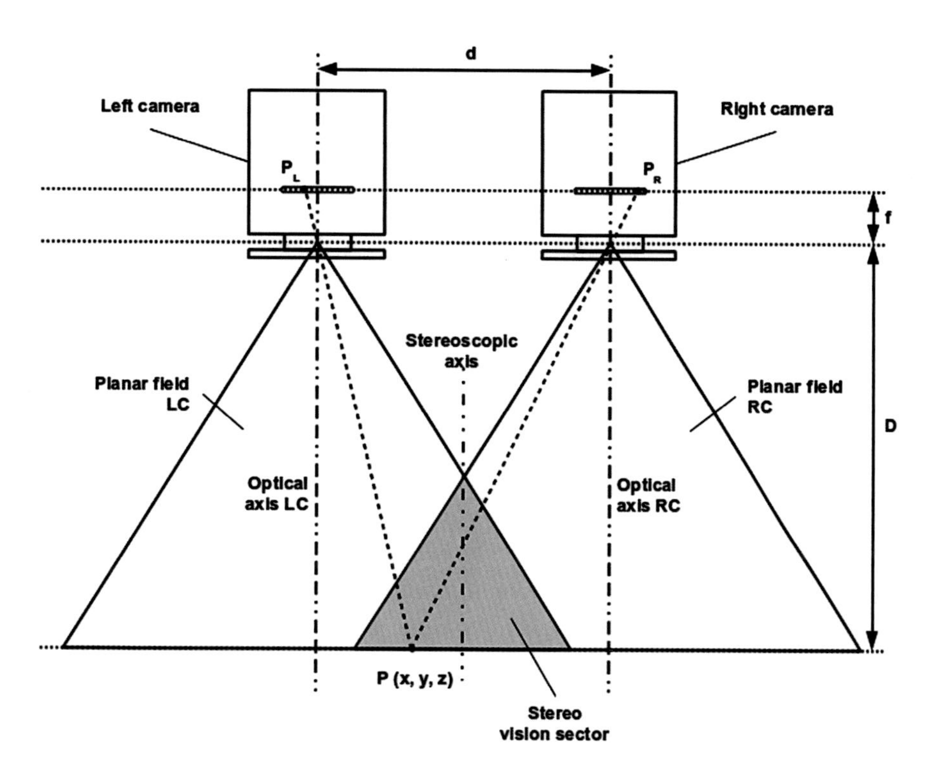

Interruption of the Light Reflected by a Retro-Reflector

The objective of this principle is to detect object intrusion through an opening with solid wall surrounding surface, which must be covered with a strip of reflective material. The sensing device must be formed by a single imaging sensor, and one or more —generally infrared— light sources for illuminating the inner wall surfaces. The illuminated reflective material is sensed by the imaging sensor and thus any non-transparent object which moves through the protected opening will be detected as a change in the reflected light intensity. Reflective materials can be designed with 1-D or 2-D pattern codes, which enable obtaining different average wall reflectivity with acceptable accuracy. The use of reflective strips also allows VBPDs to be a valid —and often more economic— solution for a variety of irregular opening shapes which are not well-suited for other ESPEs such as SLCs (Rockwell, 2012).

TOF of the Light Reflected by an Object

Surface range measurement can be directly attained using the radar TOF principle also used in 2-D scanners like SLSs. The emitter generates a laser pulse, which impinges on the object surface and the receiver detects the reflected pulse. Afterwards, the control unit electronically measures the TOF of the returning signal and its intensity. The combination of single point laser sensors, which perform point-to-point distance measurement, and scanning devices, which mechanically operate the optical camera head, are the basis for the construction of 3-D range or depth image TOF cameras (Sansoni et al., 2009). Compared to stereo cameras, TOF cameras benefit of a greater robustness respect to both color and illumination and allows for a simpler 3-D depth picture element (pixel) computation. On the other hand, unlike SLSs which are 2-D monitoring devices with high depth resolution but low speed, TOF cameras acquire depth images with high depth resolution at high frame rates with low image resolution. However, for obtaining high accuracy depth measurements for both close and far objects, special care must be taken to appropriately gain control the returning laser pulses so as to not lead to saturation for near objects or poor treatment of distant objects as a result of using low power signals which may be affected by noise. In addition, camera integration time must be also controlled in these scenes so that object motion during acquisition do not cause corrupt range data (Schmidt & Wang, 2014).

Changes from Background Patterns

This access detection approach relies on other fundamental Digital Image Processing (DIP) techniques such as motion detection and video coding (González & Woods, 2008). In this case, the VBPD control unit processes intensity changes from a background pattern which is intended to be the background of an illuminated, non-transparent body or body part whose protection is intended. This background is a static digital image of the monitored zone without any obtrusive object existing in it. For motion detection this background reference image is rapidly compared with the current image acquired by the imaging sensor in order to detect the presence of objects. This technique, widely known as background subtraction, can be also accompanied with temporal frame differencing between consecutive frames in order to make this access detection more robust. Nevertheless, this robustness may be greatly increased if both background subtraction and temporal differencing techniques are provided with shadow rejection and illumination control capabilities (Elhabian et al., 2008). In fact, this principle is an extension of the intensity difference measurement undertaken by VBPDs based on the first principle: interruption of the light reflected by a retro-reflector, as in these VBPDs background intensity is stored prior VBPD activation, and the control unit computes real-time intensity difference of the reflected light.

In general, the design and test of VBPDs must comply with the requirements of the standard IEC 61496-4: Safety of machinery – Electro-sensitive protective equipment – Part 4: Particular requirements for equipment using vision based protective devices (VBPD) (IEC 61496-4, 2007). In addition, pattern-based VBPDs must also be compliant with the recent standard extension IEC 61496-4-2: Safety of machinery – Electro-sensitive protective equipment – Part 4-2: Particular requirements for equipment using vision based protective devices (VBPD) – Additional requirements when using reference pattern techniques (VBPDPP) (IEC 61496-4-2, 2014). In the case of stereoscopic VBPDs, recent standard extension IEC 61496-4-3: Safety of machinery – Electro-sensitive protective equipment – Part 4-3: Particular requirements for equipment using vision based protective devices (VBPD) – Additional requirements when using stereo vision techniques (VBPDST) (IEC 61496-4-3, 2015) should be also considered.

Main VBPD Types

Two main types of VBPDs are typically considered based upon the kind of information extracted from the video signals. These types are known as 2-D and 3-D, and are often referred to as Video Safety Curtains (VSCs). A brief description of these devices is addressed below.

2-D VSCs

From the discussion above, 2-D camera based VSCs work upon two access detection principles: interruption of the light reflected by a retro-reflector, and changes from background patterns. Probably the most representative example of this type of VBPD is the device of Nichani et al. (2004). In this VBPD, one or more cameras are placed above a safety floor strip installed on the outer perimeter of the danger zone. Thus the area occupied by the safety floor, which is called protective zone, must be of a sufficient width, so that a 2-D camera positioned in a zenithal position over the workplace (see Figure 1) is capable of detecting any object moving on the protective zone, prior to accessing the danger zone. The safety floor strip is usually made of non-slip reflective material, so that optimal intensity difference (i.e. contrast) may be obtained from the detected image object as it is projected on the image plane, with the use of appropriate light sources.

Whereas danger zone dimensions are explicitly defined in standards ISO 13855 (ISO 13855, 2010) and ISO 13857 (ISO 13857, 2008), protective zone dimensions solely depend on VSC parameters. A schematic diagram of a typical VSC configuration is represented in Figure 3, where a worker moves towards the machine and is visualized by the camera system in three consecutively numbered frames. The longitudinal camera position relative to the hazardous machine is such that its FOV fully covers the protective zone, located within the warning zone. Camera sensor must be positioned at a distance D from the floor, which must be greater than the expected object height h. Geometrical camera parameters $OA_{\min}$ and height D are linked to the protective zone longitudinal dimension or width, FOV°. For an overhead camera such as that depicted in Figure 1 and Figure 3, the maximum value achieved for d_{PZ} in a distortionless camera is d_{PZ}. For the worker entering the danger zone (from right to left in Figure 3), only those images visualizing a projected minimum object area $d_{PZ} = 2D\tan\left(FOV^{\circ}/2\right)$ in the protective zone will be detected by the VSC. In this case, two positive object detections are achieved in frames 1 and 2. In frame 1, the projected size of the worker's foot over the safety strip is large enough to trigger machine function shutdown, which is also the case of frame 2, where the worker is now partially entering the danger zone. Finally, in frame 3 the worker is fully inside the danger zone, and thus completely outside the protective zone, so that this worker is not detected by the VSC.

Another worth noting VBPD development is the recent 2-D VSC of (Wuestefeld, 2007). This device comprises a small sized wide-angle camera, and two wide-angle light emitters located either around or at both sides of the camera, conveniently enclosed in a single housing; and a reflectance controlled stick-on tape capable of adapting itself to relatively rough surfaces. Designed to be mounted in access or passage openings, such as an opening through a wall, this device is currently commercialized by Sick (2009) and Rockwell (2012). The light sources illuminate the opening, in which inner surface the reflector tape is installed. The camera forms a picture of at least one of the reflective wall surfaces defining the opening, and directs it onto a position resolving light receiver that generates picture signals for evaluation to determine if an unauthorized project is in the opening. Light source and light receiver, the camera, as well as a control unit, are all mounted in a housing that is configured to be secured to at least one of the wall surfaces of the opening, so that it extends only negligibly into it (Wuestefeld, 2007).

Figure 3. VSC dimensional parameters: DZ = danger zone, PZ = protective zone, WZ = warning zone, $FOV^{o} = 2\arctan\left(S_{d}/2f\right)$ *= minimum projected object area*

Source: Own work.

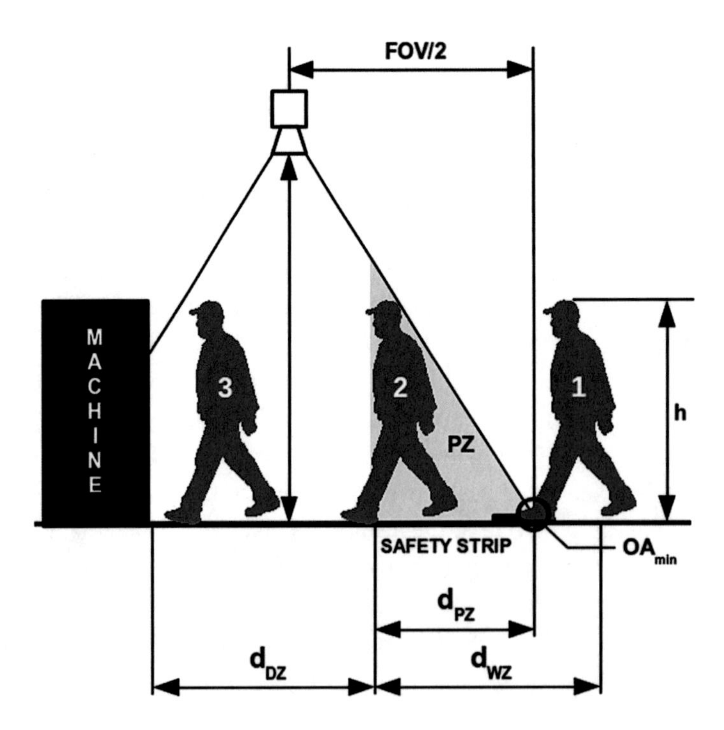

3-D VSCs

3-D VSCs have also been developed in the last two decades. The initial approach of Schatz et al. (2001), which first coined the VSC term, was designed to perform its activity with a closely mounted, double 2-D camera-based binocular system, for duplicated depth computation of objects within the stereo FOV (see Figure 2). This machine vision safety solution converts multi-camera 2-D video pixel data into 3-D point data that is used for characterization of a specific 3-D object, objects, or an area within view of a stereoscopic camera configured to provide a target protection from encroachment by another foreign object, which intrudes a nearby danger zone. Thus in this implementation, stereopsis is used in determining a 3-D location of an object with respect to cameras, or a defined reference point. A 3-D difference can then be derived and compared with a model view.

According to Figure 4, a 3-D VSC includes an image acquisition device comprising two or more video cameras, or digital cameras, arranged to view a target scene stereoscopically. The cameras pass the result-

Figure 4. Functional block diagram of a stereoscopic 3-D VSC
Source: Adapted from (Schatz et al., 2001).

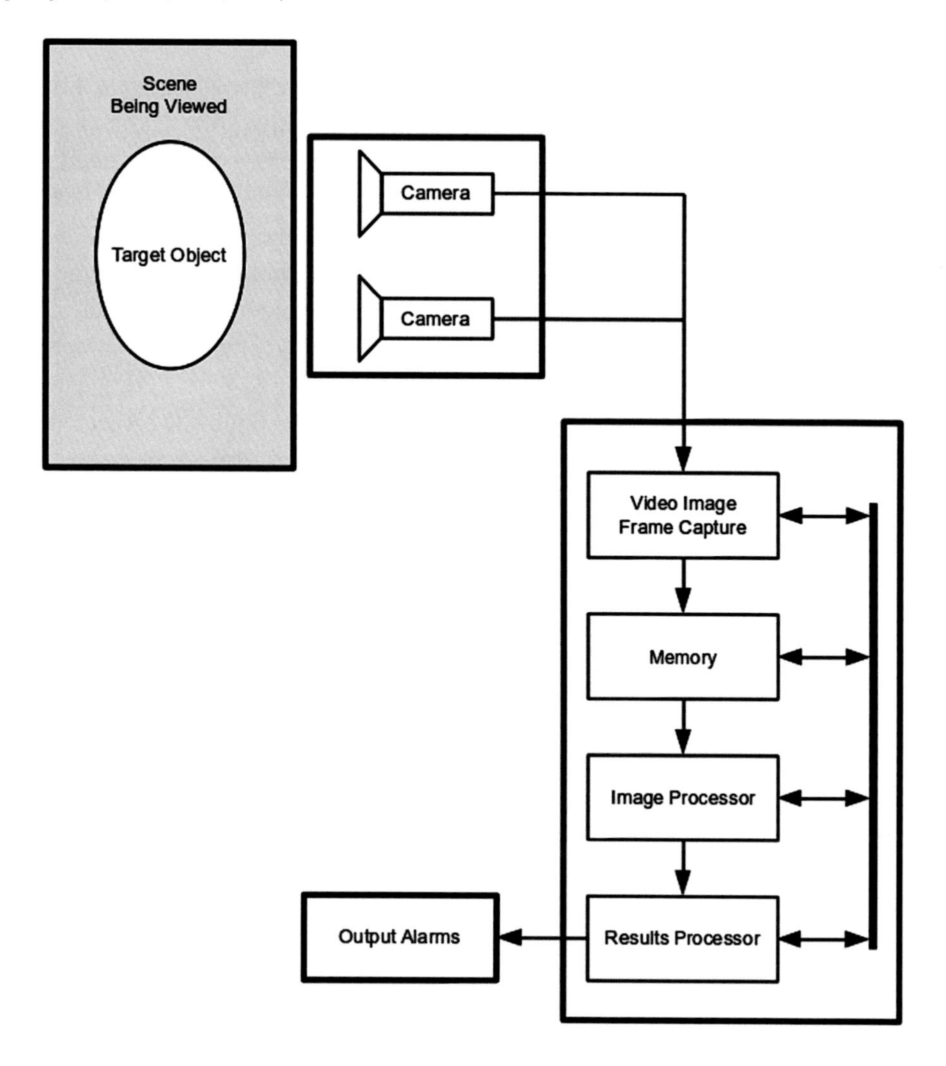

ing multiple video output signals to a computer for further processing. A computer video processor input, such as a frame grabber sub-system, is enabled for each video signal. Video images from each camera are then synchronously sampled, captured, and stored in a memory associated with a general purpose processor. The digitized image can then be stored, manipulated and further processed by a computer program associated with the computer memory, in which processing results are stored, and warning or alarm signals are sent through convenient communication ports. Since computer hardware and software enables very different implementations, it turns out that the 3-D VSC processing framework capabilities deeply affect overall system performance. Currently 2-D camera based, stereoscopic 3-D VSCs include the well-known device of (Pilz, 2015), which is derived from (Wöhler et al., 2010). The overall system is made up of various components that combine to form one safety unit. It comprises a sensing device, an analysis unit and a programmable safety and control system. The sensing device consists of three greyscale cameras, accurately mounted on a safe steel enclosure for impeding manipulation. Mounted overhead the danger zone, the camera transmits the images via high-speed cable links. The analysis unit, a high performance computer, receives and processes the image data. In doing so it generates the safety signals for the integrated programmable safety and control system, which in turn forms an interface with the actuator technology. Thus, as soon as the programmable safety and control system receives an intended alarm signal (e.g. the closest detection zone has been violated), it immediately shuts down the relevant hazardous machinery.

A different 3-D VSC approach is given in Nichani et al., (2007), where access detection is performed based upon 2-D camera acquisition of projected light patterns scattered over a target area, with a technique known as structured light. Structured light is defined as the process of illuminating an object at a known angle with a specific light pattern. This viewing angle permits analyzing a lateral projection, which is supposed to provide depth information. For example, if a line of light is generated and viewed obliquely, the distortions in the lines can be translated into height variations. Thus illuminating an object with structured light and looking at the way the light structure is changed by the object surface provides information on the 3-D shape of the object. This is a basic principle behind human depth perception, which has also been successfully applied to machine vision (Dipanda & Woo, 2005). The device works as follows. A known infrared structured lighting texture pattern is continuously projected upon a target, namely the protective zone floor of a standard VSC, and a 2-D camera receives an image of the area thus illuminated. A model image of the pattern on an empty target field is stored during an initial training step. The VSC control unit digitally interprets a camera image of the light reflected by the objects in the target area, in order to detect and characterize a pattern in the image. The pattern characterization is then processed to determine if a distortion of the characterization factors is larger than a predetermined threshold, and results in an alarm condition which is fed to the OSSD (Nichani et al., 2007). The structured lighting pattern on the target area should be implemented with a suitable monochromatic light. Typically, best performance is obtained from pattern light sources operating in the near infrared EM band, with the advantage of avoiding human interaction with the pattern.

VSC Parameters

A definition of the most relevant functional parameters of a VSC follows:

*Camera Frame Rate (*CFR*)*

The number of frames per second (fps) that the VSC camera system is capable of steadily delivering to the control unit. In Figure 3, the worker is captured in three consecutively numbered frames while moving longitudinally towards the machine on the left.

*Average Processing Time (*APT*)*

The amount of time needed by the control unit to process a single frame delivered from the camera, which is typically averaged over *CFR* consecutive frames. The information to process is highly dependent of the VSC type, where comparatively 2-D VSCs deal with the least computational burden.

*Effective Frame Rate (*EFR*)*

The number of processed frames per second that the VSC control unit is capable of steadily analyze. It is usually determined by the minimum value of *CFR* and APT^{-1}.

Maximum Detectable Object Velocity (V_{max})

For the VSC to work properly, any considered object should be detected at least once when within the protective zone. Thus, an object with instantaneous projected longitudinal length $OA_{\min}$, moving with instantaneous longitudinal velocity $L(t) \geq OA_{\min}$ across the protective zone, will be detected by the VSC if and only if $V(t)$. In the sequence depicted in Figure 3, the worker moves from right to left with longitudinal velocity $V(t) \leq V_{\max} = d_{PZ}$ *EFR* such that detection occurs in frames 1 and 2, thus in this case $V(t)$ and the VSC is properly configured.

Comparison with Other ESPEs

The recent development of VBPDs aims to overcome a series of problems associated with previously described AOPD types (see Background section). A summary of the most relevant features of SLCs, LBSDs and SLSs is given in Table 3, which highlights the most relevant issues and limitations derived from the use of these types of ESPE. A brief discussion on these follows.

SLCs and LBSDs are constructed by aligning a series of photo-transmitters and receivers in parallel to create a usually vertical *curtain* of parallel light beams for safety monitoring. However, since light travels in straight lines, each paired optical transmitter and receiver must be carefully aligned for each beam. Although mirrors may be used to bend the beams around objects, this only further increments design and calibration problems, and also reduces the safe operating range. One major disadvantage of these devices is the minimum available resolution for object detection, as determined by the inter-beam spacing. Any object smaller than this inter-beam spacing can penetrate the curtain, if between adjacent beams, without being detected. Another drawback comes from the point nature of their light sensors. Any object approaching dangerously close to the light curtain remains undetected until it physically blocks any of the beams, and thus a fast-moving intruding object might not be detected until too late, forcing the designers to physically position SLCs and LBSDs farther away from the danger zone in order

Table 3. Main features of Active Opto-electronic Protective Devices (AOPDs)

	SLC	LBSD	SLS
Minimum detectable object size	Finger, hand, body part	Human body (full)	Human body (section)
Size discrimination	Yes	No	No
Object positioning	No	No	Yes (2-D)
Workplace illumination	Dependent	Dependent	Independent
Maintenance	Dust Alignment	Dust Alignment	Dust Bumping
Protection areas	Fixed Planar (vertical)	Fixed Planar (vertical)	Fixed and Movable Planar (horizontal)
Cost	Medium	Low	High

to provide the necessary time-interval as required by the machine stopping time (Nichani et al., 2004). This additional extra space is also increased since no reflective object must be placed near the curtain, as any intrusive moving object nearby the reflector can pass the curtain without being noticed if sufficient light is reflected back to the photo-receivers. Furthermore, SLCs and LBSDs are susceptible to interference from ambient light, whether from an outside source, or reflected by a nearby object. This factor further limits the practical applications, making their use difficult in outdoor locations, in the vicinity of machines capable of emitting light bursts, such as metal welders and cutters, and near any reflective material in general. In such locations, the optical receivers may not properly sense a change in a light beam. All in all, these constraints dictates an extra space that otherwise could be used for other purposes. Finally, these devices are prone to be severely limited in cases where chips, dust, or vapors cause dispersion and attenuation of the optical beams, or where vibrations and other machine movements can cause beam misalignment (Goernemann & Stubenrauch, 2013).

More recently, the development of TOF based SLSs have made possible simultaneous object detection and positioning within a predefined nearby planar, usually horizontal area. This scanning technology uses a narrow pulsated laser beam and measures reflected light TOF to determine the position of objects within the viewing field, which encompasses a relatively large zone (e.g. ~50 m radius over 180 degrees) that can be divided into warning and danger zones. However, SLSs cannot distinguish between different sizes or characteristics of objects detected, making them unsuitable for many safety or security applications where false alarms must be minimized. In addition, SLSs typically incorporate extremely delicate moving parts, which require precision alignment. As a result, SLSs lose reliability under challenging ambient conditions. Another disadvantage of such devices is their horizontal flat detection field, which leads to multiple problems, including being susceptible to physical damage or bumping, which increases false alarms and maintenance costs. Furthermore, the protected area usually requires the use of solid objects or screens to limit its effective size, which may be a major requirement for enabling worker access to other adjacent areas (Nichani et al., 2004).

From the previous discussion, it may be demonstrated that VBPD functionality (as summarized in Table 4) enables avoiding a considerable number of problems associated with other ESPEs. However, since different technologies are being currently used to obtain its safety behavior, a brief discussion on this issue is given next. One of the major advantages of zenithal VSCs is their functional geometry. By looking top-down on a scene where intruders may enter (see Figure 2), two main advantages arise:

Table 4. Main features of Vision-Based Protective Devices (VBPDs)

	2-D VSC	3-D VSC (Stereo)	3-D VSC (Structured)
Minimum detectable object size	Finger, hand, body part, or full body	Hand, body part or full body	Hand, body part or full body
Size discrimination	Yes	Yes	Yes
Object positioning	No	Yes	Yes
Workplace illumination	Dependent	Dependent	Independent
Maintenance	Illumination	Computing hardware Illumination	Computing hardware Illumination
Protection areas	Fixed Planar (horizontal)	Fixed Volumetric (horizontal)	Fixed Volumetric (horizontal)
Cost	Low	High	Medium

1. A single camera-lighting fixture can be used, such that the whole area is uniformly lit and viewed, thus providing a more homogeneous detection capability (sensitivity) across the target area; and
2. It allows setting of precise target regions that need to be protected, which is typically done either using visible markers (or fiducials) on the floor during a setup procedure, or by a graphical user interface (gui) overlaid on the image.

This procedure is far more flexible than the required configuration available for any other existing AOPD. In addition, their relative hardware and maintenance costs are much lower.

Since any camera has a constrained FOV, in general a camera ensemble should be needed to cover the entire protective zone of a particular facility, using a suitable multi-camera FOV arrangement. This camera ensemble should allow monitoring in a projected 2-D plane the combined visual information from the different camera FOVs. The VSC then applies motion detection to this projected image by focusing in the distinctive safety strip patterns and the nearby monitored danger zone. Given their ability to monitor large protective zones, VSCs enable the possibility of object size filtering and tracking, which is considered to be a key feature used in reducing false alarm rates. However, VSCs strongly depend on multi-camera synchronization and visual data processing for obtaining a fast, reliable performance. 2-D camera based VBPDs share some drawbacks typically associated with video or closed circuit TV (CCTV) surveillance systems (Cieszynski, 2007). These video systems having motion detection sensors are also known for automatically detecting indications of malfunctions or intruders in secured areas. However, these types of known sensors are limited to the simple detection of change in the video signal caused by the perceived movement of an object, perhaps at some predefined location or region of interest (ROI).

Analog CCTV has been progressively superseded by modern digital systems. While new developments have brought clearer images, digital recording and high speed data transmission, effective security systems still rely upon proper specification and installation details, which opens a place for expertise engineering to provide tailored solutions for specific facilities. Undoubtedly, video surveillance systems are susceptible to false alarms caused by shadows coming into view that cannot be distinguished from objects. Thus, illumination control of the monitored area must be also considered. However, more problems arise from the mounting location chosen for these cameras. Typically, a non-uniform detection capability across the scene is obtained, given that perspective projections of the viewed scenario results in relative resolution changes in digitized images. Resolution may also be diminished with the use of camera lenses with radial (and tangential) distortions (González & Woods, 2008). These drawbacks may be crucial in object size discrimination applications. Therefore, while video motion detectors can be useful for general surveillance operations, the stringent requirements against false positives and false negatives do not usually permit their use for safety devices. Furthermore, in 2-D camera based video motion detectors, a low-contrast object can enter the area without triggering an alarm, which is typically based on image differencing grayscale thresholds as obtained from a subtracted background (González & Woods, 2008). Again, such systems require sufficient ambient light to uniformly illuminate the target area in order to properly view the intruding objects. However, additional lighting can cause its own problems, such as reflections that affect the workers, machines or other sensors, or cause shadows that impinge upon adjacent safety areas and cause false alarms. These and other disadvantages restrict the application of 2-D camera based video surveillance technology in VBPDs (Nichani et al., 2007).

On the other hand, locating objects in 3-D space from stereopsis requires a binocular (or trinocular) image set. In comparison, 3-D stereo VBPDs thus increase the cost and maintenance of equipment. In addition, 3-D calculations for matching and determining alarm condition is typically much more time consuming than 2-D solutions. Moreover, for an application where the camera is mounted overhead to view a target, the effective area within view is conical (i.e. the shaded area of Figure 2) and the first part of a person coming into view (i.e. the feet) is typically very close to the floor (see Figure 3), making it more difficult and error-prone to quickly detect height difference above the floor. To obtain the necessary coverage, this cone needs to be larger, the camera needs to be mounted higher from the floor, and the image resolution is thus disadvantageously diminished, which in turn further affects computed disparity accuracy. Thus with a larger cone of vision the potential false alarm rate is also increased. These disadvantages may accumulate to such an extent that the system is not reliable enough for its use in applications for protecting severe hazards, where false alarms or false positives cannot be tolerated. Some of these problems can be successfully overcome with the use of infrared projective patterns in 2-D camera based VBPD.

SOLUTIONS AND RECOMMENDATIONS

From the parameter definition of VSCs, it can be affirmed that VSC robustness (i.e. its ability to detect intruding objects without error) may be enhanced by increasing image processing speed, increasing protective zone or curtain thickness, or increasing camera $V(t) < V_{max}$, so that the time during which an intruding object or person intersects the curtain space is maximized. However, each of these elements has its own practical limitations, depending both on the hardware and the industrial environment. Better performance results are obtained for VSC implementations where the control unit has direct access to the camera output frame buffer. Typically a small video frame buffer is addressed in the control unit RAM, so that the microprocessor can read each frame and compare it to a previously recorded background pattern image, without having to wait to the frame to arrive from the camera. In this way, both camera and control unit processes can be synchronized to work at a steady *EFR*.

In multi-camera based VSCs, two architecture paradigms are typically considered: centralized and distributed. In a centralized VSC, a communication or network unit is placed between the control unit and the cameras, which must be connected to a wired, high-speed data network. Wireless camera networks are not considered a practical solution since industrial environments usually contain important sources of

EMI. In this architecture, network data delays may significantly reduce *EFR*, and the control unit processing capability must be high enough to cope with the incoming video buffers. Modern multi-threaded and multi-core processors (e.g. GPUs) are typically used for these intensive processing needs. In distributed VSCs, a dedicated control unit is hardwired to each camera. This architecture benefits of an increased overall *EFR*, since communication delay is drastically reduced and parallelization is straightforward. As a result, individual control unit processing capability needs are also reduced, which leads to more affordable solutions. However, OSSDs must be grouped together in order to connect the VSC with the MCS. This is usually achieved with the use of MSRs, or even Safety PLCs for complex camera systems.

Regardless of the camera type used, image artifacts should be also considered on the performance of VBPDs. Image artifacts are undesired effects caused by the camera sensor, the camera lens, and environmental variables. The most common artifacts are:

- Vignetting (reduction of the image brightness or saturation at the periphery, compared to the image center);
- Noise (random variation of brightness or color caused by random fluctuations of the image sensor electrical signal);
- Sensor dust in hard environmental conditions of the workplace;
- Variable reflectivity of monitored objects from e.g. different poses;
- Alternating light intensity (e.g. from fluorescent illuminators); and
- Resolution.

Many of these artifacts can be efficiently addressed in the so-called image pipeline (Ramanath et al., 2005).

FUTURE RESEARCH DIRECTIONS

Modern 3-D VBPD design trends are promoting the use of TOF depth cameras, which have experienced a noticeable drop in their costs in the last few years. Several 3-D VBPDs have been developed in recent years. Their functionality is based upon detection of differences in the target depth map. Thus any technique capable of delivering an instantaneous depth map of the scene under observation can be used as the working principle of a VBPD. As a matter of fact, robust 3-D vision overcomes many problems of 2-D vision, as depth measurement can be used to easily separate foreground objects from a teachable background. Furthermore, background teaching can also include surface reconstruction from 3-D depth data for 3-D background modelling (Saxena et al., 2008). However, frame resolution and depth noise may be a limiting issue as far as the image processing computational requirements are properly addressed.

Nonetheless, more affordable solutions may still be obtained from standard 2-D video cameras. Here the use of infrared cameras is only suited to applications where no heat, sparks and flames are produced from the machine activity, since temperature is also received by this type of camera sensor. Thus the use of combined infrared and color camera sensors is increasingly being put forward as a convenient solution for some types of machinery. Standards such as ANSI Z535.1: *Safety Colors* (ANSI, 2011) have been designated for use in warning signs and limiting strips in machinery facilities, in order to help prevent workplace accidents by training employees with uniform color messages. The color choice (e.g. black over yellow) is always based upon maximum contrast for increased readability. This same principle must be used on color based VBPDs, where several color thresholds can be used to properly distinguish between objects with otherwise much lower grayscale contrast. Recent research is providing conclusive evidence that highly chromatic colors can serve to increase or decrease the perception of lightness (for reflective materials) and brightness (for self-luminous objects) (ANSI, 2011). This effect is more dramatic in the case of colored light sources and colored reflective materials, which can be installed in VBPD protective zones.

Finally, image processing integration with camera hardware is another interesting feature currently being researched. The rapid development of high quality handheld digital camera devices such as those encountered in modern mobile phones is mainly due to improvements on the image pipeline (Ramanath et al., 2005). The main purpose of the image pipeline is to process raw image data to a final compressed image fitted to human perception, in a number of stages independently processed from one another. Pipelining is in essence a hardware processing technique employed for obtaining enhanced time performance. Typically, in an image pipeline an image is captured on a sensor. Thus, the first pipeline stage is raw data reading from the sensor. In order to reduce thermal noise effects from the sensor, optical black clamp is done to reject noise, which otherwise would lead to poor image quality. Furthermore, because sensor sensitivity is different from human perception, more pre-processing steps are usually needed (e.g. white balance, color correction, gamma correction) to obtain an enhanced image data, prior to higher-level processing tasks such as image segmentation and compression. The possibility of open hardware pipeline programming has only been made available for research in recent years (Martínez & Fernández, 2013). This clearly has allowed a path for developing modern VBPDs with enhanced image quality for machine safety applications.

CONCLUSION

This chapter has introduced the main operating features of a recently standardized type of Safeguarding Device known as Vision-Based Protective Device or VBPD. Intruding objects can be successfully detected with the use of several ESPEs, including SLCs and SLSs. Without using sensors that must be specially designed, placed, or calibrated for each different type of object to be protected, VBPDs do not rely upon any moving mechanical parts subject to the rigors of wear and tear of industrial equipment. Additionally, these devices do not require to be placed very close to, or in contact with the hazard, as would be necessary for mechanical sensors. Thus machine vision systems offer a superior approach to machine safety sensors by processing images of a scene to detect and quantify the objects being viewed. 3-D VBPDs can provide, among other things, an automated capability for performing diverse inspection, location, measurement, alignment and scanning tasks. Furthermore, properly calibrated with the aid of visual markers or fiducials on GUIs, their operation may achieve high immunity from problems caused by small contrast differences between the object and the background, and shadows casted by objects and unexpected or variable light sources, such as back-lit windows, slits or lighting sets, and even the machine operation itself. A discussion regarding several VBPD implementations has also been addressed.

REFERENCES

ANSI Z535.1 (2011). *Safety Colors*. New York: American National Standards Institute.

Caputo, A. C., Pelagagge, P. M., & Salini, P. (2013). AHP-based methodology for selecting safety devices of industrial machinery. *Safety Science*, *53*, 202–218. doi:10.1016/j.ssci.2012.10.006

Cieszynski, J. (2007). *Closed circuit television* (3rd ed.). Oxford, UK: Newnes.

Dipanda, A., & Woo, S. (2005). Towards a real-time 3D shape reconstruction using a structured light system. *Pattern Recognition*, *38*(10), 1632–1650. doi:10.1016/j.patcog.2005.01.006

Elhabian, S. Y., El-Shayed, K. M., & Ahmed, S. H. (2008). Moving object detection in spatial domain using background removal techniques - State of art. *Recent Patents on Computer Science*, *1*(1), 32–54. doi:10.2174/1874479610801010032

European Commission. (2006). Directive 2006/42/EC of the European Parliament and of the Council of 17 May 2006 on machinery. *Official Journal of the European Union, L*, *157*, 24–86.

Goernemann, O., & Stubenrauch, H. J. (2013). *Electro-sensitive protective devices (ESPE) for safe machines*. Retrieved March 15, 2016, from http://www.sick.com/group/DE/home/service/ safemachinery/standards_and_regulations/Documents/8016058_WP_ESPE_en_20130328_WEB.pdf

González, R. C., & Woods, R. E. (2008). *Digital image processing* (3rd ed.). Upper Saddle River, NJ: Prentice Hall.

IEC 61496-1. (2012). *Safety of machinery – Electro-sensitive protective equipment – Part 1: General requirements and tests*. Geneva, Switzerland: International Electrotechnical Commission.

IEC 61496-2. (2013). *Safety of machinery – Electro-sensitive protective equipment – Part 2: Particular requirements for equipment using active opto-electronic protective devices (AOPDs)*. Geneva, Switzerland: International Electrotechnical Commission.

IEC 61496-3. (2008). *Safety of machinery – Electro-sensitive protective equipment – Part 3: Particular requirements for Active Opto-electronic Protective Devices responsive to Diffuse Reflection (AOPDDR)*. Geneva, Switzerland: International Electrotechnical Commission.

IEC 61496-4-2. (2014). *Safety of machinery – Electro-sensitive protective equipment – Part 4-2: Particular requirements for equipment using vision based protective devices (VBPD) – Additional requirements when using reference pattern techniques (VBPDPP)*. Geneva, Switzerland: International Electrotechnical Commission.

IEC 61496-4. (2007). *Safety of machinery – Electro-sensitive protective equipment – Part 4: Particular requirements for equipment using vision based protective devices (VBPD)*. Geneva, Switzerland: International Electrotechnical Commission.

IEC 61496-4-3. (2015). *Safety of machinery – Electro-sensitive protective equipment – Part 4-3: Particular requirements for equipment using vision based protective devices (VBPD) – Additional requirements when using stereo vision techniques (VBPDST)*. Geneva, Switzerland: International Electrotechnical Commission.

ISO 12100. (2010). *Safety of machinery – General principles for design – Risk assessment and risk reduction*. Geneva, Switzerland: International Standards Organization.

ISO 13855. (2010). *Safety of machinery – Positioning of safeguards with respect to the approach speeds of parts of the human body*. Geneva, Switzerland: International Standards Organization.

ISO 13856-1. (2013). *Pressure-sensitive protective devices – Part 1: General principles for design and testing of pressure-sensitive mats and pressure-sensitive floors*. Geneva, Switzerland: International Standards Organization.

ISO 13857. (2008). *Safety of machinery – Safety distances to prevent hazard zones being reached by upper and lower limbs*. Geneva, Switzerland: International Standards Organization.

Martínez, A., & Fernández, E. (2013). *Learning ROS for robotics programming*. Birmingham, UK: Packt Pub.

Nichani, S., Silver, W., & Schatz, D. A. (2004). *Auto-setup of a video safety curtain system*. US Patent 6,829.371.

Nichani, S., Wolff, R., Silver, W., & Schatz, D. A. (2007). *Video safety detector with projected pattern*. US Patent 7,167,575.

Pilz GmbH & Co. KG (2015) *SafetyEYE Operating Manual*. Retrieved March 15, 2016, from https://www.pilz.com/download/open/PSENse_Operat_Man_21743-EN-18.pdf

Ramanath, R., Snyder, W. E., Yoo, Y., & Drew, M. S. (2005). Color image processing pipeline. *IEEE Signal Processing Magazine*, *22*(1), 34–43. doi:10.1109/MSP.2005.1407713

Ridley, J., & Pearce, D. (2006). *Safety with machinery* (2nd ed.). New York: Routledge.

Rockwell Automation, Inc. (Ed.). (2012). *GuardMaster SC300 hand detection safety sensor*. Retrieved March 15, 2016, from http://literature.rockwellautomation.com/idc/groups/literature/documents/pp/442l-pp001_-en-p.pdf

Sansoni, G., Trebeschi, M., & Docchio, F. (2009). State-of-the-art and applications of 3D imaging sensors in industry, cultural heritage, medicine, and criminal investigation. *Sensors (Basel, Switzerland)*, *9*(1), 568–601. doi:10.3390/s90100568 PMID:22389618

Saxena, A., Chung, S. H., & Ng, A. Y. (2008). 3-D depth reconstruction from a single still image. *International Journal of Computer Vision*, *76*(1), 53–69. doi:10.1007/s11263-007-0071-y

Schatz, D. A., Nichani, S., & Shillman, R. J. (2001). *Video safety curtain*. US Patent 6,297.844.

Schmidt, B., & Wang, L. (2014). Depth camera based collision avoidance via active robot control. *Journal of Manufacturing Systems*, *33*(4), 711–718. doi:10.1016/j.jmsy.2014.04.004

Schuon, S., Theobalt, C., Davis, J., & Thrun, S. (2008). High-quality scanning using time of-flight depth superresolution. In *Proceedings of IEEE Computer Society Conference on Computer Vision and Pattern Recognition Workshops, 2008 (CVPRW '08)* (vol. 1, pp. 1-7). Anchorage, AK: IEEE.

Sick, Inc. (Ed.). (2009). *Operating instructions. V200 Workstation Extended. V300 Workstation Extended.* Retrieved March 15, 2016, from https://www.sick.com/media/dox/3/63/763/Operating_ instructions_V200_Work_Station_Extended_V300_Work_Station_Extended_en_IM0026763.PDF

Tan, J. T. C., & Arai, T. (2011). Triple stereo vision system for safety monitoring of human-robot collaboration in cellular manufacturing. In *Proceedings of 2011 IEEE International Symposium on Assembly and Manufacturing (ISAM)* (vol. 1, pp. 1-6). Tampere, Finland: IEEE. doi:10.1109/ISAM.2011.5942335

Wöhler, C., Progscha, W., Krüger, L., Döttling, D., & Wendler, M. (2010). *Method and device for safeguarding a hazardous area.* US Patent 7,729,511.

Wuestefeld, M. (2007). *Optoelectronic Protection Device.* US Patent Application Publication US 2007/0280670.

ADDITIONAL READING

Channing, J. (Ed.). (2014). *Safety at work* (8th ed.). New York, USA: Routledge.

Koradecka, D. (Ed.). (2010). *Handbook of occupational safety and health.* Boca Raton, USA: CRC Press. doi:10.1201/EBK1439806845

Macdonald, D. M. (2004). *Practical machinery safety.* Oxford, UK: Newnes.

Rockwell Automation, Inc. (Ed.). (2011). *Safebook 4: Safety related control systems for machinery.* Retrieved March 15, 2016, from https://www.rockwellautomation.com/resources/downloads/ rockwell-automation/che/pdf/Safetyrelatedcontrolsystems_machinery.pdf

Sick, Inc. (Ed.). (2014). *Six steps to a safe machine.* Retrieved March 15, 2016, from https://mysick.com/saqqara/im0032606.pdf

KEY TERMS AND DEFINITIONS

Danger Zone: Any space within and/or around a piece of machinery in which a person can be exposed to risk of being harmed.

Fiducial (Marker): An object used in the field of view of an imaging system which is visualized in the image produced for use as a point of reference or a measure.

Harm: Physical injury or damage to health.

Hazard: Potential source of harm.

Machine Control System: Electrical means by which a set of machinery-related variable quantities is made to conform to a prescribed norm, either holding the values of the controlled quantities constant, or causing them to vary in a prescribed way.

Pipelining: A computing technique whereby the executions of computer operations are overlapped, leading to faster execution.

Risk: Severity and probability of occurrence of harm.

Stereopsis: Perception of depth and 3-D structure obtained on the basis of visual information deriving from two eyes or camera sensors with normally developed binocular vision.

Structured Light: An illumination method that uses a set of temporally encoded patterns which are sequentially projected onto a 3-dimensional scene, in order to provide a depth measurement of the illuminated objects from the camera viewpoint.

Chapter 10
A Study on Various Image Processing Techniques and Hardware Implementation Using Xilinx System Generator

Jyotsna Rani
National Institute of Technology Silchar, India

Abahan Sarkar
National Institute of Technology Silchar, India

Ram Kumar
National Institute of Technology Silchar, India

Fazal A. Talukdar
National Institute of Technology Silchar, India

ABSTRACT

This article reviews the various image processing techniques in MATLAB and also hardware implementation in FPGA using Xilinx system generator. Image processing can be termed as processing of images using mathematical operations by using various forms of signal processing techniques. The main aim of image processing is to extract important features from an image data and process it in a desired manner and to visually enhance or to statistically evaluate the desired aspect of the image. This article provides an insight into the various approaches of Digital Image processing techniques in Matlab. This article also provides an introduction to FPGA and also a step by step tutorial in handling Xilinx System Generator. The Xilinx System Generator tool is a new application in image processing and offers a friendly environment design for the processing. This tool support software simulation, but the most important is that can synthesize in FPGAs hardware, with the parallelism, robust and speed, this features are essentials in image processing. Implementation of these algorithms on a FPGA is having advantage of using large memory and embedded multipliers. Advances in FPGA technology with the development of sophisticated and efficient tools for modelling, simulation and synthesis have made FPGA a highly useful platform.

DOI: 10.4018/978-1-5225-1025-3.ch010

INTRODUCTION

Image is regarded as a relic that illustrates or records visual view point. Considering an example of a two dimensional picture, we find that it has a similar appearance to some subject, or a physical object or a person, these all provides an illustration of the specified image. Images contain numerous objects and patterns. These patterns contain valuable information required for medicine, biology, photography areas or domains. In analyzing of an image, inputs taken are images whereas outputs taken are either an image or a set of characteristics or parameters related to that of the image. The final motive of an image analysis is to acquire information from the gathered data through recognized or classified objects or classes or attributes and then take action accordingly to the acquired information.

BACKGROUNDS

Image is regarded as an artifact that depicts or records visual perception. Let us consider an example of a two dimensional picture, we find that it has a similar appearance to some subject, or a physical object or a person, these all provides a depiction of the specified image. Images contain plenty of objects and patterns. These patterns are actually attached with valuable information which are required for medicine, biology, photography areas or domains. In analyzing of an image, inputs taken are images whereas outputs taken are either an image or a set of characteristics or parameters related to that of the image. The final motive of an image analysis is to acquire information from the gathered data through recognized or classified objects or classes or attributes and then take action accordingly to the acquired information. For a digital image representation, an image may be represented as a two-dimensional function where x co-ordinate and y co-ordinate are spatial or planar co-ordinates, and the amplitude of the particular function at any pair of co-ordinates are regarded as the intensity of the image at that particular point.

A new descriptive name is introduced which is regarded as gray level which refers to the intensity of monochromatic images. In general, Color images are formed by a specified combinational mixture of individual images. For example, in the RGB color system, a color image is composed of three individual monochromatic images. They are denoted as the red (R), green (G) and blue (B) for primary images. An image may be continuous with respect to the spatial or planar co-ordinates. Its intensity may be continuous in terms of amplitude of the particular image. To convert such an image into a digitalized from we need to digitalize the co-ordinates, as well as the amplitude too. The process of getting the co-ordinate values digitalized are termed as sampling and the amplitude values is termed as quantization. Thus, when the spatial and planar co-ordinates along with the amplitude values of the particular two dimensional functions are all finite and discrete quantities, we regard the image as a digital image (Hasan et al., 2010).

In general a digital image can be categorized into four groups or types. They are:

- **Gray Scale Image:** A gray scale image is a two dimensional matrix barcode which encodes the raw received information. Its values are represented as the shades of gray having integer values ranging from [0, 255] or [0, 65535] respectively.
- **Binary Images:** A binary image is a digital image. It has only two possible values allotted to each and every specified pixel. The two colors generally taken into consideration for a binary image are black and white. But it is not compulsory to choose the two fixed colors. Any two colors can be taken into consideration in order to represent a binary image. The color used for the targeted

object in the image is the foreground color and the rest of the image is the background color. A binary image can be referred as a logical set of arrays consisting of 0s and 1s.

- **RGB Images:** RGB Images are the combination of Red (R), Green (G), and Blue (B) color in a specified proportion. The RGB color mode is regarded as an add-on color model in which the light of all the three colors are added together in every possible ways as per requirement to generate a broad array of colors.
- Indexed images (Gonzalez et al., 2002).

IMAGE PROCESSING

Image processing can be termed as processing of images using mathematical operations by using various forms of signal processing techniques. The input taken is an image, or a series of images, or a video, such as a photograph or video frame; the output of image processing may be either an image or a set of characteristics or parameters related to the image. Most image-processing techniques involve treating the image as a two-dimensional signal and applying standard signal-processing techniques to it. Images are also processed as three-dimensional signals where the third-dimension being time or the z-axis.

Closely related to image processing are computer graphics and computer vision. In computer graphics, images are manually made from physical models of objects, environments, and lighting, instead of being acquired (via imaging devices such as cameras) from natural scenes, as in most animated movies. Computer vision, on the other hand, is often considered high-level image processing out of which a machine/computer/software intends to decipher the physical contents of an image or a sequence of images (e.g., videos or 3D full-body magnetic resonance scans).

In modern sciences and technologies, images also gain much broader scopes due to the ever growing importance of scientific visualization (of often large-scale complex scientific/experimental data). Examples include microarray data in genetic research, or real-time multi-asset portfolio trading in finance.

Image Processing Toolbox Product Description

Image Processing Toolbox presents an extensive set of reference-standard methods, design fow, functions, and apps for image processing, reasoning, visualization, and algorithm development. We can perform various processing techniques such as:

- Image analysis,
- Image segmentation,
- Image enhancement,
- Noise reduction,
- Geometric transformations,
- Image registration and many other digital image processing techniques.

Moreover, many toolbox functions support multicore -processors, GPUs, and C-code generation. Image Processing Toolbox presents a varied set of image types, including:

- High dynamic range,
- Gigapixel resolution,
- Embedded ICC profile, and
- Tomographic.

Visualization functions and apps let you explore images and videos, examine a specific region of pixels, adjust color and contrast, create contours or histograms, and manipulate regions of interest (ROIs). The toolbox supports workflows for processing, displaying, and navigating large images.

Fundamental Steps Involved in Digital Image Processing

There are 6 fundamental steps which are involved in the overall method of digital image processing, as seen below in Figure 1.

Image Acquisition

Image acquisition can be simply described as procurement of an image from a valid source. In terms of image processing, it can be described as the action of retrieving an image from some source, usually a hardware-based source, so it can be passed through whatever processes need to occur afterward. Perform-

Figure 1. Fundamentals of image processing

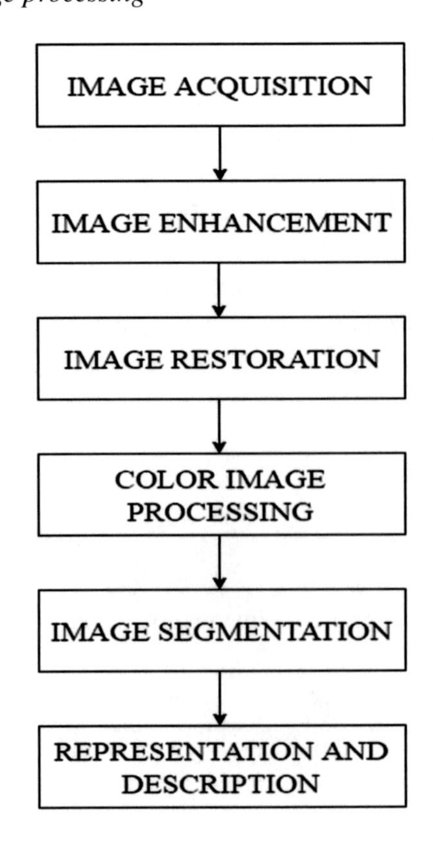

ing image acquisition in image processing is always the first step in the workflow sequence of digital image processing as we know without an image, no processing is possible. The image that is procured is completely unprocessed. This is the result of whatever hardware was used to generate it, which can be very important in some fields to have a proper and stable baseline from which to work. The main aim of this processing is to have a source of input that operates within a particular controlled and measured guidelines that the same image can if necessary, be almost perfectly reproduced under the same necessary conditions so abnormal factors are easier to trace, locate and be removed.

Image Enhancement

In terms of computer graphics, the process of enhancing the quality of a digitally stored image by manipulating the image with any required software is terms as the process of Image Enhancement. In short, it is the process of making an image lighter or darker, or to increase or decrease contrast. Various advanced image enhancement software supports many filters for manipulating images in various ways. Programs specialized for image enhancements are generally regarded as image editors.

Image Restoration

Image Restoration is defined as the process of taking a corrupt or a noisy image, thereby retrieving the clean, original image. Corruption may generate in many forms such as motion blur, noise and camera miss-focus. Image restoration is performed by reversing the process that blurred the image and such is performed by imaging a point source and use the point source image, which is called the Point Spread Function (PSF) to restore the image information lost to the blurring process. Image restoration is different from image enhancement in that the previous one is designed to emphasize features of the image that make the image more pleasing to the observer, but not necessarily to produce realistic data from a scientific point of view. Image enhancement techniques (like contrast stretching or de-blurring by a nearest neighbor procedure) provided by imaging packages use no a priori model of the process that created the image.

Color Image Processing and Compression

Image compression may be lossy or lossless. Lossless compression is preferred for archival purposes and often for medical imaging, technical drawings, clip art, or comics. Lossy compression methods, especially when used at low bit rates, introduce compression artifacts. Lossy methods are especially suitable for natural images such as photographs in applications where minor (sometimes imperceptible) loss of fidelity is acceptable to achieve a substantial reduction in bit rate. The lossy compression that producible differences may be called visually lossless.

Image Segmentation

Image segmentation is the process of partitioning a digital image into multiple segments (sets of pixels, also known as super pixels). The goal of segmentation is to simplify change the representation of an image into something that is more meaningful and easier to analyze. Image segmentation is typically used to locate objects and boundaries in images. More precisely, image segmentation is the process of assigning a label to every pixel in an image such that pixels with the same label share certain characteristics.

Representation and Description

Finally the Image obtained after processing is represented with proper description after processing.

IMAGE PROCESSING USING THE SYSTEM GENERATOR

FPGAs have demonstrated many benefits in applications involving large and independent data sets as wells as the acquisition of the digital data itself. One of the most vital and fundamental use of an FPGA is digital signal processing (DSP). DSPs are enforced using ASICs usually. Presently, many are now being replaced by FPGA's due to drastic decrease in cost and re-configurability. According to Moore's law, the number of transistors in a dense integrated circuit has doubled approximately in every two years. The potential use of FPGAs have also expanded as the chip density as increased accordingly as per Moore's law. The newest FPGA boards are equipped with a variety of I/O ports, AD/DA converters as well as soft or hard core microprocessors. These all have facilitated the device to act as a system on chip. Moreover, software and hardware co-designs are turning out to be increasingly popular as a solution to various other complex and in-depth computations using FPGA's (Hasan et al., 2010).

The main use of FPGAs are designing applications which require high speed as well as parallel data processing, generally in digital image processing. Image processing in terms of computer graphics is alteration of various pixels in various desired ways within a digital image to present in several requirements accordingly. Increase in application areas of signal processing in present era in accordance to growth of powerful and low-cost processing chips has led to dramatic growth of new applications such as multimedia delivery and hand-held communications delivery. Image processing which has a strong mathematical basis is one of the most important areas to be focused on. Image enhancement technique is widely implemented on many vital fields such as:

- Medical,
- Video surveillance,
- Various robotics applications, and
- Target recognition,

to increase the quality of the given image. FPGAs are increasingly used in modern imaging applications such as:

- Filtering,
- Medical imaging,
- Wireless communications and
- Image compressions,

due to the requirement to process the image in real time which less processing time and implementation in hardware level offering parallelism (Ownby et al.,2003). Digital image processing has varied applications in wide variety of fields of human endeavor. There are number of well-defined processes which go to make up a typical image application. Acquisition, Enhancement, Restoration, Segmentation and Analysis are the steps needed by just about every application which involves image processing (Bin Othman et al., 2010). Once images are inside the computer system, or more specifically, once they are read inside a program, the images are nothing but matrices hence, all the operations that can be applied to matrices should theoretically be applicable to the images as well. Image arithmetic is the implementation of standard arithmetic operations, such as addition, subtraction, multiplication, and division for images. Image arithmetic has many uses in image processing, both as a preliminary step in more complex operations and by itself.

DSP functions are implemented on two primary platforms such as Digital Signal Processors (DSPs) and FPGAs. FPGA is a form of highly manipulative hardware set whereas DSPs is a specialized form of microprocessors. Mostly engineers prefer FPGA over DSP due to its massive parallel processing capabilities inherent to FPGA and time to market make it the better choice. As mentioned FPGAs can be configured in hardware, FPGAs offer complete hardware Customization while used and implemented on various DSP applications. System Generator is a DSP design tool from Xilinx that enables the use of the MathWorks model-based design environment Simulink for FPGA design.

It is a system level modeling tool in which designs are encapsulated in the DSP friendly Simulink modeling environment using Xilinx specific Blockset al.l of the downstream FPGA implementation steps including synthesis and place and route are automatically performed to generate an FPGA programming file. System Generator provides many features such as System Resource Estimation to take full advantage of the FPGA resources, Hardware Co-Simulation and accelerated simulation through hardware in the loop co-simulation; which give many orders of simulation performance increase. This objective lead to the use of Xilinx System Generator (XSG), a tool with a high- level graphical interface under the Matlab, Simulink based blocks which makes it very easy to handle with respect to other software for hardware description (Harinarayan et al., 2011). Rout et al. (2013) presented the information regarding FPGA implementation for various image processing algorithm using the most efficient tool called Xilinx System Generator for Matlab. Gupta et al. (2015) reported the conceptual description of hardware and software simulation for image processing using Xilinx System Generator (XSG).

Xilinx System Generator

System Generator is a DSP design tool from Xilinx that facilitates the use of the MathWorks model-based Simulink design environment for FPGA design. The design tools enable the design processes by obscuring the technical knowledge necessary for FPGA a Register Transfer Level (RTL) design. Xilinx System Generator.

As a substitute, a design is created using the perceptive visual environment within Simulink that uses various specific block sets expedite the development. Moreover, System Generator can help in performing various the FPGA implementation steps such as synthesis, mapping, and place and route to generate the FPGA executable file.

The Gateway In and Gateway Out blocks in FPGA define the input and output from the FPGAs block set. Outside of this region, Simulink are entitled to send customized signals and data into the simulated FPGA to scrutinize the results. The System Generator block determines the type of FPGA board to be used, moreover providing several additional operations such as clock speed, compilation type and analysis.

Comprising of a vast library function set of over 90 DSP building blocks, System Generator provides faster prototyping and design from a high-level programming stand point. Some blocks such as the M-code and Black box provides a platform for direct programming in MATLAB M-code, C code, and Verilog to simplify integration with existing projects or customized block behavior. System Generator projects can also easily be placed directly onto the FPGA as an executable bit stream file as well as generating Verilog code for additional optimizations or integration with existing projects within the Xilinx ISE environment.

METHODOLOGY OF PROPOSED HARDWARE IMPLEMENTATION OF IMAGE PROCESSING METHOD

The image processing method needs to be carried out in hardware concerning to meet the real time applications. FPGA implementation can be achieved using prototyping environment using Matlab/Simulink and Xilinx System Generator tool. The input in Simulink block set is given as image source and image viewer and output image can be viewed on image viewer block set. Image pre-processing and image post-processing units are common for all the image processing applications which are implemented using Simulink Blocksets.

Image Pre-Processing Unit

Image pre-processing in MATLAB helps in providing input to FPGA. It acts as a specific test vector array which is applicable for FPGA Bit stream compilation using system generator.

- **Resize:** Set Input dimensions for an image and interpolation i.e. bicubic it helps in preserving fine detail in an image.
- **Convert 2-D to 1-D:** Converts the image into single array of pixels.
- **Frame Conversion and Buffer:** It incorporates setting sampling mode along with buffering of data.

The model based design used for image pre-processing is shown as follows in Figure 2 and Figure 3:

Image Post-Processing Unit

Image post processing provides recreating image from 1D array. Post-processing unit uses the following steps (Figure 4 and Figure 5).

- **Data Type Conversion:** It transforms image signal into unsigned integer format.
- **Buffer:** Transforms scalar samples to frame output at a much lower sampling rate.
- **Convert 1D to 2D:** Convert 1D image signal to 2D image matrix.
- **Video Viewer:** It provides with the display of output image back on the monitor.

Figure 2. Pre-processing block flow chart

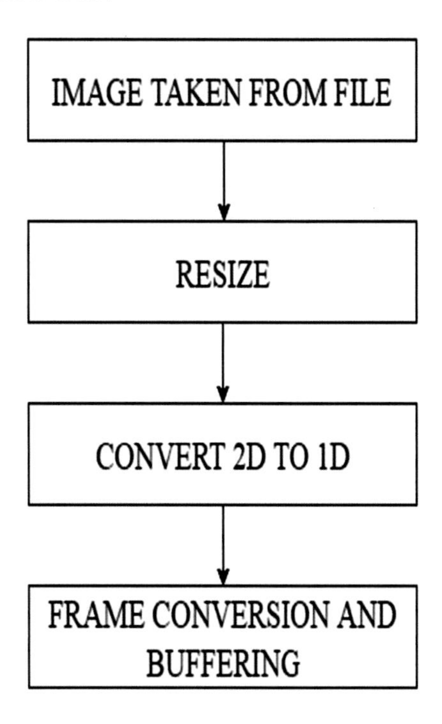

Figure 3. Pre-processing block

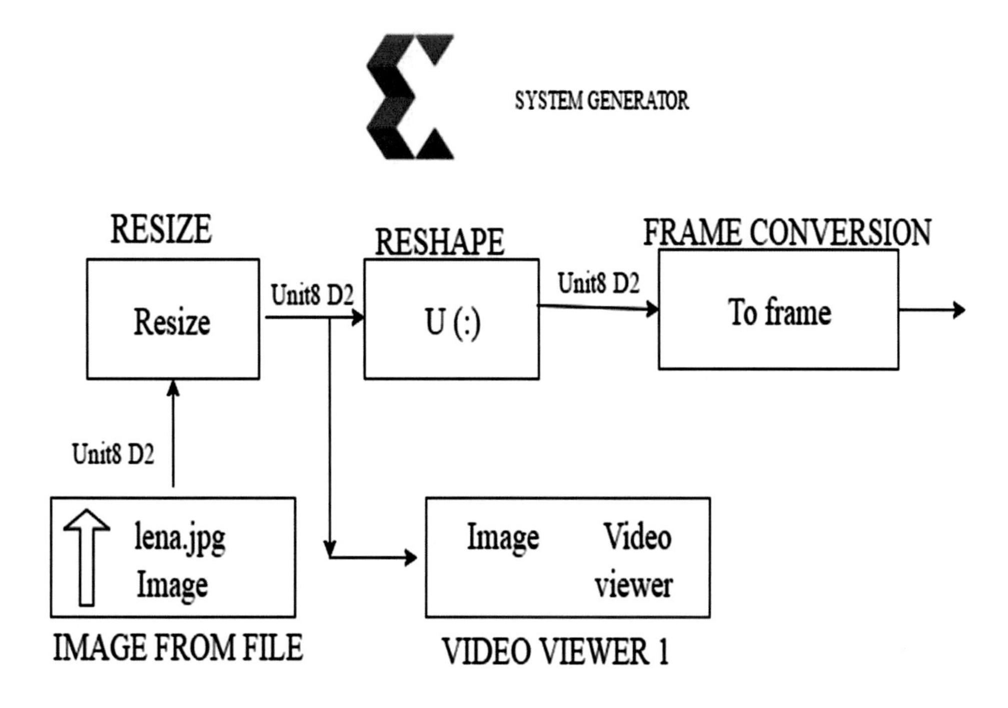

Figure 4. Post-processing block flow chart

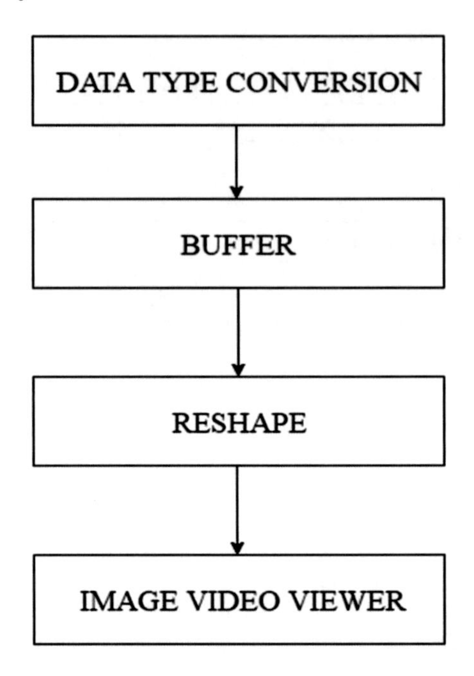

Figure 5. Post-processing block

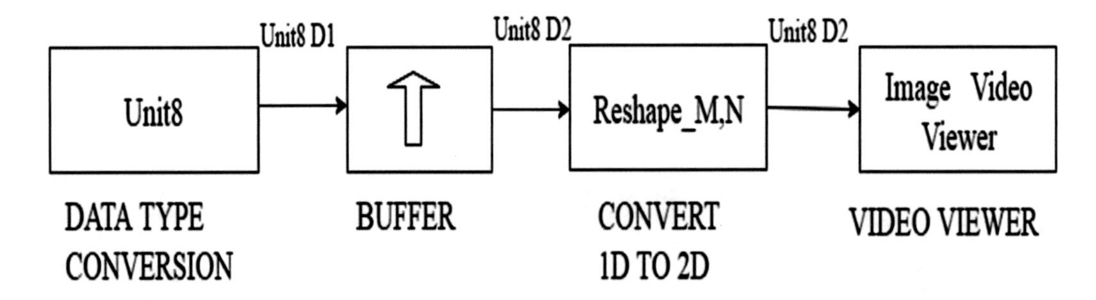

XILINX MODELS FOR OPERATIONS OF IMAGE PROCESSING

The development of various required models are based on algorithms used for Image Processing. Once the FPGA boundaries have been constructed using the Gateway blocks, the DSP design set can be implemented using blocks from the Xilinx DSP block set.

Various Algorithms for Negative Image

Various Algorithms to obtain negative images has been implemented using different operations and outputs are obtained accordingly.

Image Negative Using XOR Operation

The Exclusive OR function sets bits that are the same in each operand to 0 and bits that are different in operands to 1. All pixels of a certain value can be found by applying XOR function. A negative image is produced by XOR with 255 to the image. Figure 6 and Figure 7, given below, show the results of an image with XSG bloc.

Image Negative Using NOT Operation

To obtain the negative of an image, we have to set the values in the image matrix to double precision or we can say to invert all pixels of matrix. This is accomplished in order to invert the color matrix so we can achieve this by using the gate NOT which is available in Xilinx system generator. Figure 8 and Figure 9 show the results of an image with XSG blocks.

Figure 6. Algorithm for XOR operation

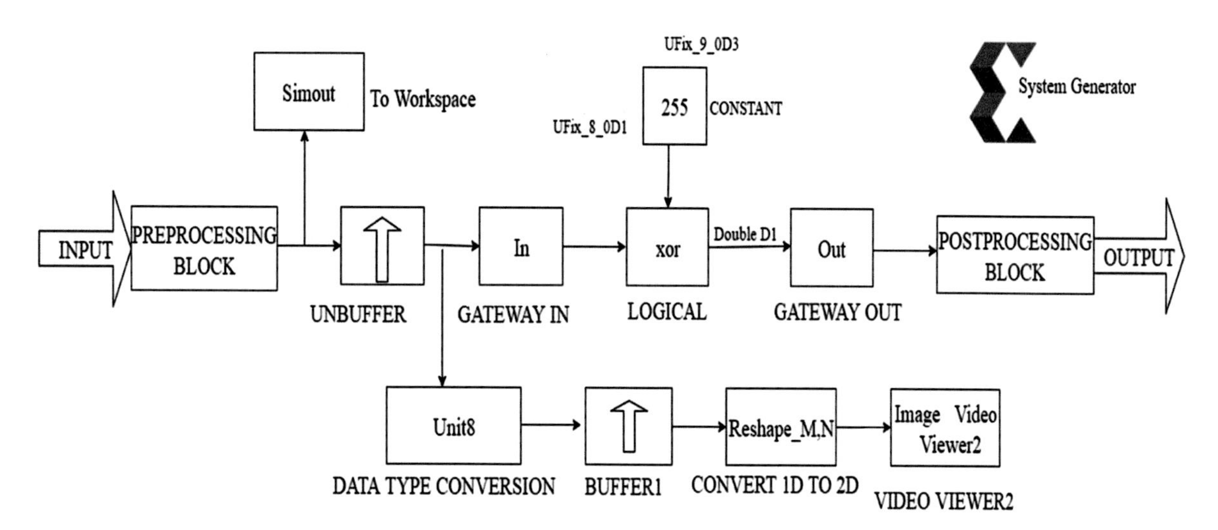

Figure 7. Original image and the negative image obtained as output

Figure 8. Algorithm for NOT operation

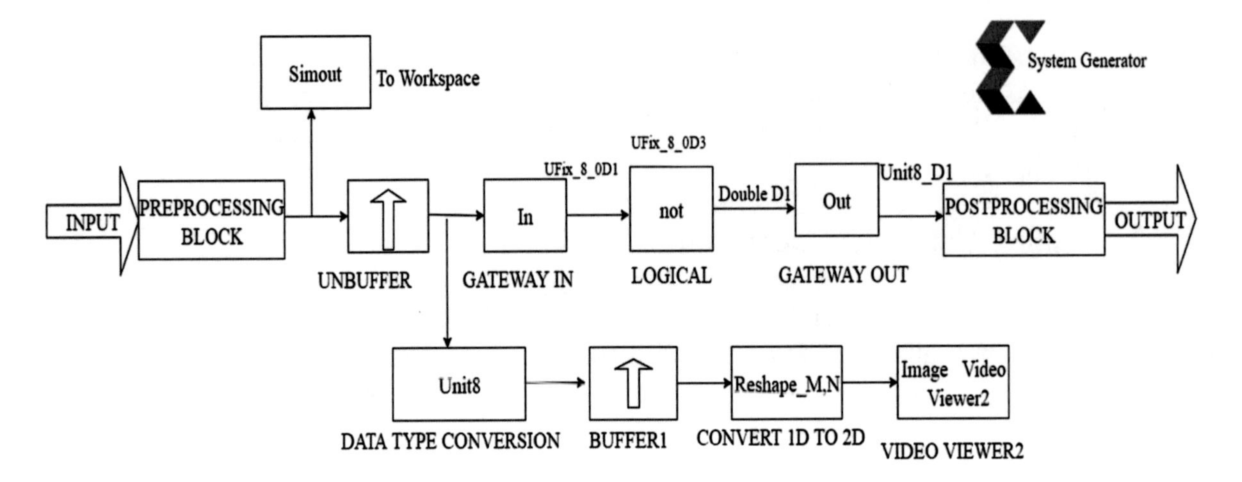

Figure 9. Original image and the negative image obtained as output

Negative Using Add Sub Block

The negative transform exchanges dark values for light values and vice versa accordingly (See Figures 10 and 11).

Algorithm for Image Enhancement

In image enhancement technique, we have demonstrated how an image can be enhanced by adding a constant to each pixel values. Image filtering can also be accomplished using model based design different filtering architecture can be defined and Xilinx block can be created. Figures 12 and 13 show the simulation model along with its result.

Figure 10. Algorithm used for Addition-Subtraction Block

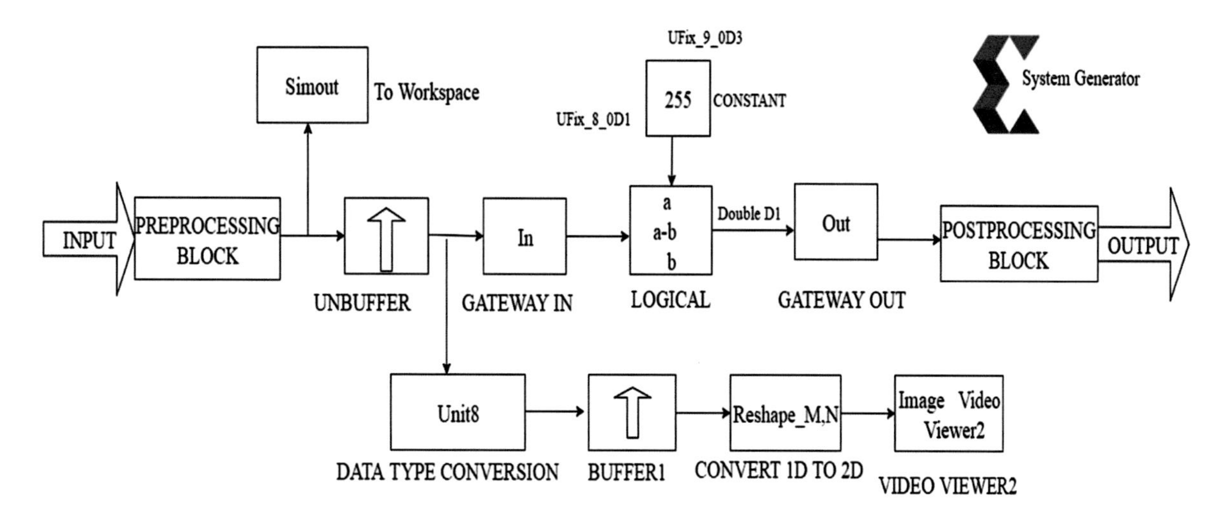

Figure 11. Original image and the negative image obtained as output

Figure 12. Algorithm for Image Enhancement Operation

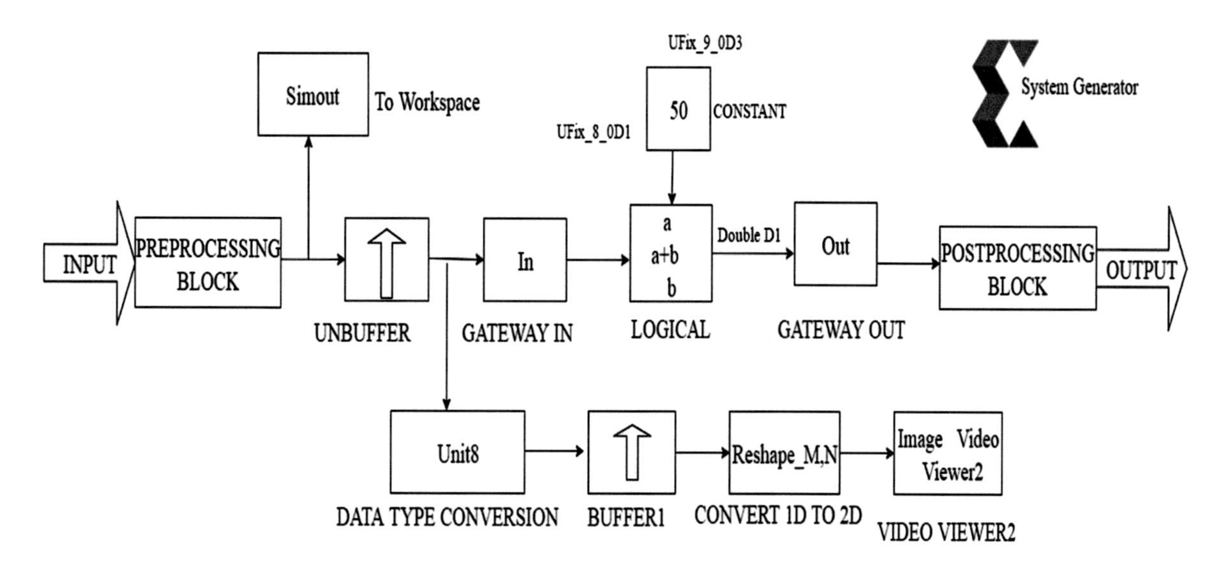

Figure 13. Original image and the obtained output

HARDWARE CO-SIMULATION

Now once the hardware board is installed, the starting point for hardware co-simulation is the System Generator model or subsystem we would like to run in hardware. A model can be co-simulated, given it provides all the requirements of the underlying hardware board. This model must include a System Generator token which defines the process of compilation into hardware.

Choosing a Compilation Target

This is done following these steps:

- **Part:** Defines the FPGA part to be used.
- **Synthesis Tool:** Specifies the tool in order to synthesize the design.
- **Hardware Description Language:** Specifies the HDL language to be used for compilation commonly called as Verilog.
- **Create Test Bench:** Instructs System Generator to generate a HDL test bench.
- Design is synthesized and implemented.

Selecting a Clocking Tab

- **FPGA Clock Period (ns):** Defines the period in nanoseconds of the system clock
- **Clock Pin Location:** Defines the pin location for the hardware clock.
- The code generator is entreating by pressing the Generate button in the System Generator token dialog box.

CONCLUSION

Digital Image Processing can be termed as a pictorial representation for human interpretation and analysis. It involves processing an altering a digital image in a desired manner and then obtaining the image in a readable format. Digital image processing are very important in various fields such as document imaging, signature verification, biometrics, face detection and recognition and various other fields. The Xilinx System Generator tool is a latest application in the field of digital image processing. It offers a friendly environment design for the processing where the units are designed by blocks. This tool support software simulation, but the most important is that can synthesize in FPGAs hardware which has many advanced features such as parallelism, robust and speed which are very essential in the field of digital image processing. Advantage of FPGA is that it has large memory and embedded multipliers. Advances in FPGA technology with the development of sophisticated and efficient tools for modelling, simulation and synthesis have made FPGA a highly vital and friendly platform. This article provides an introductory overview to FPGA and along with a step by step tutorial to the beginners in handling Xilinx System Generator.

ACKNOWLEDGMENT

We acknowledge NIT Silchar which provides such a good platform to work in this area.

REFERENCES

Gonzalez, R. C., & Woods, R. E. (2002). *Digital image processing* (2nd ed.). Upper Saddle River, NJ: Prentice hall.

Gupta, A., Vaishnav, H., & Garg, H. (2015). Image processing using Xilinx System Generator (XSG) in FPGA. *IJRSI, 2*(9).

Harinarayan, R., Pannerselvam, R., Ali, M. M., & Tripathi, D. K. (2011, March). Feature extraction of Digital Aerial Images by FPGA based implementation of edge detection algorithms. In *Emerging Trends in Electrical and Computer Technology (ICETECT), 2011 International Conference on* (pp. 631-635). IEEE. doi:10.1109/ICETECT.2011.5760194

Hasan, S., Yakovlev, A., & Boussakta, S. (2010, July). Performance efficient FPGA implementation of parallel 2-D MRI image filtering algorithms using Xilinx system generator. In *Communication Systems Networks and Digital Signal Processing (CSNDSP), 2010 7th International Symposium on* (pp. 765-769). IEEE.

Othman, D. M. F. B., Abdullah, N., & Rusli, N. A. B. A. (2010, October). An overview of MRI brain classification using FPGA implementation. In *Industrial Electronics & Applications (ISIEA), 2010 IEEE Symposium on* (pp. 623-628). IEEE. doi:10.1109/ISIEA.2010.5679389

Ownby, M., & Mahmoud, W. H. (2003, March). A design methodology for implementing DSP with Xilinx System Generator for Matlab. In Southeastern Symposium on System Theory (Vol. 35, pp. 404-408).

Raut, N. P., & Gokhale, A. V. (2013). FPGA implementation for image processing algorithms using xilinx system generator. *IOSR Journal of VLSI and Signal Processing, 2*(4).

KEY TERMS AND DEFINITIONS

Field Programmable Gate Array (FPGA): Field Programmable Gate Arrays (FPGAs) are semiconductor devices that are based around a matrix of configurable logic blocks (CLBs) connected by programmable interconnects.

Image Processing: Refers to the processing of images using some specific mathematical operations or some form of signal processing for which the input is an image and the obtained output of image processing may be either an image or a set of characteristics or parameters which refers to the image.

Xilinx System Generator: System Generator is a Digital Signal Processor's (DSP's) design tool from Xilinx that enables the use of the MathWorks model-based design environment Simulink for FPGA design.

Chapter 11
New Redundant Manipulator Robot with Six Degrees of Freedom Controlled with Visual Feedback

Claudio Urrea
Universidad de Santiago de Chile, Chile

Héctor Araya
Universidad de Santiago de Chile, Chile

ABSTRACT

The design and implementation stages of a redundant robotized manipulator with six Degrees Of Freedom (DOF), controlled with visual feedback by means of computational software, is presented. The various disciplines involved in the design and implementation of the manipulator robot are highlighted in their electric as well as mechanical aspects. Then, the kinematics equations that govern the position and orientation of each link of the manipulator robot are determined. The programming of an artificial vision system and of an interface that control the manipulator robot is designed and implemented. Likewise, the type of position control applied to each joint is explained, making a distinction according to the assigned task. Finally, functional mechanical and electric tests to validate the correct operation of each of the systems of the manipulator robot and the whole robotized system are carried out.

INTRODUCTION

Currently, robotics is possibly immersed in all industrial fields because of the countless applications that it can be given (Siqueira & Terra, 2011; Ben-Gharbia et al., 2014). These systems have improved the productivity and quality of manufactured products and their use has extended from the automobile industry (General Motors: Unimate in 1960) (Hunt, 1983) to the aerospace industry (National Aeronautics and Space Administration: Curiosity in 2012) (Kaufman, 2012).

Regarding Degrees Of Freedom (DOF), robots are classified in two categories:

DOI: 10.4018/978-1-5225-1025-3.ch011

- Redundant and
- Non-redundant.

The robot developed in this study corresponds to a redundant robot, since it possess a high number of DOFs (equal or superior to 6). Industrial robots, which are commercialized globally, are non-redundant and built with enough DOFs as to carry out specific tasks within their physical limitations. Having different types of manipulator robots focused on only one activity is impractical and expensive. In addition, the extensive range of applications has therefore required flexibilizing the work space of the robots as well as improving their position, accuracy and orientation. These characteristics can be achieved by increasing their degrees of freedom, i.e., providing them with redundancy.

However, all these activities would not be possible without an adequate design of the robot and of its technical control. Fulfilling this requires the knowledge and study of a mathematical model and of a certain class of "intelligence" that can direct the manipulator robot to perform the assigned tasks. Specifically, the control of the manipulator robot is based on an artificial vision system, but the technique used is not the same for all the joints. In the case of the first three rotational joints, this visual feedback system filters images and is capable of determining the position of a laser beam that is placed in the manipulator robot's end-effector. In the case of prismatic joints, position is controlled by limit switches, while for the rest of the joints there is no position feedback, but a functioning routine with sequential movements. Image processing is key for achieving a good control of a manipulator robot. For this purpose, there is a variety of techniques, among which morphological operations and filtering may be underlined.

Using the basic laws of physics that govern the robot's dynamics, it is possible to derive a mathematical model that represents its behavior, and through appropriate programming tools, develop an environmental simulation to subject it to different tests such as, for example, following trajectories (Angulo & Avilés, 1989; Selig, 1992; Iñigo & Vidal, 2004; Torres et al., 2002; Craig, 2006).

On the other hand, there are many kinds and configurations of robots with different Degrees Of Freedom (DOF) (Siciliano & Khatib, 2008; Urrea & Coltters, 2015; Urrea & Kern, 2016).

To make the design and implementation of a Selective Compliant Assembly Robot Arm (SCARA) type manipulator robot with six DOF and an RRRPRP (R: Revolution and P: Prismatic) with visual feedback, and in addition get this robotized system to develop correctly an application in real time, it is necessary to be conversant with several subjects, among them robotics fundamentals, servomotor and microcontroller operation, the use of artificial vision, computer programming, systems control, etc.

RELATED WORK

The applications of the manipulator robots are varied and flexible, even more when the extension of the environment or the location of the elements that the robot should manipulate are uncertain. Occasionally, tasks are undermined by other factors, such as dark environments or bad weather. In contrast to other studies, the visual feedback system used does not comprise obstacle avoidance, nor does integrate a system that helps to counteract the effects of ambient noise. The main difference is that it has been implemented in a redundant robotic design. Most research works with non-redundant robots, which restricts movement and results in a decrease in the volume of work.

Control of manipulator robots based on vision is widely used in a number of situations. However, many questions remain unsolved regarding the control of this type of manipulator robots in real time. For example, image processing is not exact, image recognition is not fast and the manipulator robot's motion is not smooth. These problems negatively affect the performance of the robot. In view of these inconveniences, some works present analyses of several algorithms for manipulator robot's processing, which in its control system are based on artificial vision. Some improvements in the visual manipulator robot control system are shown in order to obtain better effective images and a faster response in the real system's response time (Guo, Ju, & Yao, 2009).

Likewise, in some cases the manipulator robots need to operate in complex environments, characterized by noise and/or bad weather, which impedes them to reach uncertain locations. Wang et al.'s (2008) study proposes a RRRRRP configuration technique for a 6-DOF manipulator robot, which integrates binocular and stereo vision to help scanning the objective and environment.

The operation of nuclear power plants is regulated by very strict safety standards. In a regular inspection cycle of the current in steam generators, approximately 100 out of 6000 tubes need to be tested and, thus, this may lead to omissions and identification errors in the collected data. Therefore, Birgmajer et al. (2006) developed an artificial vision system to guide a manipulator robot directly to the tube network. Likewise, this system supervises the robot movements on the tube network and verifies the correct positioning of the current probes inside the desired tube. Of course, this eliminates the need of manual operation and reduces probe wearing.

SERVOMOTORS

Servomotors are electric motors that possess the capacity of being controlled for both speed and position. Their motor has a speed reducer, a torque multiplier and a circuit that controls the system. The rotation angle is 180° commonly, but it can be modified so as to the rotation is 360°.

To control a servomotor, a specific frequency and duration pulse should be sent. All servos have three cables, two for power (Vcc and Gnd), and one for sending the servo control pulse train, which makes the internal differential circuit place the servo axis in the indicated position, depending on the pulse width.

Servomotors are divided in analog and digital. Digital servomotors possess, as well as analog servos, a potentiometer for position feedback and a control electronics embedded in the servo. The difference between these two types lies in the control board, which, in the case of digital servomotors, has a microprocessor in charge of analyzing the signal, processing it and controlling the motor. This enables reducing the response dead zone, allowing for faster responses, e.g. for a same time lag, a digital servomotor can receive 5 or 6 times more pulses than an analog servomotor.

MICROCONTROLLERS

A microcontroller is an integrated circuit that includes three functional units, namely a Microprocessor (CPU), Memory and I/O (Input/Output) Units, and that allows carrying out logical processes. These processes are programed within the microcontroller in a particular language, and are introduced in it by a programmer.

Figure 1. Comparison between the quantity of pulses that an analog servomotor and a digital servomotor can receive

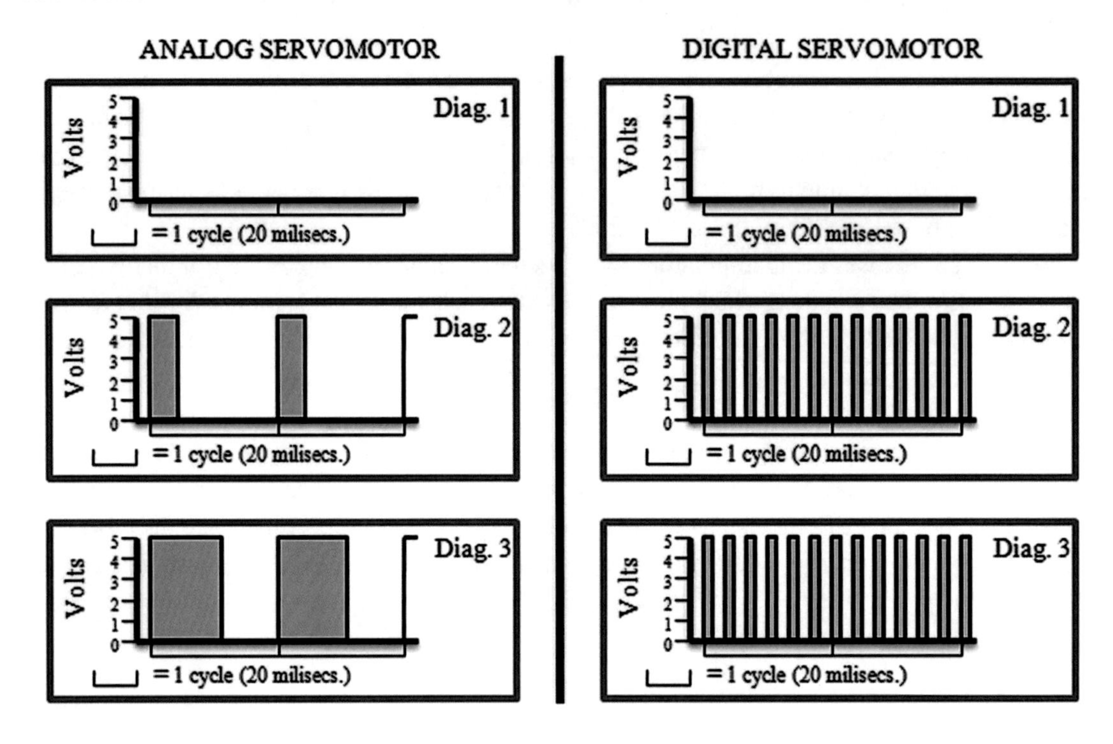

Since microcontrollers are programmable, they can be used in all types of projects, due to its wide variety of features such as:

- Analog or digital input and output ports,
- Serial communication ports,
- A/D converters,
- Timers,
- Meters,
- Comparators,
- Synchronous serial port,
- Different types of memory (ram, rom, eprom, eeprom, flash).

This versatility makes microcontrollers useful tools for electronics in any field, particularly in robotics, because allows one to obtain the pulse trains necessary to operate servomotors, In addition, since it has serial communication ports, sending information to and from a PC serial port becomes easy.

ARTIFICIAL VISION

As a tool, artificial vision allows establishing a relation between the three-dimensional world and bidimensional views taken from some device that permits that artificial vision. Thanks to this, a reconstruction of a three-dimensional space can be made –from its views– and on the other hand, -for a desired position- carry out a representation of a projection of a three-dimensional scene on a two-dimensional surface. Also, artificial vision can be used to get more detailed information on the images by digitizing them. At present, digital image processing is widely used and the areas in which it is applied are so varied that it is convenient to group them according to their origin. The main energy source of real images corresponds to the electromagnetic energy spectrum (Figure 2). The spectral bands are grouped by photons, depending on the energy level, in this way getting a spectrum that goes from gamma rays (high energy) to radio waves (low energy). Bands are shaded to convey that electromagnetic energy spectrum bands are not different, but that move smoothly from one to another.

To illustrate this, some uses of digital image processing are described according to the band used.

Gamma-Ray Band

The most extended uses for gamma rays are nuclear medicine and astronomical observations.

X-Ray Band

X-rays are among the oldest sources of electromagnetic energy used for images. Their most widely known use is medical diagnosis, but they are also used in astronomy and in the circuit boards for the failure detection industry. Another medical application of X-rays is angiography.

Ultraviolet-Ray Band

The applications of ultraviolet light are varied. They include lithography, industrial inspection, microscopy, lasers, biological imaging, and astronomical observation.

Visible and Infrared Band

Visible band uses are by far the most common applications. Infrared band is used in conjunction with visible bands and, thus, they are in the same group. Their uses range from pharmaceutical and micro-inspection to material characterization, astronomy, remote sensing, industrial and law enforcement uses.

Figure 2. Electromagnetic energy spectrum

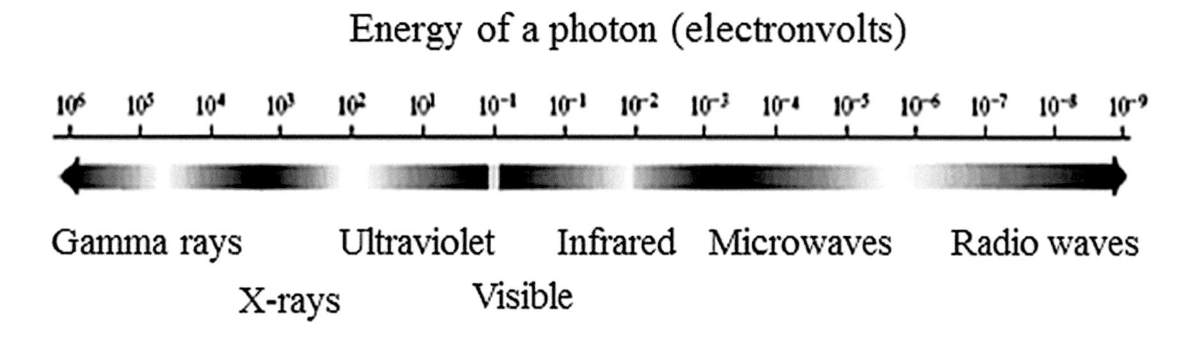

Microwave Band

The principal application of the microwave band is the radar, although this band has other uses such as fingerprints and notes verification and automatic number plate recognition.

Radio Wave Band

As in the case of gamma rays, the most widespread application of radio wave bands is in the medical and astronomical field. In medicine, these bands are used to obtain images by magnetic resonance.

The visible band was chosen for this work, because the images are acquired by a WebCam with a resolution of 640x480 pixels.

IMAGE PROCESSING

Image processing is a technique used with the aim of improving the aspect or the quality of an image, in order to enable the information search. The images must be processed according to the application in which they will be used. There are many image processing techniques, among which filtering and morphological operations should be mentioned.

Filters

The filtering process is based on a set of techniques and consists in getting, from an initial image, a final image that is more adequate for a specific application. Some of the main objectives of the use of filters are:

- To eliminate noise. Exclude those pixels whose intensity level is too different from that of its neighbors and whose origin can be in the acquisition as well as in the transmission of the image.
- To highlight the edges. Give prominence to the edges of the figures in an image.
- To detect edges. Identify pixels where sudden changes take place in the intensity function.
- To detect objects. Identify the shape of an object that has a particular color in the image.
- To smooth the image. Reduce the number of intensity variations between neighboring pixels.

Therefore, the filters are considered as operators, at the pixel level, of a digital image. The filtering process can be performed either in the space or in the frequency domain.

Morphological Operations

Morphological operations are based on set theory operations. They simplify images conserving the main shape characteristics of the objects, and they also extract the useful components in the representation and description of the shape of the regions. They are used mainly to:

- Preprocess images (noise suppression, shape simplification, etc.).
- Highlight the structure of the objects (extracting skeleton, detecting objects, convex hull, enlarge, reduce, etc.).

- Describe objects (area, perimeter, etc.).

The fundamental morphological operations are dilation and erosion; starting from them, the opening and closure operations are composed, which are closely related to the representation of shapes, the decomposition and extraction of primitives.

Image Acquisition and Representation

Visual processing begins with image acquisition using a capture device. There are numerous capture devices depending on the type of lighting, sensors and image scanning mechanisms. The most popular devices are those based on E Charge Coupled Devices (CCD) and with Complementary Metal-Oxide-Semiconductor (CMOS) chips. CCDs are chips that integrate a photo-detector matrix. Each photo-detector (photosite) corresponds to a pixel. When a photon reaches the semiconductor, the latter releases electrons and each cell acts as a well that accumulates the released electrons and, thus, the number of electrons is proportional to the light intensity received by the semiconductor.

Another capture device is the *Active Pixel Sensor* (APS), which is based on the CMOS technology that, in turn, depends on silicon semiconductors and functions in a similar way. The difference with CCD is that, in APS, each pixel incorporates an electric signal amplifier and an A/D converter, which allows pixels to be simultaneously exposed and read. Other advantages of APS are lower electricity consumption, simultaneous reading of a larger number of pixels, scarce or inexistent "Blooming", greater flexibility in reading other construction topologies, different types of combinable pixels, and high frequency of

Figure 3. Image of the composition of a CCD sensor

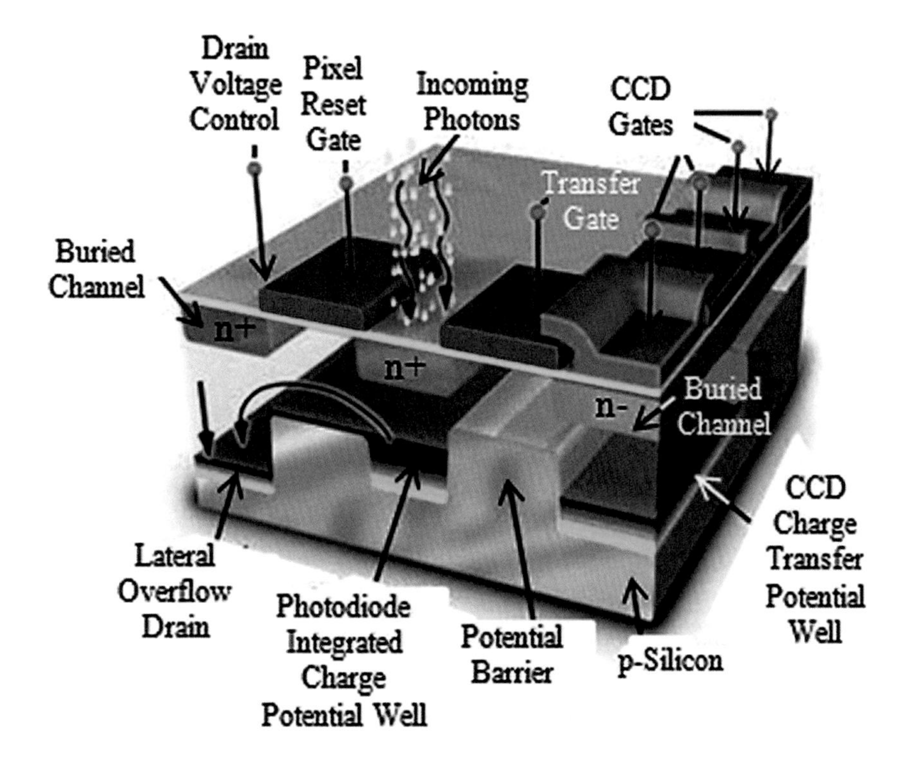

images. However, some of the APS disadvantages are a smaller light receiving surface per pixel and less pixel uniformity due to higher Fixed Pattern Noise (FPN) (Ohta, 2007).

Both devices use independent color detectors, so a red, green and blue microfilter is used in a typical arrangement (Bayer filter) on the photosites. Each square formed by four pixels has a red, a blue and two green filters, since the human eye is more sensitive to green color.

Many others capture devices are more used on non-luminous images —e.g. ultrasound or magnetic resonance images— than the abovementioned.

Once an image has been captured with the device, it must be digitized to process it computationally. This task is currently performed by the same capture devices, which include image digitization. When

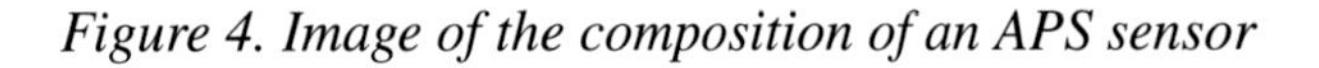

Figure 4. Image of the composition of an APS sensor

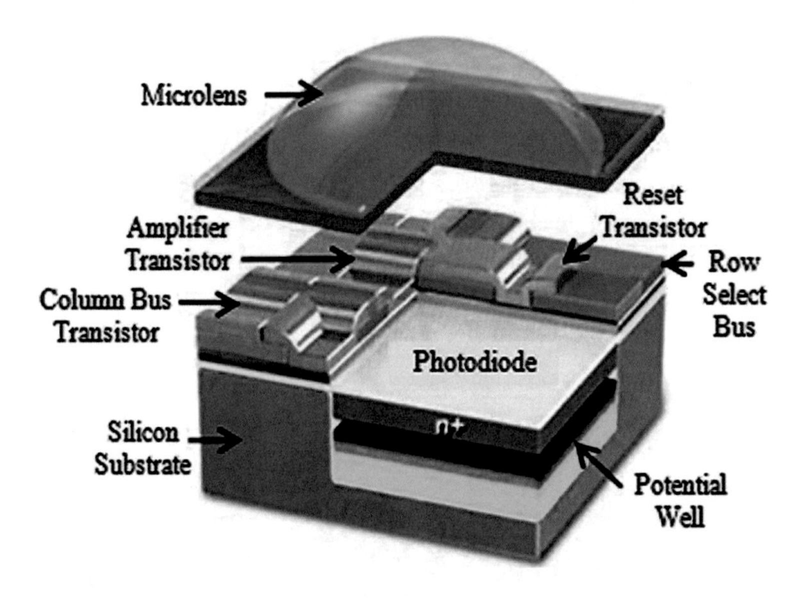

Figure 5. Typical microfilter arrangement, called Bayer mosaic

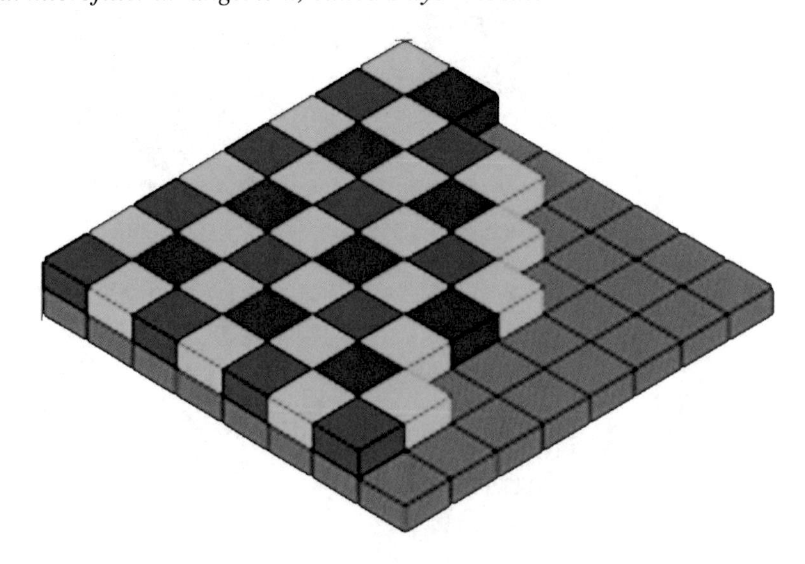

the image has been already digitized, is important to know its storage format. There are various storage formats for the images, such as BMP, TIFF, GIF, JPEG, etc. The differences between them lie in the depth levels admitted and the kind of compression, among other properties. In the depth levels there are differences that define the kind of image. For example, if the information per pixel is on the order of 1 bit, then we are dealing with a black-and-white image, if the information per pixel is 1 byte, we are dealing with an image in a gray scale or with a palette of 256 colors. Also, if the information per pixel is 3 bytes, then we have an RGB image, which has three information bands per pixel: one for red (R), one for green (G), and one for blue (B).

BASICS OF INDUSTRIAL ROBOTICS APPLIED TO THE ROBOTIZED MANIPULATOR

The manipulator robots' design is divided in two categories: general purpose and exclusive work. General purpose manipulator robots, particularly, possess 6 or more DOF, which are necessary to allow the free positioning and placing of the end of the robot (end effector), making the robot capable of undertaking any kind of work. Although manipulator robots with more than 7-DOF are able to perform tasks that are more complex. It is also true that, depending of the task provided, their structures may generate control problems due to the risk of collisions. Whereas five or less Degrees Of Freedom manipulator robots are of the exclusive work type, i.e. designed with specific structures to execute particular tasks, which constrain the posture and work volume of these robots. For these reasons, the study of manipulator robots —and particularly of 6-DOF manipulator robots— becomes a practical and essential issue.

The manipulator robot that is designed presents a type RRRPRP configuration, i.e., the first three DOF are revolution (R), then it has a prismatic joint (P), and then the end-effector has a gripper with a wrist that can turn (R) and open or close (P). This particular configuration allows the manipulator robot to work on the x-y plane with mechanical redundancy. The kinematic study of this robot is now presented.

Solution of the Direct Kinematic Problem of the Robotized Manipulator

In what follows only the first five DOF that have an incidence on the calculation of the robot's direct kinematics are considered (Urrea & Kern, 2016). Therefore, the gripper's opening/closing prismatic joint is not considered. The assignment of the standard Denavit-Hartenberg coordinate system is specified in Figure 6.

Finally, the homogeneous transformation matrix obtained is:

$$ {}^{0}T_{5} = \begin{bmatrix} c_{123-5} & s_{123-5} & 0 & l_1c_1 + l_2c_{12} + l_3c_{123} \\ s_{123-5} & -c_{123-5} & 0 & l_1s_1 + l_2s_{12} + l_3s_{123} \\ 0 & 0 & -1 & l_0 - d_4 - l_5 \\ 0 & 0 & 0 & 1 \end{bmatrix} \quad (1) $$

where:

Figure 6. Side view of the manipulator robot

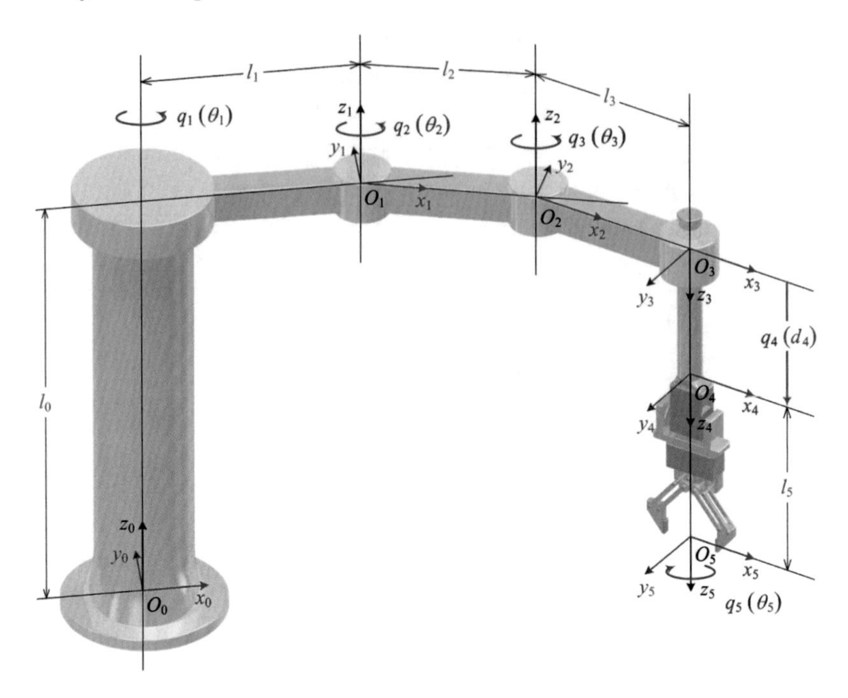

$$\sin\theta_1 = s_1\ ,\quad \sin(\theta_1+\theta_2) = s_{12}\ ,\quad \sin(\theta_1+\theta_2+\theta_3) = s_{123}\ ,\quad \sin(\theta_1+\theta_2+\theta_3-\theta_5) = s_{123-5}$$

$$\cos\theta_1 = c_1\ ,\quad \cos(\theta_1+\theta_2) = c_{12}\ ,\quad \cos(\theta_1+\theta_2+\theta_3) = c_{123}\ ,\quad \cos(\theta_1+\theta_2+\theta_3-\theta_5) = c_{123-5}$$

l_0: Length of the fixed base.
l_1: Length of the first link.
l_2: Length of the second link.
l_3: Length of the third link.
d_4: Variable length of the prismatic joint.
l_5: Length of the fifth link.

DESIGN OF THE ROBOTIZED MANIPULATOR

The robot presented here has fundamental differences with respect to a classical SCARA robot; among others, there is a kinematic redundancy because it has an additional link to work on the *x-y* Cartesian plane, allowing it to be tolerant to some mechanical faults. Moreover, to carry out at every instant the position and orientation feedback of its links, use is made of an artificial vision system. This allows the robot to locate with no difficulty the objects that will be manipulated in its work space. For the prismatic joint, a discrete position control is designed, and range of motion limits are included to ensure stopping electromechanically the movement at both ends of its displacement. In the open kinematic chain of the manipulator robot, after the prismatic joint there is a gripper that has rotational motion, and also a prismatic joint that corresponds to the opening/closing of the gripper (Figure 8a).

In the manipulator robot's physical design, it is established that the base should be a 50 cm high, 10 cm diameter and 3 or 4 mm of thickness cylindrical metal tube, since it has to support the weight of all the links and joints. In addition, its lower part needs a flange to be fixed to a supporting surface.

Then, the distance between the joints' axes should be 15 cm, and, therefore, the lengths of the links may vary up to 15 cm. Furthermore, they have to be made of a 50-mm-per-side and 2.5-mm-thickness squared aluminum tube in order to bear the weight and the inertia of the movements.

One of the most important elements in the mechanical design of a manipulator robot are joints. Prismatic joints were devised to move the end-effector about 15cm away from the highest position, about 23 cm away from the work surface, to the lowest position, 7cm away from the work surface. This joint is designed in such a way that the vertical movement is carried out using a rack and pinion connected to the servomotor axis, which transforms rotational motion into linear motion. The end-effector is composed of the gripper and a rotational joint in the wrist, thus, possessing two degrees of freedom, wrist rotation and the gripper tightening movement.

The robot's first rotational joint is conceived considering that it must support the total weight of the rest of the robotic arm (Figure 8b), and allow the free rotation of the servomotor. The same criterion is used for the second and third joints (Figure 8c). To this end, joints should be made of 2-mm-thickness iron. The design comprises a UC 205 and its respective PL-205 support in the first joint and two UC 201 bearings with PFL-20 supports in the second and third joint, respectively. Bearings are used in joints to enable the movement of the servomotor axis, avoiding overload and prolonging the servomotor useful life. However, this generates a greater load on the manipulator robot, because of the weight of these joints and bearings.

Another important element to the manipulator robot's implementation is the supporting table to which the manipulator robot is attached. This table has as a cover the working grid, a 1-m^2 iron-and-wood surface, which is also mobile, since the laboratory already possess one fixed 1-m^2 supporting table and a new table would cause a lack-of-space issue. In order to solve this problem, it is determined that the

Figure 7. Gripper used in the manipulator robot, which includes two servomotors

Figure 8. a) Gripper. b) First rotational joint. c) Second rotational joint

manipulator robot table has wheels and a structure that permits pushing it to the desks and placing it on them. The metal structure is made up of 5-cm-per-side square tubes.

DEVELOPMENT OF THE VISUAL FEEDBACK SYSTEM

An artificial vision system is complex because there are various factors that have a large effect on its performance, which can be divided into:

- An intrinsic part of the hardware of the image capture device. These are associated with data acquisition, such as lighting, focus, perspective, resolution, pixel size, noise level, etc.
- Factors associated with image processing, such as algorithms, filters, morphological operations, programming, etc.

A visual feedback system allows reinforce the performance of a manipulator robot, correcting on line possible lags in the links' movements, which may be caused by external disturbance. In particular, in this research this system gives the manipulator robot the position of its end-effector as well as that of an object to be manipulated in a work grid. The problem focuses on the algorithms to be implemented in the system, which should be capable of providing this information by means of the recognition of specific parts or of colors.

The proper lighting of the artificial vision system's work area is fundamental to obtain good information, since this influences directly the value that is assigned to each element of an image's digital representation. This effect can be clearly seen when algorithms for color filtering are developed.

Implementation of the Visual Feedback System

The visual feedback system consists of three elements: a web camera (webcam), a work grid, and an objective (Figure 9). The function of the webcam is to capture images to be processed by a PC; this procedure consists of two fundamental tasks. The first task is accomplished before the automatic image acquisition, and consists in linking the image of the work grid with the real work grid, by mapping the frames captured by the camera and the later assignment of the values of the real distances for each captured square of the work grid. The second task, accomplished during the automatic image acquisition process, consists in filtering the images, and detecting and locating the objective and the feedback signal on the work grid. With this information the system is capable of controlling the position of the manipulator robot. The work grid fulfills the function of linking the actual position of the objective with its relative position in the image, because each element of the image of the grid is assigned a real value from the work grid, and in this way the grid represents an x-y Cartesian plane. The manipulator robot is located at the center of this Cartesian plane (0,0), oriented over the positive axis of component x. On the y axis, the manipulator robot covers both the positive and the negative side. Finally, the objective fulfills the function of indicating the location of an object to be manipulated, and it has two purposes: one is to indicate the position of the end-effector, giving the position to which the manipulator robot's gripper must be displaced; and the second is to localize the object that will be manipulated. For both purposes the use of color filters is fundamental, because it is the fastest way to identify the objectives.

The webcam used has a resolution of 640x480 pixels; it is placed 70 cms from the base of the manipulator robot and approximately 30 cms above the grid. At that distance, the work space is limited to 30 cm to the sides and around 40 cm forward. With this configuration some 280 elements of the grid are detected. Figure 10 shows the view caught by the camera.

Once the required hardware has been installed, the graphic interface and the algorithms are developed to allow the detection of the grid and the objectives.

Graphic Interface Development and Programming

The graphic interface is the tool that allows the user to interact with the manipulator robot, and it was designed with the Graphical User Interface Design Editor (GUIDE) tool of the MATLAB software. From the functional standpoint it presents two modes of operation: the manual and the automatic modes. Graphically, as shown in Figure 11, it has two windows (axes1 and axes2) which show, depending on the chosen function: the images taken by the webcam for the calibration, the processed image of the work grid with the different elements that constitute it in different colors, the filtered image to detect the objective, and a graph with the top view of the positioning of the first three rotational links. Also,

Figure 9. Scheme of the visual feedback system

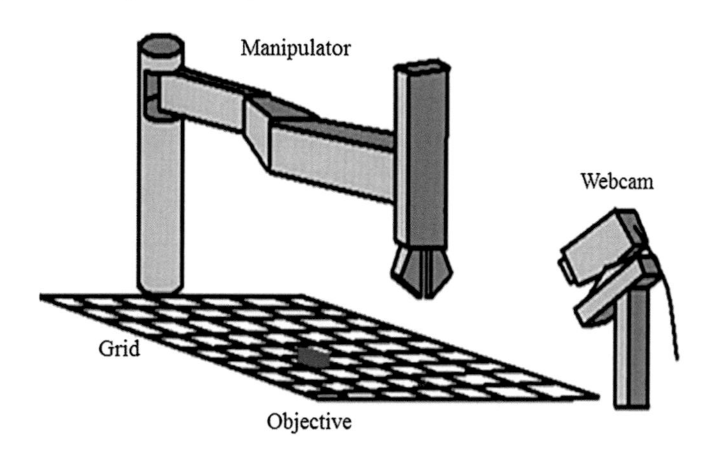

Figure 10. View of the webcam

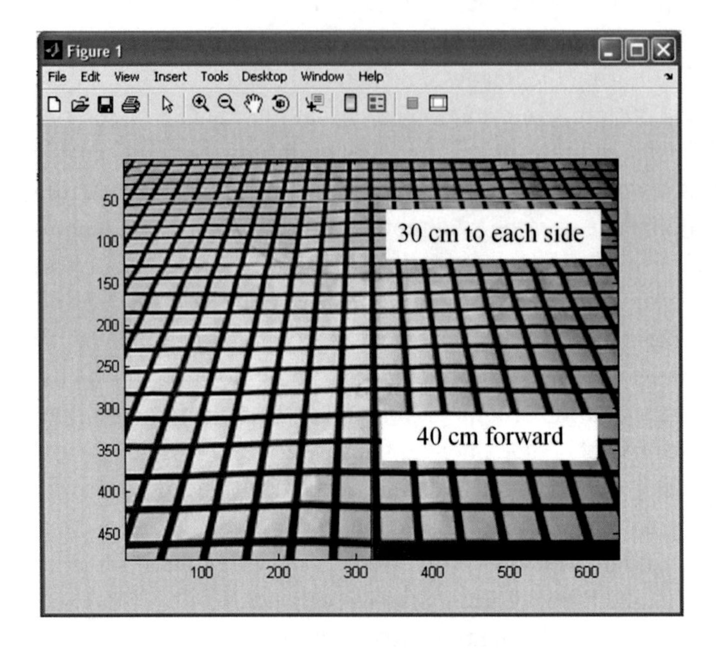

the interface shows the x-y coordinates, measured in centimeters, of the objective and of the feedback; the values that are calculated automatically, by means of a computer program, for the angles of the rotational joints $(\theta_1, \theta_2, \theta_3)$; the values of the Pulse Width Modulation (PWM) pulses that are sent to the servomotors of the rotational joints (pulse1, pulse2 and pulse3), and the calculated lengths of the links of the rotational joints (link1, link2 and link3).

Manual Mode

When the manipulator robot is not operating in automatic mode, a human operator can operate it freely through the manual mode of the interface, which consists of a number of controls:

- Prismatic and Gripper Manual Control.
    - It is divided into three panels —one for each joint— that separate the command buttons that allow the manual control of the prismatic joint, the rotation of the gripper's wrist, and the opening/closing of the gripper. As manual control is being used, care must be taken not to damage the manipulator robot. The command buttons on the prismatic joint allow the lifting, lowering and detention of this joint, while the buttons on the opening/ closing of the gripper permit the opening and closing of this piece, and those on the gripper's wrist enable wrist rotation.
- Joints Manual Control.
    - It allows positioning the first three rotational joints of the manipulator robot. It consists in entering values from 600 to 2400 microseconds in each edit box, with 1500 microseconds as the center for positioning. In addition, it is also possible to enter the time required for servomotors to position in milliseconds.

Figure 11. Graphic interface. Location of the windows, coordinates chart, and values table.

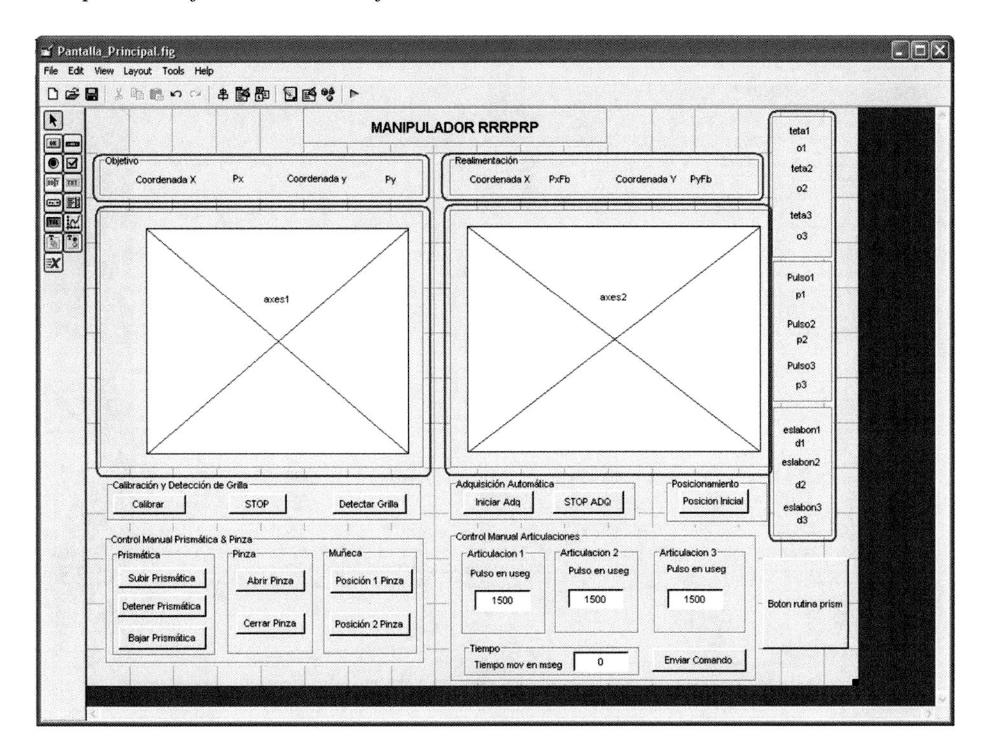

- Prismatic Routine Button.
  - It has the function of automating the movement of the prismatic joint and of the gripper, and performing it as a pre-established routine that consists in opening the gripper, lowering the prismatic joint, closing the gripper, and finally raising the gripper.
- Positioning Button.
  - It has the function of sending the command that takes the manipulator robot to its initial position, leaving it ready to operate in automatic mode.

Automatic Acquisition Mode

In this mode, the manipulator robot detects the objective in the grid and automatically positions the gripper over it. Then the prismatic routine starts, allowing the gripper to open, GO down, close, take the object and rise, and then deposit it in a programmed position. But for this to take place properly, the grid calibration and detection steps must be taken first.

Calibration

The calibration button allows visualizing the manipulator robot and the work grid. From this the webcam is oriented so that the lines that appear in the image coincide with the grid lines on the base of the manipulator robot. Once the camera's position and orientation have been calibrated, the Grid Detection button must be pressed.

Figure 12. Graphic interface. Location of the manual acquisition mode controls.

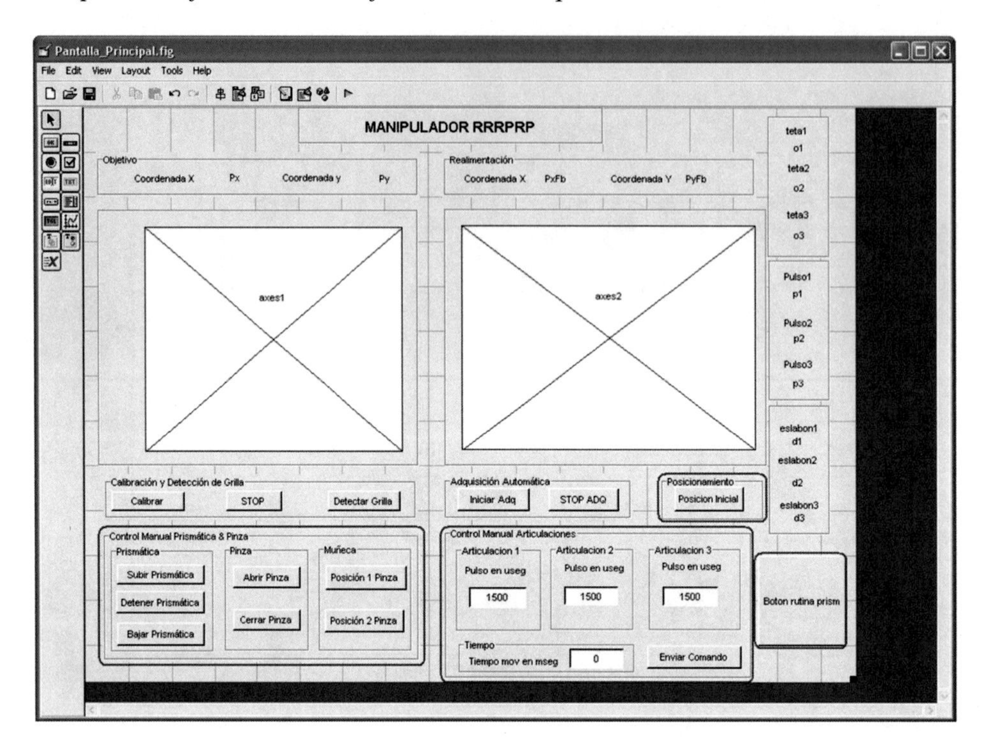

Figure 13. Graphic interface. Location of the automatic acquisition mode controls.

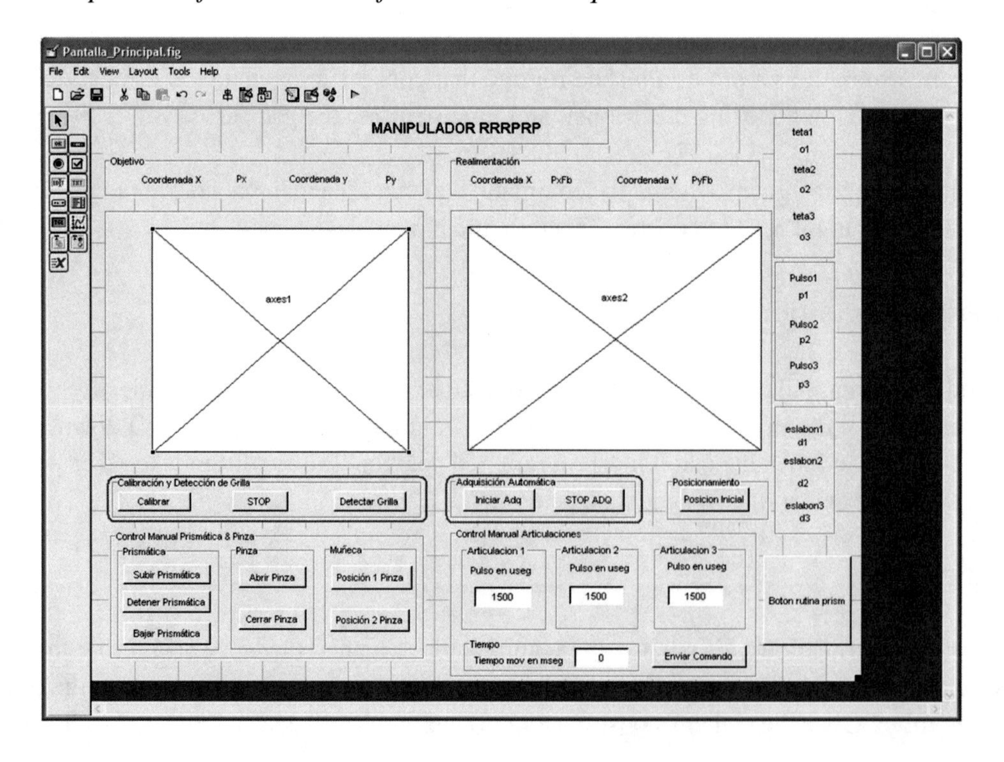

Grid Detection

It allows assigning to each element of the image of the work grid a real position in the x-y Cartesian plane. The process begins with the adjustment of the parameters of the webcam and getting an image of the grid; then the image is binarized using a threshold. The binarized image is eroded, using a square structural element, to eliminate the noise and possible errors in the capture of the image or in the binary conversion. The edges of the image are eliminated to avoid generating errors in the algorithms that allow the search for neighborhoods in the elements near the edges of the images. Some elements are eliminated, leaving only the work space without considering the elements that are behind the manipulator robot, i.e., on the blue line in Figure 14, to later number the quantity of elements. The total image is divided in half by a red line obtained in the calibration, after which these two images are labeled. The labeling consists in generating a matrix identical to the one entered; each of whose pixels is numbered, according to the connectivity existing between them. Therefore, the algorithm labels all the elements formed by connected pixels, each element being identified by a number. Figure 15 illustrates the result of the process described above, where after the grid's squares have been labeled, they are given a different color when the values of the squares of that grid are different, and they are also assigned an *x-e* individual coordinate that corresponds to the center of each square.

Automatic Acquisition

It allows starting the automated process of image capture, filtering, positioning calculation, and manipulator robot positioning. The acquisition of images is continuous because it uses the webcam as a video camera, that is, frame-by-frame images are obtained. The processing of the images is done frame by frame. The first step of the processing corresponds to filtering the frame in RGB format, captured by the camera. This filtration is done for two specific colors:

- Green, which represents the objective, and
- Bright red, which corresponds to the light beam emitted by a laser to carry out the system's feedback.

The position delivered by the red beam of the laser, mounted on the end-effector, is used as visual feedback information for the control system. This allows correcting the position of the manipulator robot's links every time it is required.

CONTROL SYSTEM OF THE ROBOTIZED MANIPULATOR

The design of the manipulator robot's control system is fundamental, because the correct fulfillment of the tasks that the robot must perform depend on it. The kind of control is not the same for all the manipulator robot's joints. Since the system used for obtaining the controller's feedback information is an artificial vision one, the control used will be separated by requested task, and divided into:

- Position control of the first three joints.
- Control of the prismatic joint and the gripper.

Figure 14. Calibration process of the webcam

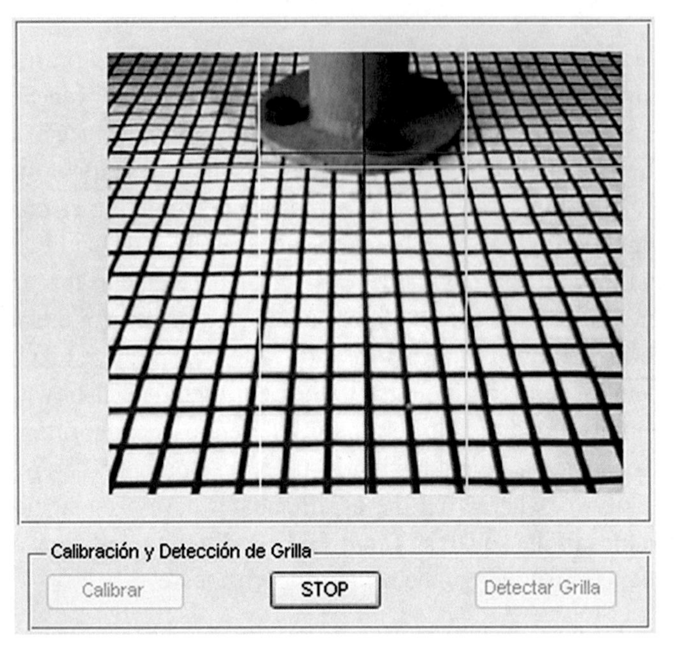

Figure 15. Graphic interface. Detection process of the work grid.

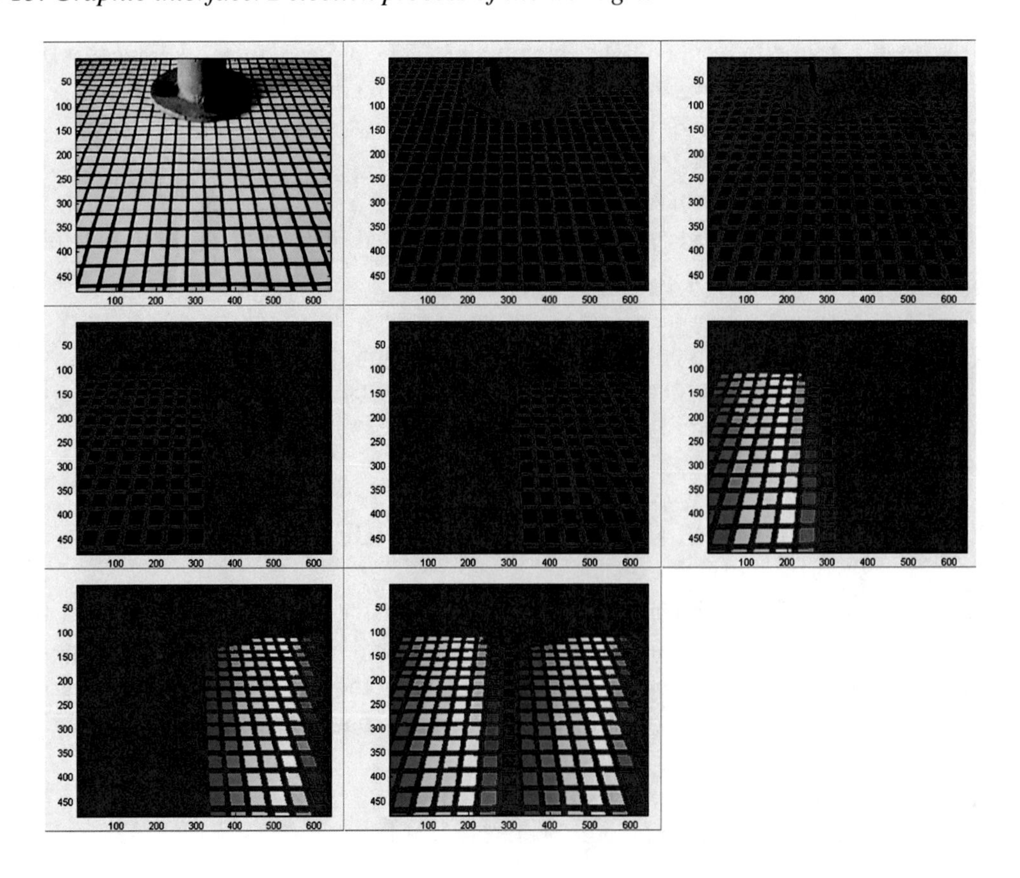

Figure 16. Graphic interface. Operation in automatic acquisition mode.

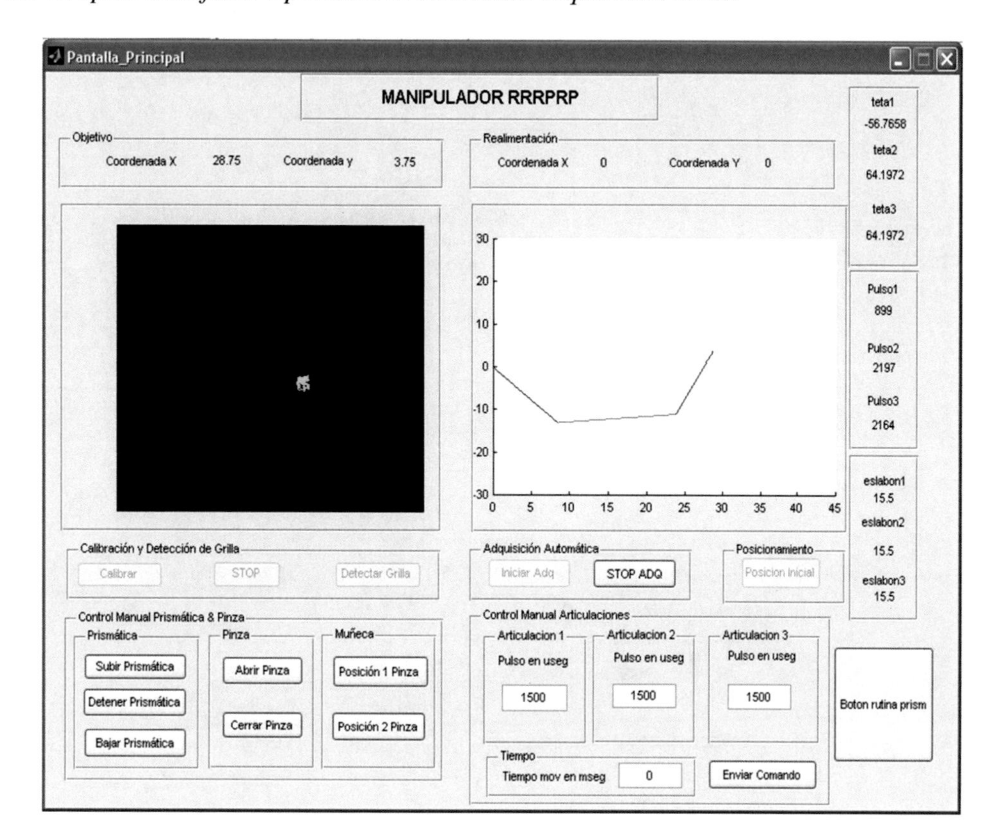

Position Control of the First Three Joints

Position control that is applied to the first rotational joints and can be performed in open loop or closed loop, depending on whether the visual feedback system does or does not detect correctly the red light beam of the laser mounted on the end-effector. The visual feedback system uses the same technique to filter the image and get the position delivered by the laser that is used to get the position of the objective. Once this position is delivered by the laser, which is the actual position of the end-effector, it is compared with the calculated position; the difference between these two positions is used to correct the calculated position. The flowchart of the operating logic of the robot's first three joints is shown in Figure 17.

The MATLAB processing is conducted frame by frame. Then, after obtaining the XY coordinates for the center of the objective's base and for the red beam, the angles that the different joints should have to allow the positioning of the end-effector in the objective's position are calculated through the equations obtained from the inverse kinematics study. In case that the difference between the objective and the laser red beam coordinates is not zero, the end-effector's position is corrected.

It should be noted that the manipulator robot design has also considered constructive restrictions, such as the servomotors limited rotation (up to 180°) and the specific length of the links, among others.

Once the angles that the joints should adopt have been calculated by means of the servomotors, this information is sent to the SSC-32 controller card through serial communication.

Figure 17. Logical operation flowchart-artificial vision system

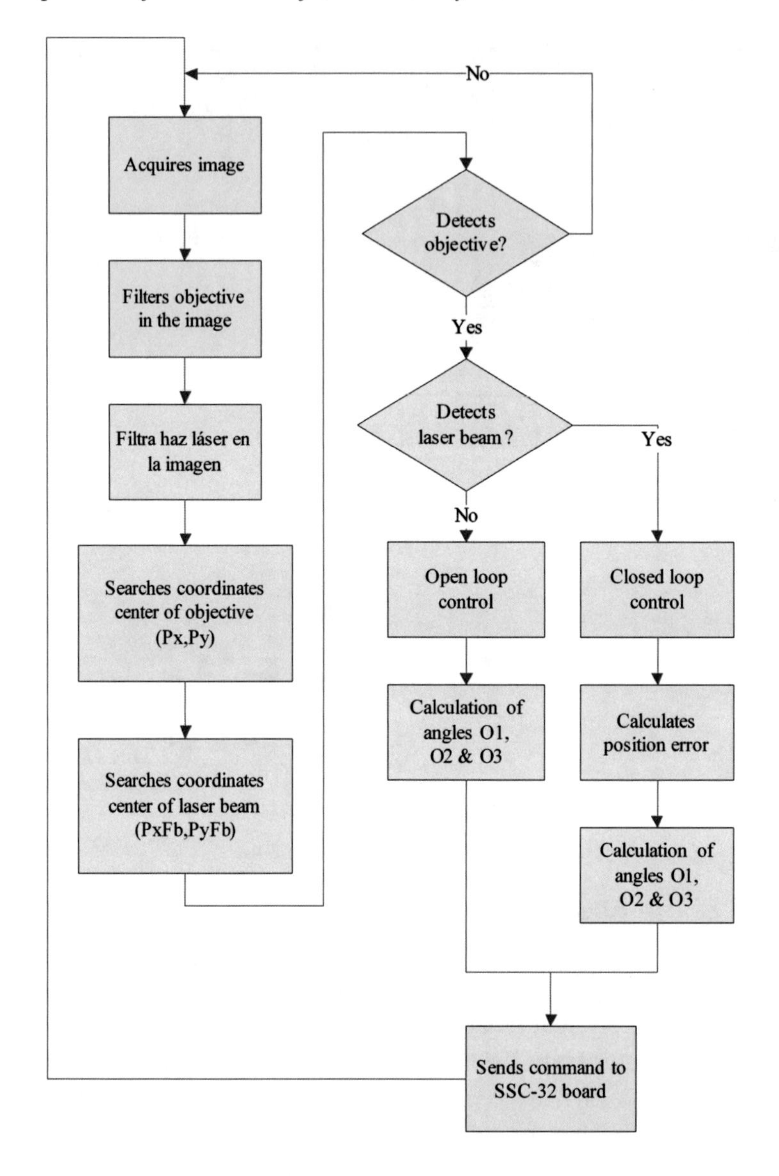

Control of the Prismatic Joint and the Gripper

The position of the prismatic joint corresponding to the fourth link is obtained by a computer routine in a discrete manner between the prismatic joint's up and down extremes. The other DOFs of the manipulator robot do not require additional position feedback, but use is made of the feedback of each servomotor of the gripper, allowing the turning of the wrist and the opening/closing of the gripper. The flowchart of the operating logic of the robot's last three joints is shown in Figure 18.

Figure 18. Logical operation flowchart- prismatic joint

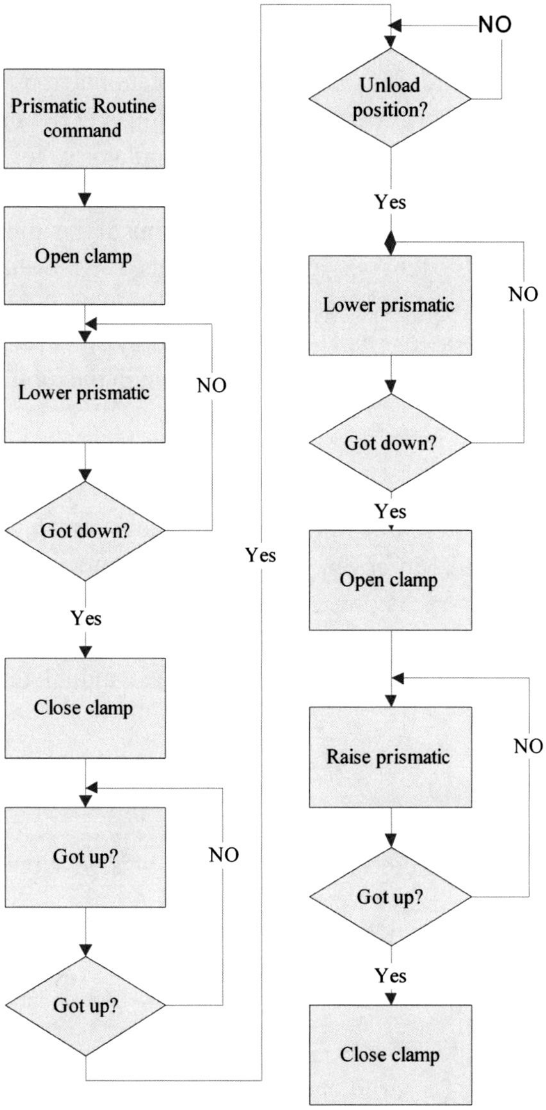

FUNCTIONAL TESTING OF THE ROBOTIZED MANIPULATOR

The implemented robot manipulator incorporates diverse systems, such as communication, mechanical, electronic and electric. Additionally, there is a system, the artificial vision system, which is fundamental to the robot operation in automatic mode, albeit not being part of it. Now, to verify the proper functioning of a robot manipulator, functional tests must be conducted in each of the robot systems as well as in the systems' operation as a whole.

Mechanical Tests

Functional tests for mechanical systems involved in the motion of the manipulator robot were carried out as the construction moved forward. This is the case for rotational and prismatic joints, which were tested one by one when constructed. However, the real interaction between the joint's motions and inertia in the implemented configuration was tested when the manipulator robot was completely assembled. Once assembled, motion tests were carried out using the *LynxTerm* software, supplied with the SSC-32 controller card. By means of this software, the proper functioning of the joints was tested one by one and as a set. The inertia present in the system was also tested, since the movement of one or two joints — while the rest of the joints are not controlled— modifies the position of these. In addition, once the robot was assembled, the 0° of the servomotors had to be calibrated by software, because the 0° mechanical position of the servomotor did not coincide with the 0° position in the joint, due to a construction error.

Electrical Tests

Once constructed, modular connection cards were checked so they do not have wiring misplacements that might cause a power supply shortcut. Moreover, when the manipulator robot was completely assembled, wiring conductivity was verified in each connector, ensuring that the signal successfully traveled from the manipulator robot's base to the end-effector card.

The power supply unit was assembled and tested prior to the manipulator robot construction in order to test the servomotors.

Artificial Vision System Tests

Tests on the artificial vision system were carried out as this was developed, in order to continuously debug the programs and to improve the software performance.

- The green color filtering and the laser red beam ranges were tested and modified to cover wide ranges of luminosity.
- The HMI implemented with MATLAB GUIDE was tested and modified so it does not present problems or errors when using a serial object for communicating with the SSC-32 card, or with the video object when communicating with the webcam.
- Comments on the .m MATLAB files were registered for the manipulator robot future use and modification.
- Different additional programs were created for specific tasks, such as color filtering, communications testing, calibration, collecting graphs associated with solutions of the manipulator robot, work space as well as functions that, in the end, were not used in the programming or that were replaced by more efficient ones.

CONCLUSION

The developed artificial vision system allows getting correctly, through images, the position of objects located on the robot's Cartesian work plane. The acquisition, frame by frame, of these images allows positioning in real time the links of the manipulator robot. However, this is restricted by the camera speed of frame acquisition (fps: frames per second). Additionally, it allows correcting any positioning error generated by the inertia of the mechanical movement, since the system sends the same positioning command in the frame that follows, making the servomotor correct its position, in case this was wrong.

The artificial vision system is capable of identifying successfully the squares that compose the robot's work grid, as well as the object and the feedback signal, in this way validating the correct positioning of the end-effector.

The equations obtained from the inverse kinematics study were validated by means of their implementation in the automatic control system of the manipulator robot, thus, achieving the correct position in which the manipulator robot should be placed for a specific objective. Only the kinematic equations were determined in the case of the links with the same length.

The redundant robotized manipulator with RRRPRP configuration, feedback by means of the artificial vision system, functioned correctly. In spite of this, due to its nature, the system is susceptible to blooming and reflections on the acrylic surface, generated by other sources of light. The same factor affects the objective colors and feedback signal recognition, so a good and correct lighting is fundamental in this case. A good camera positioning is also crucial during calibration. Otherwise, the system will not operate correctly in automatic mode. The web camera's position limits the workspace of the manipulator robot, but allows improving the acquisition of the grid frames that are closest to the manipulator robot's base.

DISCUSSION

Despite the correct functioning of the manipulator robot, making improvements and modifications is intended. Free connections were established in the electronic cards, particularly for the signals of the rotational joints' limit switches and for the position analog signal obtained by servomotor intervention. Also, the vision system can also be improved or replaced so the manipulator robot is able to execute a different task. The main idea of equipping the Robotics Laboratory with a robust, modular and programmable manipulator robot is to develop research projects and to potentiate the Robotics study in the Departamento de Ingeniería Eléctrica de la Universidad de Santiago de Chile. From this point of view, and considering the overactuated configuration and the other mechanical and electric characteristics of the manipulator robot, as well as the artificial vision system that controls the robot, this study results might be used for:

- Research and development of fault correction software for overactuated manipulator robots.
- Research and development of a motion control in joints with high inertia software.
- Research and development of a trajectory generation software.
- Research and development of a physical proprietary interface of the Universidad de Santiago to carry out the speed control of a manipulator robot servomotors.
- Development of a software for different applications associated with artificial vision, such as unique and multiple object detection and manipulation by form, color, volume, etc.

ACKNOWLEDGMENT

This work has been supported by Proyectos Basales and the Vicerrectoría de Investigación, Desarrollo e Innovación (VRIDEI), Universidad de Santiago de Chile, Chile.

REFERENCES

Angulo, J.M., & Avilés, R. (1989). *Curso de robótica*. Madrid, España: Paraninfo.

Ben-Gharbia, K., Maciejewski, A., & Roberts, R. (2014). A kinematic analysis and evaluation of planar robots designed from optimally fault-tolerant Jacobians. *IEEE Transactions on Robotics*, *30*(2), 516–524. doi:10.1109/TRO.2013.2291615

Birgmajer, B., Kovacic, Z., & Postruzin, Z. (2006). Integrated visión system for supervisión and guidance of a steam generator tuve inspection manipulator. In *Proceedings of the 2006 IEEE International Conference on Control Applications*. Munich, Germany: IEEE.

Craig, J. J. (2006). *Robótica, tercera edición*. Naucalpan de Juárez, México: Pearson Educación.

Guo, D., Ju, H., & Yao, Y. (2009). *Research of manipulator motion planning algorithm based on vision*. Paper presented at 2009 Sixth International Conference on Fuzzy Systems and Knowledge Discovery, Chengdu, China. doi:10.1109/FSKD.2009.664

Hunt, V. D. (1983). *Industrial robotics handbook* (1st ed.). South Norwalk, CT: Industrial Press Inc.

Iñigo, R., & Vidal, E. (2004). *Robots industriales manipuladores*. Ciudad de México, México: Alfaomega.

Kaufman, M. (2012). *Mars landing 2012: Inside the NASA curiosity mission*. Washington, DC: National Geographic Society.

Selig, J. M. (1992). *Introductory robotics*. London: Prentice Hall.

Siciliano, B., & Khatib, O. (2008). *Springer handbook of robotics*. Berlin, Germany: Springer-Verlag. doi:10.1007/978-3-540-30301-5

Siqueira, A., Terra, M., & Bergerman, M. (2011). *Robust control of robots: Fault tolerant approaches*. London: Springer London. doi:10.1007/978-0-85729-898-0

Torres, F., Pomares, J., Gil, P., Puente, S., & Aracil, R. (2002). *Robots y sistemas sensoriales, segunda edición*. Madrid, España: Pearson Educación.

Urrea, C., & Coltters, J. P. (2015). Design and implementation of a graphic 3D simulator for the study of control techniques applied to cooperative robots. *International Journal of Control Automation and Systems*, *13*(6), 1476–1485. doi:10.1007/s12555-014-0278-y

Urrea, C., & Kern, J. (2016). Trajectory tracking control of a real redundant manipulator of the SCARA type. *Journal of Electrical Engineering & Technology*, *11*(1), 215–226. doi:10.5370/JEET.2016.11.1.215

Wang, H., Zou, X., Liu, C., Lu, J., & Liu, T. (2008). *Study on behavior simulation for picking manipulator in virtual environment based on binocular stereo vision*. Paper presented at System Simulation and Scientific Computing, 2008. ICSC 2008. Asia Simulation Conference, Beijing, China.

Chapter 12
Insilico Approach for Epitope Prediction toward Novel Vaccine Delivery System Design

P. Raja Rajeswari
K. L. University, India

S. Viswanadha Raju
JNTUH, India

Amira S. Ashour
Tanta University, Egypt

Nilanjan Dey
Techno India College of Technology – Kolkata, India

ABSTRACT

Vaccines build a defense mechanism against the disease causing agents through the immune system stimulation and disease agents' imitation. Some of the vaccines contain a part of the disease causing agents that are either weakened or dead. Along with using vaccines with viral infections, it can be used against the various types of cancers for both therapy and prevention. The use of cancer's vaccines in cancer therapies is called immunotherapy. It can be done either by specific cancer vaccine or universal cancer vaccine that contains tumor antigens, which stimulate the immune system. This in turn initiates various mechanisms that terminate tumor cells and prevents recurrence of these tumors. The present work proposed an Insilico approach in epitope prediction and analysis of antigenecity and Immunogenecity of Haemophilus influenzae strains. It was interested with the design of a novel vaccine delivery system with better adjuvancity, where vaccine adjuvant is significant for the improvement of the antigens' immunogenicity that present in the vaccines. The conducted insilico approaches selected the best strain target proteins, m strain selection, epitope prediction, antigenicity and immunogenicity prediction of target proteins to find out the best targets.

DOI: 10.4018/978-1-5225-1025-3.ch012

INTRODUCTION

Haemophilus influenza is a gram negative bacterium that causes meningitis and acute respiratory infection, mainly in children (Turk DC,1984; Booy et al., 1997). Invasive disease initiated by Haemophilus Influenzae type b (Hib) is one of the foremost transferable diseases for young children (Rahman et al., 2008). Vaccination can be provided even for individuals at increased risk for Hib ailment who are already suffering from HIV infection, and immunoglobulin deficiency (Briere et al., 2014). Haemophilus Influenza is the most significant bacterial infections cause. Additionally, it origins an extensive range of other severe infections described by bloodstream invasion with the association of organ systems other than the central nervous system (CNS). Despite the effective antimicrobials availability, Hib is considered to be a substantial mortality and morbidity cause.

Haemophilus influenza is commensal bacteria to human upper respiratory tract (Wolf et al., 2007). There is an urgent need for a vaccine of this pathogen and still Hib disease post vaccination period cases are still reported (Bajanca et al., 2004; G-Kushnir et al., 2012; Galil et al., 1999). The constant increase of antibiotic resistant strains of these bacteria is one cause for evolving novel vaccines. The H. influenzae resistance to β-lactam antibiotics is an accumulative difficulty. The ampicillin resistance in this organism changes from 10% to 60% based on the terrestrial region (Bae et al., 2010).

Vaccination has a significant role to prevent influenza infection (Lottenbach et al., 2004). However, existing influenza vaccines have several limitations, such as

- The high production time,
- The limited vaccine capacity,
- The lack of knowledge in population,
- Dependence on egg-based production,
- Regulatory approval procedure,
- Limited worldwide vaccine availability,
- Inadequate effectiveness in aging and unprimed residents, and
- The cross-reactivity lack by present vaccines (Allan et al., 2005; Blain et al., 2014).

The possible solutions for such limitations includes:

- The virus culture-based construction,
- Synthetic vaccines,
- Recombination antigens,
- Increased steadiness of vaccines,
- Increase the immunogenicity by increasing antigen dose,
- T-cells vaccines, and
- Cross reactive vaccines (Barenkamp et al., 1981; Foxwell et al., 1998).

The present status and the production developments of the influenza vaccines attract the researches focus to study the biotechnical aspects like production rather than clinical outcomes (Florea et al., 1998).

Over the past few years several promising strategies of vaccines were introduced to prevent infections and cancer diseases. These include recombinant viral vectors, Deoxyribonucleic acid (DNA) vaccines; delivery system based on lipid including virosomes and liposomes. Though, these may suffer from some restrictions such as side effects linked with adjuvants, and weak immunogenicity, contemporary progresses in vaccine technology have delivered additional insights for controlling vaccine design.

Recently, based on recombinant viruses, an enormous number of delivery systems have emerged for vaccine purposes and for gene therapy. A realistic outer-membrane proteins number in 'H.influenzae' was investigated. Utmost of them emanated from isolates of 'H.influenzae' type b (Verez-Bencomo et al., 2004). Analyzing both encapsulated/non-encapsulated 'H.influenzae' strains exhibited protein mechanisms to encompass up to 36 proteins, where 6 signify the major protei content (Briere et al., 2009).

Some of the vaccines contain a part of the disease causing agents that are either weakened or dead. Apart from using vaccines only for viral infections, utilizing the same against Cancers both as therapeutic and preventative has captured huge interest. The use of Cancer vaccines in cancer therapies is called immunotherapy which is done either by specific cancer vaccine or universal cancer vaccine which contain tumor antigens that stimulate the immune system which in turn initiate various mechanisms that terminate tumor cells and prevents recurrence of these tumors.

Current influenza vaccines convince neutralizing antibodies alongside the viral membrane surface proteins hemagglutinin (HA) and neuraminidase (NA). Due to drift of HA, antigenic shift, and NA genes, defusing antibodies produced by influenza vaccines deficiency cross-reactivity in contradiction of non-matching influenza strains. Although, seasonal modifications to the vaccine strains are completed to manage this delinquent, it is not as suitable as a possible cross-protective influenza vaccine. Therefore, the alternative correlates identification of protection (CoPs) alongside influenza is an imperative stage to the cross-reactive influenza vaccines improvement. Three Hib outer membrane proteins act to fulfill this necessity. Data related to the P6 protein specify that this protein has at least one surface epitope, which is common to all the pathogen strains (Loeb et al., 1980). The novel vaccine targets in Haemophilus influenza outer membrane proteins have focused primarily on their antigenic potentials (Briere et al., 2009). Nevertheless, one significant criterion that should be met is the successful vaccine of Hib outer membrane protein if the protein retains surface-unprotected and antibody available antigenic causes that are common to most if not all strains of this pathogen (Loeb et al., 1980).

The existing influenza vaccines aforesaid limitations may be determined through the new technologies employment of in the influenza production domain and vaccine formulation. Original antigens frequently necessitate new construction techniques that have their rewards and drawbacks. Furthermore, these original antigens often are essential to be formulated with excipients and adjuvants.

Consequently, the current work proposed a method for the antigen delivery systems with prominence on versatile immunogenic vaccine delivery candidates. In addition, the advances in immunological, formulation and production features for promising novel influenza vaccine antigens are presented. Moreover, their impact for solving the influenza vaccines limitations is discussed.

BACKGROUND

The novel vaccine targets in Haemophilus influenza outer membrane proteins are focused primarily on their antigenic capacities as probable vaccine candidates (Briere et al., 2009). Conversely, one essential criterion to be met is the effective vaccine of Hib outer membrane protein if the protein retains surface-exposed.

Epitope Prediction

The epitope (antigenic determinant) is considered to be a part of antigen recognized by the immune system. It is a non-self-protein/sequence resulting from the host that termed as epitopes. The body part which identifies epitope is called a paratope. Prediction of antigenic epitopes on protein surface is significant for vaccines design and immune diagnostic reagents. It can be represented as a protein surface regions estimation that is especially predictable by antibodies. Consequently, the surface of an antigen/a foreign material can be predicted for further use in particular drug design for haemophilus influenza.

The Immune Epitope Database (IEDB) analysis tool is used for Epitope prediction. It provides an easy search of the tentative data symbolizing antibody and T cell epitopes deliberate in humans, non-human primates, and animal type. Epitopes elaborated in infectious disease, autoimmunity, allergy, and transplant (Bui et al., 2007; Nielsen et al., 2003). Also, the IEDB hosts used to assist the analysis and prediction of B cell and T cell epitopes.

Antigenicity

Antigenicity is the capability of a chemical structure and antigen to bind precisely with certain products group that has adaptive immunity. The facility to cause the antibodies production is known as the. The host sensitivity degree and its ability to produce antibodies is called immunogenicity. The IEDB Antigens 3.0 tool is used to predict the antigenicity in the present work. Through this tool, the antigens are searched upon by the organism name or by the antigen name. Antigen is a natural source from which an epitope is derived. The IEDB contains epitope data related to infectious diseases. The organism from which epitope are derived includes virus, bacteria or fungi.

Immunogenicity

The immunogenicity is the capability to convince a specific immune response. It is also known as the capability of a specific substance including as an antigen or epitope to stimulate an immune reaction in the human/animal body. It can be categories into

1. Wanted, and
2. Unwanted categories (Peters et al., 2003; Stranzl et al., 2004).

Where:

- **Wanted Immunogenicity:** It is related to vaccines where the injection of an antigen (the vaccines) has to lead to an immune response against the pathogen.
- **Unwanted Immunogenicity:** It is occurred when organism mounts an immune response against an antigen. Unwanted immunogenicity is strongly linked with therapeutic proteins. A fraction of the patients treated with those drugs mount anti-drug-antibodies

The antigen sequence of Haemophilus influenzae Type b is given as input to the Immune Epitope Database (IEDB) analysis tool. Afterward, the antigenecity score is calculated as a performance measurement of the area under a receiver operator characteristics (ROC) curve. The Analytical ultracentrifugation (AUC) on a protein base is calculated. This guarantees the estimate of all residues in a protein as only nonepitopes or only epitopes has an AUC of 0.5 analogous to a random expectation. Each method performance was calculated as the average AUC, average sensitivity, and average specificity for the 25 antigen collections.

PERFORMANCE MEASUREMENTS

Epitope Prediction Using Log-Odds Ratios

For epitope residues prediction, log-odds ratios were employed in arrangement with a scoring function, which sums the amino acids ratios in the spatial locality around each residue to provide a log-odds ratio score for each residue in an assumed protein. Enthused by Andersen et al. (2006) and Sweredoski et al. (2008), a scoring function is defined in the present work that declines weight on log-odds ratios as a distance function. The function used in (Sweredoski et al., 2008) included five distance thresholds to progressively decline the weight on log-odds ratios that set to 8, 10, 12, 14, and 16 Å.

However, in the current work, simpler design for this function is proposed with a single distance threshold and additionally a smoothing window size w is included. This parameter was established based on the sequence-based predictions optimization (w=9) in Andersen, et al. (2006), and approved by Sweredoski et al. (2008). The proximity sum (PS) function is given by:

$$PS\left(r, w, k_{ps}\right) = \sum_{i} \beta_{i}.ls\left(r_{i}, w\right) \tag{1}$$

$$\beta_{i} = 0.8\left(1 - \left(d_{i}/k_{ps}\right)\right) + 0.2 \tag{2}$$

Here, r is the query residue that used to compute the log-odds ratio score (PS), $ls(r_i,w)$ is the log-odds ratio value of r_i, where r_i is any residue within k_{ps} distance from r, sequentially averaged over a window of w residues and d_i is the distance between r and r_i. The minimal weight was set to 0.2 in order to guarantee that log-odds ratios are involved in the neighborhood sphere influence the final score. This weight value has been provided successful results in similar previous scoring functions (Turk, 1984). A 2-dimentional grid search were carried out to determine the optimal parameters set using the grids: *w*={1,3.....11}, *kps*={4,6.....28 }A°. *w={1,3...11}*, k_{ps}*={4,6...28 Å}*. The distance between two resides were considered as the distance between C_α atoms.

Epitope Prediction Using Surface Measurements

For epitope estimation, five dissimilar surface measures are designed from the protein structure. They are trained and tested due to their capability to predict B-cell epitope.

1. Full sphere neighbor count (FS) (Andersen et al., 2006),
2. Upper half-sphere neighbor count (UHS), and
3. Half-sphere exposure as described (HSE) (Verez-Bencomo et al., 2004) that was beforehand employed for B-cell prediction.

In the present work, the residue is classified as neighbor to the query residue if the C_α - C_α distance were below k_{sur}. The extensively used relative surface accessibility (RSA) was tested (Kabsch et al., 1983). Moreover, a fusion of the neighbor count and the RSA (Ta) is performed by significant neighbor residues as residues holding any atom within *T* distance of any atom in the query residue. The neighbor count in the upper and lower half-spheres was designed using the structural bio-python module established by Hamelryck et al. (2003).

Additionally, the surface accessibility is measured by the DSSP using the standard 4 Å probe. Then, the RSA are attained by dividing the surface accessibility with the maximum surface accessibility, deliberated from the peptide GGXGG, where X is the amino acid in question. The optimal sphere radius k_{sur}, for the FS, UHS, RSA, and HSE, and the distance threshold *T* for Ta was estimated by a grid search using the grids; ksur = {4,6.....28 A°.} and T={4,6....28 A° *}* k_{sur}*={4,6...28 Å}* and *T={4,6...28 Å}.*

Combining Log-Odds and Surface Measures

The log-odds ratio scores are includes the tested surface procedures to provide an overall estimate score. The scores were weighted using:

$$DS\left(r,\alpha\right) = -\alpha.SS\left(r\right) + \left(1-\alpha\right).PS\left(r\right) \quad (3)$$

where, *SS* and *PS* are surface scores and the log-odds ratio scores; respectively. The parameter values obtained to optimize the prediction power of the surface measures and log-odds ratio scores individually with the five training sets are used as inputs. In addition, the optimal values of α are obtained by the grid search using the grid: α*={0.005,0.010...1.0}*. Since, the RSA scores numerical values were much lower than the log-odds ratio scores, thus the RSA values are multiplied by 10 for optimization curve smoothing.

MAIN FOCUS OF THE CHAPTER

Vaccine adjuvant have their importance to improve immunogenicity of antigens present in vaccines. They are mostly used to achieve more specific effects. Some in silico approaches are conducted to select best strain target proteins, m strain selection, epitope prediction, antigenicity and immunogenicity prediction of target proteins are done to find out the best targets for further studies. Vaccination has a significant role to prevent influenza infection (Lottenbach et al., 2004). However, existing influenza vaccines have several limitations, such as

- The high production time,
- The limited vaccine capacity,
- The lack of knowledge in population.,
- Dependence on egg-based production,
- Regulatory approval procedure,
- Limited worldwide vaccine availability,
- Restricted efficiency in aging and unprimed residents, and
- The cross-reactivity lack by existing vaccines (Allan et al., 2005; Blain et al., 2014).

The possible solutions for such limitations includes

- The culture-based virus production,
- Synthetic vaccines,
- Recombination antigens increase the immunogenicity by increasing antigen dose,
- T-cells vaccines and cross reactive vaccines (Barenkamp et al., 1981; Foxwell et al., 1998).

The progresses in the influenza vaccines production attract the researches focus to study the biotechnical aspects like production rather than clinical outcomes (Florea et al., 1998).

Haemophilus influenzae Type b(Hib) is a foremost reason of severe bacterial infection in the early childhood. Although 6 different serotypes exist, it is now recognized that over the 99% of the typable strains which cause invasive disease in humans are of type b (Hib). Over the past few years several promising strategies of vaccines were introduced to prevent infections and cancer diseases. These include recombinant viral vectors, Deoxyribonucleic acid (DNA) vaccines; lipid based delivery system.

Recently, based on recombinant viruses, an enormous number of delivery systems have emerged for vaccine purposes and for gene therapy. Most of outer-membrane proteins occur due to isolation of 'H.influenzae' type b (Verez-Bencomo et al., 2004). Analyzing both encapsulated/non-encapsulated 'H.influenzae' strains exhibited protein components to comprise up to 36 proteins (Briere et al., 2009).

PROPOSED SYSTEM

With emphasizing on adaptable immunogenic vaccine delivery candidates, the current work proposed a method for the antigen delivery systems. In addition, the advances in immunological, formulation and production characteristics for the novel influenza vaccine antigens are presented.

SOLUTIONS AND RECOMMENDATIONS (RESULTS AND DISCUSSION)

Vaccine adjuvant have significant role to improve the immunogenicity of antigens presented in the vaccines. Thus, in the proposed approach, some silico schemes are conducted to select the best strain target proteins, train selection, epitope prediction, antigenicity and immunogenicity prediction of the target proteins to realize the best targets for further studies. Table 1 depicted a list of antigens of Haemophilus influenzae Type b with the calculated values of the antigenicity score and the immunogenicity score.

Table 1 depicts that antigenicity score and immunogenicity score for the Haemophilus influenzae Type b, Hap (1.0024,0.064), Adhesion (1.0050,0.064), Lipoprotein(1.0050,0.064), Tb2(1.000,0.64).

The Antigenicity Score for Some of the Antigens

- *The graph projects the Antigenicity of Antigen Hap.* The numbers of the residues in Hap are 522, the Antigenic propensity is 1.0024. There are 14 antigenic determinants in the sequence of the graph projects the antigenicity of antigen Hap. The sequence number is taken in the x-axis and the average antigenic propensity is taken in the y-axis.
- *The graph projects the Antigenicity of Antigen Adhesion.* The numbers of the residues in Adhesion are 1390 and the Antigenic propensity is 1.0050. There are 16 antigenic determinants in the sequence of the graph projects the antigenicity of antigen adhesion, where the sequence number is taken in the x-axis and the average antigenic propensity is taken in the y-axis.
- *The graph projects the Antigenicity of Antigen Tbp2(transferrin-binding protein 2 precursor).* The numbers of the residues in Adhesion are 629, and the Antigenic propensity is 1.0000. There are 19 antigenic determinants in the sequence of the graph projects the antigenicity of antigen Tbp2.

Consequently, the proposed approach realizes the best disease causing target antigens by applying epitope prediction for binding site analysis among the predicted antigens. The results establish that the best antigenicity and immunogenecity scores using the propensity values for antigenicity and Immunogencity values as well as the ranking of immunogenicity. Thus, the best targets of HPV type b strains are selected. Among the all selected antigens, Tbp2 (transferrin-binding protein 2 precursor) (1.0000), Hap(1.0024), Adhesion (1.0050), are showing the best Antigencity and also showing best Immunogenecity. Antigen Hap (immunogencity score (IM score) 0.64 for 48 residues; P-value -1.22023e), Lipoprotein (IM score

Table 1. The antigenicity score and immunogenicity score for the Haemophilus influenzae Type b

Haemophilus Influenzae Type b [Antigens]	Type	Antigenicity Score	Immunogenicity Score
LPS	Antigen	**1.0050**	**0.064**
D15	Protective antigen	1.0135	0.064
Hap from H. influenzae	Protective antigen	**1.0024**	**0.064**
Adhesion	Protective antigen	**1.0050**	**0.064**
NucA	Protective antigen	1.0211	0.035
Omp26	Protective antigen	1.0171	0.035
OmpP1	Protective antigen	1.0234	0.035
OmpP2	Protective antigen	1.0186	0.053
OmpP5	Protective antigen	1.0283	0.064
Lipoprotein	Protective antigen	**1.0050**	**0.064**
Tbp2	Protective antigen	**1.0000**	**0.064**

0.64 for 32 residues P-value - 0.000615951), IgG (IM score 0.64 for 36 residues; P-value – 0.00123556) are showing best results we can consider them for them as best targets.

CONCLUSION

Cancer vaccines in cancer therapies is called immunotherapy which is done either by specific cancer vaccine or universal cancer vaccine that contains tumor antigens. They stimulate the immune system which in turn initiates various mechanisms that terminate tumor cells and prevent recurrence of these tumors. The proposed approach identified the best antigens, where these target antigens can be powerful for further studies as well as they can be employed in the development of new drugs which can bitterly interact with selected targets. The antigens Tbp2 (1.0000), Hap(1.0024), Adhesion(1.0050) are showing best Antigencity and also showing best Immunogenecity Hap (immunogencity score(IM score) 0.64 for 48 residues ; P-value -1.22023e), Lipoprotein(IM score 0.64 for 32 residues P-value - 0.000615951), IgG (IM score 0.64 for 36 residues ; P-value – 0.00123556) are showing best results we can consider them for them as best targets.

REFERENCES

Allan, I., Loeb, M. R., & Moxon, E. R. (1987). Limited genetic diversity of Haemophilus influenzae (type b). *Microbial Pathogenesis*, *2*(2), 139–145. doi:10.1016/0882-4010(87)90105-7 PMID:3509858

Bae, S., Lee, J., Lee, J., Kim, E., Lee, S., Yu, J., & Kang, Y. (2010). Antimicrobial resistance in Haemophilus influenzae respiratory tract isolates in Korea: Results of a nationwide acute respiratory infections surveillance. *Antimicrobial Agents and Chemotherapy*, *54*(1), 65–71. doi:10.1128/AAC.00966-09 PMID:19884366

Bajanca, P., & Caniça, M. (2004). Emergence of nonencapsulated and encapsulated non-b-type invasive Haemophilus influenzae isolates in Portugal (19892001). *Journal of Clinical Microbiology*, *42*(2), 807–810. doi:10.1128/JCM.42.2.807-810.2004 PMID:14766857

Barenkamp, S. J., Munson, R. S., & Granoff, D. M. (1981). Subtyping isolates of Haemophilus influenzae type b by outer-membrane protein profiles. *The Journal of Infectious Diseases*, *143*(5), 668–676. doi:10.1093/infdis/143.5.668 PMID:6972422

Blain, A., MacNeil, J., Wang, X., Bennett, N., Farley, M. M., Harrison, L. H., . . . Reingold, A. (2014, September). Invasive Haemophilus influenzae Disease in Adults≥ 65 Years, United States, 2011. In Open forum infectious diseases (Vol. 1, No. 2). Oxford University Press.

Booy, R., Heath, P. T., Slack, M. P., Begg, N., & Moxon, E. R. (1997). Vaccine failures after primary immunisation with Haemophilus influenzae type-b conjugate vaccine without booster. *Lancet*, *349*(9060), 1197–1202. doi:10.1016/S0140-6736(96)06392-1 PMID:9130940

Briere, E. C., Jackson, M., Shah, S. G., Cohn, A. C., Anderson, R. D., MacNeil, J. R., & Messonnier, N. E. et al. (2012). Haemophilus influenzae type b disease and vaccine booster dose deferral, United States, 1998–2009. *Pediatrics*, *130*(3), 414–420. doi:10.1542/peds.2012-0266 PMID:22869828

Briere, E. C., Rubin, L., Moro, P. L., Cohn, A., Clark, T., & Messonnier, N. (2014). Prevention and control of haemophilus influenzae type b disease: Recommendations of the Advisory Committee on Immunization Practices (ACIP). *MMWR. Recommendations and Reports*, *63*(RR-01), 1–14. PMID:24572654

Bui, H. H., Sidney, J., Li, W., Fusseder, N., & Sette, A. (2007). Development of an epitope conservancy analysis tool to facilitate the design of epitope-based diagnostics and vaccines. *BMC Bioinformatics*, *8*(1), 1. doi:10.1186/1471-2105-8-361 PMID:17897458

Florea, L., Halld, B., Kohlbacher, O., Schwartz, R., Hoffman, S., & Istrail, S. (2003, August). Epitope prediction algorithms for peptide-based vaccine design. In *Bioinformatics Conference, 2003. CSB 2003. Proceedings of the 2003 IEEE* (pp. 17-26). IEEE. doi:10.1109/CSB.2003.1227293

Foxwell, A. R., Kyd, J. M., & Cripps, A. W. (1998). Nontypeable Haemophilus influenzae: Pathogenesis and prevention. *Microbiology and Molecular Biology Reviews*, *62*(2), 294–308. PMID:9618443

Galil, K., Singleton, R., Levine, O. S., Fitzgerald, M. A., Bulkow, L., Getty, M., & Parkinson, A. et al. (1999). Reemergence of invasive Haemophilus influenzae type b disease in a well-vaccinated population in remote Alaska. *The Journal of Infectious Diseases*, *179*(1), 101–106. doi:10.1086/314569 PMID:9841828

Greenberg-Kushnir, N., Haskin, O., Yarden-Bilavsky, H., Amir, J., & Bilavsky, E. (2012). Haemophilus influenzae type b meningitis in the short period after vaccination: a reminder of the phenomenon of apparent vaccine failure. *Case reports in infectious diseases, 2012.*

Hamelryck, T. (2005). An amino acid has two sides: A new 2D measure provides a different view of solvent exposure. *Proteins: Structure, Function, and Bioinformatics*, *59*(1), 38–48. doi:10.1002/prot.20379 PMID:15688434

Hamelryck, T., & Manderick, B. (2003). PDB file parser and structure class implemented in Python. *Bioinformatics (Oxford, England)*, *19*(17), 2308–2310. doi:10.1093/bioinformatics/btg299 PMID:14630660

Haste Andersen, P., Nielsen, M., & Lund, O. (2006). Prediction of residues in discontinuous B-cell epitopes using protein 3D structures. *Protein Science*, *15*(11), 2558–2567. doi:10.1110/ps.062405906 PMID:17001032

Kabsch, W., & Sander, C. (1983). Dictionary of protein secondary structure: Pattern recognition of hydrogen-bonded and geometrical features. *Biopolymers*, *22*(12), 2577–2637. doi:10.1002/bip.360221211 PMID:6667333

Loeb, M. R., & Smith, D. H. (1980). Outer membrane protein composition in disease isolates of Haemophilus influenzae: Pathogenic and epidemiological implications. *Infection and Immunity*, *30*(3), 709–717. PMID:6971807

Lottenbach, K. R., Granoff, D. M., Barenkamp, S. J., Powers, D. C., Kennedy, D., Irby-Moore, S., & Mink, C. M. et al. (2004). Safety and immunogenicity of Haemophilus influenzae type B polysaccharide or conjugate vaccines in an elderly adult population. *Journal of the American Geriatrics Society*, *52*(11), 1883–1887. doi:10.1111/j.1532-5415.2004.52511.x PMID:15507066

Nielsen, M., Lundegaard, C., Worning, P., Lauemøller, S. L., Lamberth, K., Buus, S., & Lund, O. et al. (2003). Reliable prediction of T-cell epitopes using neural networks with novel sequence representations. *Protein Science*, *12*(5), 1007–1017. doi:10.1110/ps.0239403 PMID:12717023

Peters, B., Bulik, S., Tampe, R., Van Endert, P. M., & Holzhütter, H. G. (2003). Identifying MHC class I epitopes by predicting the TAP transport efficiency of epitope precursors. *Journal of Immunology (Baltimore, MD.: 1950)*, *171*(4), 1741–1749. doi:10.4049/jimmunol.171.4.1741 PMID:12902473

Rahman, M., Hossain, S., Baqui, A. H., Shoma, S., Rashid, H., Nahar, N., & Khatun, F. et al. (2008). Haemophilus influenzae type-b and non-b-type invasive diseases in urban children (< 5years) of Bangladesh: Implications for therapy and vaccination. *The Journal of Infection*, *56*(3), 191–196. doi:10.1016/j.jinf.2007.12.008 PMID:18280571

Stranzl, T., Larsen, M. V., Lundegaard, C., & Nielsen, M. (2010). NetCTLpan: Pan-specific MHC class I pathway epitope predictions. *Immunogenetics*, *62*(6), 357–368. doi:10.1007/s00251-010-0441-4 PMID:20379710

Sweredoski, M. J., & Baldi, P. (2008). PEPITO: Improved discontinuous B-cell epitope prediction using multiple distance thresholds and half sphere exposure. *Bioinformatics (Oxford, England)*, *24*(12), 1459–1460. doi:10.1093/bioinformatics/btn199 PMID:18443018

Turk, D. C. (1984). The pathogenicity of Haemophilus influenzae. *Journal of Medical Microbiology*, *18*(1), 1–16. doi:10.1099/00222615-18-1-1 PMID:6146721

Verez-Bencomo, V., Fernandez-Santana, V., Hardy, E., Toledo, M. E., Rodríguez, M. C., Heynngnezz, L., & Villar, A. et al. (2004). A synthetic conjugate polysaccharide vaccine against Haemophilus influenzae type b. *Science*, *305*(5683), 522–525. doi:10.1126/science.1095209 PMID:15273395

Wolf, J., & Daley, A. J. (2007). Microbiological aspects of bacterial lower respiratory tract illness in children: Typical pathogens. *Paediatric Respiratory Reviews*, *8*(3), 204–211. doi:10.1016/j.prrv.2007.08.002 PMID:17868918

KEY TERMS AND DEFINITIONS

Antigenecity: The capability to cause the antibodies production. The antigenicity degree of a substance is determined by the amount and kind of the substance.

Antigens: A toxin substance that induces an immune response in the body.

Epitope: On the antigen surface, a localized region that accomplished of causing an immune response is called epitope.

Epitope Mimicry: Mimicry is the theoretical opportunity that sequence resemblances between foreign and self-peptides are adequate to result in the cross-activation of autoreactive T or B cells by pathogen-derived peptides.

Immunogenicity: Immunogenicity is the capability of a specific substance to aggravate an immune response in the human or animal body.

Pathogen: A pathogen is something that causes disease, such as a virus.

Related References

To continue our tradition of advancing academic research, we have compiled a list of recommended IGI Global readings. These references will provide additional information and guidance to further enrich your knowledge and assist you with your own research and future publications.

Adeyemo, O. (2013). The nationwide health information network: A biometric approach to prevent medical identity theft. In *User-driven healthcare: Concepts, methodologies, tools, and applications* (pp. 1636–1649). Hershey, PA: Medical Information Science Reference; doi:10.4018/978-1-4666-2770-3.ch081

Adler, M., & Henman, P. (2009). Justice beyond the courts: The implications of computerisation for procedural justice in social security. In A. Martínez & P. Abat (Eds.), *E-justice: Using information communication technologies in the court system* (pp. 65–86). Hershey, PA: Information Science Reference; doi:10.4018/978-1-59904-998-4.ch005

Aflalo, E., & Gabay, E. (2013). An information system for coping with student dropout. In L. Tomei (Ed.), *Learning tools and teaching approaches through ICT advancements* (pp. 176–187). Hershey, PA: Information Science Reference; doi:10.4018/978-1-4666-2017-9.ch016

Ahmed, M. A., Janssen, M., & van den Hoven, J. (2012). Value sensitive transfer (VST) of systems among countries: Towards a framework. *International Journal of Electronic Government Research, 8*(1), 26–42. doi:10.4018/jegr.2012010102

Aikins, S. K. (2008). Issues and trends in internet-based citizen participation. In G. Garson & M. Khosrow-Pour (Eds.), *Handbook of research on public information technology* (pp. 31–40). Hershey, PA: Information Science Reference; doi:10.4018/978-1-59904-857-4.ch004

Aikins, S. K. (2009). A comparative study of municipal adoption of internet-based citizen participation. In C. Reddick (Ed.), *Handbook of research on strategies for local e-government adoption and implementation: Comparative studies* (pp. 206–230). Hershey, PA: Information Science Reference; doi:10.4018/978-1-60566-282-4.ch011

Aikins, S. K. (2012). Improving e-government project management: Best practices and critical success factors. In *Digital democracy: Concepts, methodologies, tools, and applications* (pp. 1314–1332). Hershey, PA: Information Science Reference; doi:10.4018/978-1-4666-1740-7.ch065

Akabawi, M. S. (2011). Ghabbour group ERP deployment: Learning from past technology failures. In E. Business Research and Case Center (Ed.), Cases on business and management in the MENA region: New trends and opportunities (pp. 177-203). Hershey, PA: Business Science Reference. doi:10.4018/978-1-60960-583-4.ch012

Akabawi, M. S. (2013). Ghabbour group ERP deployment: Learning from past technology failures. In *Industrial engineering: Concepts, methodologies, tools, and applications* (pp. 933–958). Hershey, PA: Engineering Science Reference; doi:10.4018/978-1-4666-1945-6.ch051

Akbulut, A. Y., & Motwani, J. (2008). Integration and information sharing in e-government. In G. Putnik & M. Cruz-Cunha (Eds.), *Encyclopedia of networked and virtual organizations* (pp. 729–734). Hershey, PA: Information Science Reference; doi:10.4018/978-1-59904-885-7.ch096

Akers, E. J. (2008). Technology diffusion in public administration. In G. Garson & M. Khosrow-Pour (Eds.), *Handbook of research on public information technology* (pp. 339–348). Hershey, PA: Information Science Reference; doi:10.4018/978-1-59904-857-4.ch033

Al-Shafi, S. (2008). Free wireless internet park services: An investigation of technology adoption in Qatar from a citizens' perspective. *Journal of Cases on Information Technology*, *10*(3), 21–34. doi:10.4018/jcit.2008070103

Al-Shafi, S., & Weerakkody, V. (2009). Implementing free wi-fi in public parks: An empirical study in Qatar. *International Journal of Electronic Government Research*, *5*(3), 21–35. doi:10.4018/jegr.2009070102

Aladwani, A. M. (2002). Organizational actions, computer attitudes and end-user satisfaction in public organizations: An empirical study. In C. Snodgrass & E. Szewczak (Eds.), *Human factors in information systems* (pp. 153–168). Hershey, PA: IRM Press; doi:10.4018/978-1-931777-10-0.ch012

Aladwani, A. M. (2002). Organizational actions, computer attitudes, and end-user satisfaction in public organizations: An empirical study. *Journal of Organizational and End User Computing*, *14*(1), 42–49. doi:10.4018/joeuc.2002010104

Allen, B., Juillet, L., Paquet, G., & Roy, J. (2005). E-government and private-public partnerships: Relational challenges and strategic directions. In M. Khosrow-Pour (Ed.), *Practicing e-government: A global perspective* (pp. 364–382). Hershey, PA: Idea Group Publishing; doi:10.4018/978-1-59140-637-2.ch016

Alshawaf, A., & Knalil, O. E. (2008). IS success factors and IS organizational impact: Does ownership type matter in Kuwait? *International Journal of Enterprise Information Systems*, *4*(2), 13–33. doi:10.4018/jeis.2008040102

Ambali, A. R. (2009). Digital divide and its implication on Malaysian e-government: Policy initiatives. In H. Rahman (Ed.), *Social and political implications of data mining: Knowledge management in e-government* (pp. 267–287). Hershey, PA: Information Science Reference; doi:10.4018/978-1-60566-230-5.ch016

Amoretti, F. (2007). Digital international governance. In A. Anttiroiko & M. Malkia (Eds.), *Encyclopedia of digital government* (pp. 365–370). Hershey, PA: Information Science Reference; doi:10.4018/978-1-59140-789-8.ch056

Amoretti, F. (2008). Digital international governance. In A. Anttiroiko (Ed.), *Electronic government: Concepts, methodologies, tools, and applications* (pp. 688–696). Hershey, PA: Information Science Reference; doi:10.4018/978-1-59904-947-2.ch058

Amoretti, F. (2008). E-government at supranational level in the European Union. In A. Anttiroiko (Ed.), *Electronic government: Concepts, methodologies, tools, and applications* (pp. 1047–1055). Hershey, PA: Information Science Reference; doi:10.4018/978-1-59904-947-2.ch079

Amoretti, F. (2008). E-government regimes. In A. Anttiroiko (Ed.), *Electronic government: Concepts, methodologies, tools, and applications* (pp. 3846–3856). Hershey, PA: Information Science Reference; doi:10.4018/978-1-59904-947-2.ch280

Amoretti, F. (2009). Electronic constitution: A Braudelian perspective. In F. Amoretti (Ed.), *Electronic constitution: Social, cultural, and political implications* (pp. 1–19). Hershey, PA: Information Science Reference; doi:10.4018/978-1-60566-254-1.ch001

Amoretti, F., & Musella, F. (2009). Institutional isomorphism and new technologies. In M. Khosrow-Pour (Ed.), *Encyclopedia of information science and technology* (2nd ed., pp. 2066–2071). Hershey, PA: Information Science Reference; doi:10.4018/978-1-60566-026-4.ch325

Andersen, K. V., & Henriksen, H. Z. (2007). E-government research: Capabilities, interaction, orientation, and values. In D. Norris (Ed.), *Current issues and trends in e-government research* (pp. 269–288). Hershey, PA: CyberTech Publishing; doi:10.4018/978-1-59904-283-1.ch013

Anderson, K. V., & Henriksen, H. Z. (2005). The first leg of e-government research: Domains and application areas 1998-2003. *International Journal of Electronic Government Research, 1*(4), 26–44. doi:10.4018/jegr.2005100102

Anttiroiko, A. (2009). Democratic e-governance. In M. Khosrow-Pour (Ed.), *Encyclopedia of information science and technology* (2nd ed., pp. 990–995). Hershey, PA: Information Science Reference; doi:10.4018/978-1-60566-026-4.ch158

Association, I. R. (2010). Networking and telecommunications: Concepts, methodologies, tools and applications (Vols. 1–3). Hershey, PA: IGI Global; doi:10.4018/978-1-60566-986-1

Association, I. R. (2010). Web-based education: Concepts, methodologies, tools and applications (Vols. 1–3). Hershey, PA: IGI Global; doi:10.4018/978-1-61520-963-7

Baker, P. M., Bell, A., & Moon, N. W. (2009). Accessibility issues in municipal wireless networks. In C. Reddick (Ed.), *Handbook of research on strategies for local e-government adoption and implementation: Comparative studies* (pp. 569–588). Hershey, PA: Information Science Reference; doi:10.4018/978-1-60566-282-4.ch030

Becker, S. A., Keimer, R., & Muth, T. (2010). A case on university and community collaboration: The sci-tech entrepreneurial training services (ETS) program. In S. Becker & R. Niebuhr (Eds.), *Cases on technology innovation: Entrepreneurial successes and pitfalls* (pp. 68–90). Hershey, PA: Business Science Reference; doi:10.4018/978-1-61520-609-4.ch003

Becker, S. A., Keimer, R., & Muth, T. (2012). A case on university and community collaboration: The sci-tech entrepreneurial training services (ETS) program. In Regional development: Concepts, methodologies, tools, and applications (pp. 947-969). Hershey, PA: Information Science Reference. doi:10.4018/978-1-4666-0882-5.ch507

Bernardi, R. (2012). Information technology and resistance to public sector reforms: A case study in Kenya. In T. Papadopoulos & P. Kanellis (Eds.), *Public sector reform using information technologies: Transforming policy into practice* (pp. 59–78). Hershey, PA: Information Science Reference; doi:10.4018/978-1-60960-839-2.ch004

Bernardi, R. (2013). Information technology and resistance to public sector reforms: A case study in Kenya. In *User-driven healthcare: Concepts, methodologies, tools, and applications* (pp. 14–33). Hershey, PA: Medical Information Science Reference; doi:10.4018/978-1-4666-2770-3.ch002

Bolívar, M. P., Pérez, M. D., & Hernández, A. M. (2012). Municipal e-government services in emerging economies: The Latin-American and Caribbean experiences. In Y. Chen & P. Chu (Eds.), *Electronic governance and cross-boundary collaboration: Innovations and advancing tools* (pp. 198–226). Hershey, PA: Information Science Reference; doi:10.4018/978-1-60960-753-1.ch011

Borycki, E. M., & Kushniruk, A. W. (2010). Use of clinical simulations to evaluate the impact of health information systems and ubiquitous computing devices upon health professional work. In S. Mohammed & J. Fiaidhi (Eds.), *Ubiquitous health and medical informatics: The ubiquity 2.0 trend and beyond* (pp. 552–573). Hershey, PA: Medical Information Science Reference; doi:10.4018/978-1-61520-777-0.ch026

Borycki, E. M., & Kushniruk, A. W. (2011). Use of clinical simulations to evaluate the impact of health information systems and ubiquitous computing devices upon health professional work. In *Clinical technologies: Concepts, methodologies, tools and applications* (pp. 532–553). Hershey, PA: Medical Information Science Reference; doi:10.4018/978-1-60960-561-2.ch220

Buchan, J. (2011). Developing a dynamic and responsive online learning environment: A case study of a large Australian university. In B. Czerkawski (Ed.), *Free and open source software for e-learning: Issues, successes and challenges* (pp. 92–109). Hershey, PA: Information Science Reference; doi:10.4018/978-1-61520-917-0.ch006

Buenger, A. W. (2008). Digital convergence and cybersecurity policy. In G. Garson & M. Khosrow-Pour (Eds.), *Handbook of research on public information technology* (pp. 395–405). Hershey, PA: Information Science Reference; doi:10.4018/978-1-59904-857-4.ch038

Burn, J. M., & Loch, K. D. (2002). The societal impact of world wide web - Key challenges for the 21st century. In A. Salehnia (Ed.), *Ethical issues of information systems* (pp. 88–106). Hershey, PA: IRM Press; doi:10.4018/978-1-931777-15-5.ch007

Burn, J. M., & Loch, K. D. (2003). The societal impact of the world wide web-Key challenges for the 21st century. In M. Khosrow-Pour (Ed.), *Advanced topics in information resources management* (Vol. 2, pp. 32–51). Hershey, PA: Idea Group Publishing; doi:10.4018/978-1-59140-062-2.ch002

Bwalya, K. J., Du Plessis, T., & Rensleigh, C. (2012). The "quicksilver initiatives" as a framework for e-government strategy design in developing economies. In K. Bwalya & S. Zulu (Eds.), *Handbook of research on e-government in emerging economies: Adoption, e-participation, and legal frameworks* (pp. 605–623). Hershey, PA: Information Science Reference; doi:10.4018/978-1-4666-0324-0.ch031

Cabotaje, C. E., & Alampay, E. A. (2013). Social media and citizen engagement: Two cases from the Philippines. In S. Saeed & C. Reddick (Eds.), *Human-centered system design for electronic governance* (pp. 225–238). Hershey, PA: Information Science Reference; doi:10.4018/978-1-4666-3640-8.ch013

Camillo, A., Di Pietro, L., Di Virgilio, F., & Franco, M. (2013). Work-groups conflict at PetroTech-Italy, S.R.L.: The influence of culture on conflict dynamics. In B. Christiansen, E. Turkina, & N. Williams (Eds.), *Cultural and technological influences on global business* (pp. 272–289). Hershey, PA: Business Science Reference; doi:10.4018/978-1-4666-3966-9.ch015

Capra, E., Francalanci, C., & Marinoni, C. (2008). Soft success factors for m-government. In A. Anttiroiko (Ed.), *Electronic government: Concepts, methodologies, tools, and applications* (pp. 1213–1233). Hershey, PA: Information Science Reference; doi:10.4018/978-1-59904-947-2.ch089

Cartelli, A. (2009). The implementation of practices with ICT as a new teaching-learning paradigm. In A. Cartelli & M. Palma (Eds.), *Encyclopedia of information communication technology* (pp. 413–417). Hershey, PA: Information Science Reference; doi:10.4018/978-1-59904-845-1.ch055

Charalabidis, Y., Lampathaki, F., & Askounis, D. (2010). Investigating the landscape in national interoperability frameworks. *International Journal of E-Services and Mobile Applications*, *2*(4), 28–41. doi:10.4018/jesma.2010100103

Charalabidis, Y., Lampathaki, F., & Askounis, D. (2012). Investigating the landscape in national interoperability frameworks. In A. Scupola (Ed.), *Innovative mobile platform developments for electronic services design and delivery* (pp. 218–231). Hershey, PA: Business Science Reference; doi:10.4018/978-1-4666-1568-7.ch013

Chen, I. (2005). Distance education associations. In C. Howard, J. Boettcher, L. Justice, K. Schenk, P. Rogers, & G. Berg (Eds.), *Encyclopedia of distance learning* (pp. 599–612). Hershey, PA: Information Science Reference; doi:10.4018/978-1-59140-555-9.ch087

Chen, I. (2008). Distance education associations. In L. Tomei (Ed.), *Online and distance learning: Concepts, methodologies, tools, and applications* (pp. 562–579). Hershey, PA: Information Science Reference; doi:10.4018/978-1-59904-935-9.ch048

Chen, Y. (2008). Managing IT outsourcing for digital government. In A. Anttiroiko (Ed.), *Electronic government: Concepts, methodologies, tools, and applications* (pp. 3107–3114). Hershey, PA: Information Science Reference; doi:10.4018/978-1-59904-947-2.ch229

Chen, Y., & Dimitrova, D. V. (2006). Electronic government and online engagement: Citizen interaction with government via web portals. *International Journal of Electronic Government Research*, *2*(1), 54–76. doi:10.4018/jegr.2006010104

Chen, Y., & Knepper, R. (2005). Digital government development strategies: Lessons for policy makers from a comparative perspective. In W. Huang, K. Siau, & K. Wei (Eds.), *Electronic government strategies and implementation* (pp. 394–420). Hershey, PA: Idea Group Publishing; doi:10.4018/978-1-59140-348-7.ch017

Chen, Y., & Knepper, R. (2008). Digital government development strategies: Lessons for policy makers from a comparative perspective. In H. Rahman (Ed.), *Developing successful ICT strategies: Competitive advantages in a global knowledge-driven society* (pp. 334–356). Hershey, PA: Information Science Reference; doi:10.4018/978-1-59904-654-9.ch017

Cherian, E. J., & Ryan, T. W. (2014). Incongruent needs: Why differences in the iron-triangle of priorities make health information technology adoption and use difficult. In C. El Morr (Ed.), *Research perspectives on the role of informatics in health policy and management* (pp. 209–221). Hershey, PA: Medical Information Science Reference; doi:10.4018/978-1-4666-4321-5.ch012

Cho, H. J., & Hwang, S. (2010). Government 2.0 in Korea: Focusing on e-participation services. In C. Reddick (Ed.), *Politics, democracy and e-government: Participation and service delivery* (pp. 94–114). Hershey, PA: Information Science Reference; doi:10.4018/978-1-61520-933-0.ch006

Chorus, C., & Timmermans, H. (2010). Ubiquitous travel environments and travel control strategies: Prospects and challenges. In M. Wachowicz (Ed.), *Movement-aware applications for sustainable mobility: Technologies and approaches* (pp. 30–51). Hershey, PA: Information Science Reference; doi:10.4018/978-1-61520-769-5.ch003

Chuanshen, R. (2007). E-government construction and China's administrative litigation act. In A. Anttiroiko & M. Malkia (Eds.), *Encyclopedia of digital government* (pp. 507–510). Hershey, PA: Information Science Reference; doi:10.4018/978-1-59140-789-8.ch077

Ciaghi, A., & Villafiorita, A. (2012). Law modeling and BPR for public administration improvement. In K. Bwalya & S. Zulu (Eds.), *Handbook of research on e-government in emerging economies: Adoption, e-participation, and legal frameworks* (pp. 391–410). Hershey, PA: Information Science Reference; doi:10.4018/978-1-4666-0324-0.ch019

Ciaramitaro, B. L., & Skrocki, M. (2012). mHealth: Mobile healthcare. In B. Ciaramitaro (Ed.), Mobile technology consumption: Opportunities and challenges (pp. 99-109). Hershey, PA: Information Science Reference. doi:10.4018/978-1-61350-150-4.ch007

Comite, U. (2012). Innovative processes and managerial effectiveness of e-procurement in healthcare. In A. Manoharan & M. Holzer (Eds.), *Active citizen participation in e-government: A global perspective* (pp. 206–229). Hershey, PA: Information Science Reference; doi:10.4018/978-1-4666-0116-1.ch011

Cordella, A. (2013). E-government success: How to account for ICT, administrative rationalization, and institutional change. In J. Gil-Garcia (Ed.), *E-government success factors and measures: Theories, concepts, and methodologies* (pp. 40–51). Hershey, PA: Information Science Reference; doi:10.4018/978-1-4666-4058-0.ch003

Cropf, R. A. (2009). ICT and e-democracy. In M. Khosrow-Pour (Ed.), *Encyclopedia of information science and technology* (2nd ed., pp. 1789–1793). Hershey, PA: Information Science Reference; doi:10.4018/978-1-60566-026-4.ch281

Cropf, R. A. (2009). The virtual public sphere. In M. Pagani (Ed.), *Encyclopedia of multimedia technology and networking* (2nd ed., pp. 1525–1530). Hershey, PA: Information Science Reference; doi:10.4018/978-1-60566-014-1.ch206

D'Abundo, M. L. (2013). Electronic health record implementation in the United States healthcare industry: Making the process of change manageable. In V. Wang (Ed.), *Handbook of research on technologies for improving the 21st century workforce: Tools for lifelong learning* (pp. 272–286). Hershey, PA: Information Science Publishing; doi:10.4018/978-1-4666-2181-7.ch018

Damurski, L. (2012). E-participation in urban planning: Online tools for citizen engagement in Poland and in Germany. *International Journal of E-Planning Research, 1*(3), 40–67. doi:10.4018/ijepr.2012070103

de Almeida, M. O. (2007). E-government strategy in Brazil: Increasing transparency and efficiency through e-government procurement. In M. Gascó-Hernandez (Ed.), *Latin America online: Cases, successes and pitfalls* (pp. 34–82). Hershey, PA: IRM Press; doi:10.4018/978-1-59140-974-8.ch002

de Juana Espinosa, S. (2008). Empirical study of the municipalitites' motivations for adopting online presence. In A. Anttiroiko (Ed.), *Electronic government: Concepts, methodologies, tools, and applications* (pp. 3593–3608). Hershey, PA: Information Science Reference; doi:10.4018/978-1-59904-947-2.ch262

de Souza Dias, D. (2002). Motivation for using information technology. In C. Snodgrass & E. Szewczak (Eds.), *Human factors in information systems* (pp. 55–60). Hershey, PA: IRM Press; doi:10.4018/978-1-931777-10-0.ch005

Demediuk, P. (2006). Government procurement ICT's impact on the sustainability of SMEs and regional communities. In S. Marshall, W. Taylor, & X. Yu (Eds.), *Encyclopedia of developing regional communities with information and communication technology* (pp. 321–324). Hershey, PA: Information Science Reference; doi:10.4018/978-1-59140-575-7.ch056

Devonshire, E., Forsyth, H., Reid, S., & Simpson, J. M. (2013). The challenges and opportunities of online postgraduate coursework programs in a traditional university context. In B. Tynan, J. Willems, & R. James (Eds.), *Outlooks and opportunities in blended and distance learning* (pp. 353–368). Hershey, PA: Information Science Reference; doi:10.4018/978-1-4666-4205-8.ch026

Di Cerbo, F., Scotto, M., Sillitti, A., Succi, G., & Vernazza, T. (2007). Toward a GNU/Linux distribution for corporate environments. In S. Sowe, I. Stamelos, & I. Samoladas (Eds.), *Emerging free and open source software practices* (pp. 215–236). Hershey, PA: Idea Group Publishing; doi:10.4018/978-1-59904-210-7.ch010

Diesner, J., & Carley, K. M. (2005). Revealing social structure from texts: Meta-matrix text analysis as a novel method for network text analysis. In V. Narayanan & D. Armstrong (Eds.), *Causal mapping for research in information technology* (pp. 81–108). Hershey, PA: Idea Group Publishing; doi:10.4018/978-1-59140-396-8.ch004

Dologite, D. G., Mockler, R. J., Bai, Q., & Viszhanyo, P. F. (2006). IS change agents in practice in a US-Chinese joint venture. In M. Hunter & F. Tan (Eds.), *Advanced topics in global information management* (Vol. 5, pp. 331–352). Hershey, PA: Idea Group Publishing; doi:10.4018/978-1-59140-923-6.ch015

Drnevich, P., Brush, T. H., & Luckock, G. T. (2011). Process and structural implications for IT-enabled outsourcing. *International Journal of Strategic Information Technology and Applications*, *2*(4), 30–43. doi:10.4018/jsita.2011100103

Dwivedi, A. N. (2009). Handbook of research on information technology management and clinical data administration in healthcare (Vols. 1–2). Hershey, PA: IGI Global; doi:10.4018/978-1-60566-356-2

Elbeltagi, I., McBride, N., & Hardaker, G. (2006). Evaluating the factors affecting DSS usage by senior managers in local authorities in Egypt. In M. Hunter & F. Tan (Eds.), *Advanced topics in global information management* (Vol. 5, pp. 283–307). Hershey, PA: Idea Group Publishing; doi:10.4018/978-1-59140-923-6.ch013

Eom, S., & Fountain, J. E. (2013). Enhancing information services through public-private partnerships: Information technology knowledge transfer underlying structures to develop shared services in the U.S. and Korea. In J. Gil-Garcia (Ed.), *E-government success around the world: Cases, empirical studies, and practical recommendations* (pp. 15–40). Hershey, PA: Information Science Reference; doi:10.4018/978-1-4666-4173-0.ch002

Esteves, T., Leuenberger, D., & Van Leuven, N. (2012). Reaching citizen 2.0: How government uses social media to send public messages during times of calm and times of crisis. In K. Kloby & M. D'Agostino (Eds.), *Citizen 2.0: Public and governmental interaction through web 2.0 technologies* (pp. 250–268). Hershey, PA: Information Science Reference; doi:10.4018/978-1-4666-0318-9.ch013

Estevez, E., Fillottrani, P., Janowski, T., & Ojo, A. (2012). Government information sharing: A framework for policy formulation. In Y. Chen & P. Chu (Eds.), *Electronic governance and cross-boundary collaboration: Innovations and advancing tools* (pp. 23–55). Hershey, PA: Information Science Reference; doi:10.4018/978-1-60960-753-1.ch002

Ezz, I. E. (2008). E-governement emerging trends: Organizational challenges. In A. Anttiroiko (Ed.), *Electronic government: Concepts, methodologies, tools, and applications* (pp. 3721–3737). Hershey, PA: Information Science Reference; doi:10.4018/978-1-59904-947-2.ch269

Fabri, M. (2009). The Italian style of e-justice in a comparative perspective. In A. Martínez & P. Abat (Eds.), *E-justice: Using information communication technologies in the court system* (pp. 1–19). Hershey, PA: Information Science Reference; doi:10.4018/978-1-59904-998-4.ch001

Fagbe, T., & Adekola, O. D. (2010). Workplace safety and personnel well-being: The impact of information technology. *International Journal of Green Computing*, *1*(1), 28–33. doi:10.4018/jgc.2010010103

Fagbe, T., & Adekola, O. D. (2011). Workplace safety and personnel well-being: The impact of information technology. In *Global business: Concepts, methodologies, tools and applications* (pp. 1438–1444). Hershey, PA: Business Science Reference; doi:10.4018/978-1-60960-587-2.ch509

Farmer, L. (2008). Affective collaborative instruction with librarians. In S. Kelsey & K. St.Amant (Eds.), *Handbook of research on computer mediated communication* (pp. 15–24). Hershey, PA: Information Science Reference; doi:10.4018/978-1-59904-863-5.ch002

Favier, L., & Mekhantar, J. (2007). Use of OSS by local e-administration: The French situation. In K. St.Amant & B. Still (Eds.), *Handbook of research on open source software: Technological, economic, and social perspectives* (pp. 428–444). Hershey, PA: Information Science Reference; doi:10.4018/978-1-59140-999-1.ch033

Fernando, S. (2009). Issues of e-learning in third world countries. In M. Khosrow-Pour (Ed.), *Encyclopedia of information science and technology* (2nd ed., pp. 2273–2277). Hershey, PA: Information Science Reference; doi:10.4018/978-1-60566-026-4.ch360

Filho, J. R., & dos Santos Junior, J. R. (2009). Local e-government in Brazil: Poor interaction and local politics as usual. In C. Reddick (Ed.), *Handbook of research on strategies for local e-government adoption and implementation: Comparative studies* (pp. 863–878). Hershey, PA: Information Science Reference; doi:10.4018/978-1-60566-282-4.ch045

Fletcher, P. D. (2004). Portals and policy: Implications of electronic access to U.S. federal government information services. In A. Pavlichev & G. Garson (Eds.), *Digital government: Principles and best practices* (pp. 52–62). Hershey, PA: Idea Group Publishing; doi:10.4018/978-1-59140-122-3.ch004

Fletcher, P. D. (2008). Portals and policy: Implications of electronic access to U.S. federal government information services. In A. Anttiroiko (Ed.), *Electronic government: Concepts, methodologies, tools, and applications* (pp. 3970–3979). Hershey, PA: Information Science Reference; doi:10.4018/978-1-59904-947-2.ch289

Forlano, L. (2004). The emergence of digital government: International perspectives. In A. Pavlichev & G. Garson (Eds.), *Digital government: Principles and best practices* (pp. 34–51). Hershey, PA: Idea Group Publishing; doi:10.4018/978-1-59140-122-3.ch003

Franzel, J. M., & Coursey, D. H. (2004). Government web portals: Management issues and the approaches of five states. In A. Pavlichev & G. Garson (Eds.), *Digital government: Principles and best practices* (pp. 63–77). Hershey, PA: Idea Group Publishing; doi:10.4018/978-1-59140-122-3.ch005

Gaivéo, J. M. (2013). Security of ICTs supporting healthcare activities. In M. Cruz-Cunha, I. Miranda, & P. Gonçalves (Eds.), *Handbook of research on ICTs for human-centered healthcare and social care services* (pp. 208–228). Hershey, PA: Medical Information Science Reference; doi:10.4018/978-1-4666-3986-7.ch011

Garson, G. D. (1999). *Information technology and computer applications in public administration: Issues and trends*. Hershey, PA: IGI Global; doi:10.4018/978-1-87828-952-0

Garson, G. D. (2003). Toward an information technology research agenda for public administration. In G. Garson (Ed.), *Public information technology: Policy and management issues* (pp. 331–357). Hershey, PA: Idea Group Publishing; doi:10.4018/978-1-59140-060-8.ch014

Garson, G. D. (2004). The promise of digital government. In A. Pavlichev & G. Garson (Eds.), *Digital government: Principles and best practices* (pp. 2–15). Hershey, PA: Idea Group Publishing; doi:10.4018/978-1-59140-122-3.ch001

Garson, G. D. (2007). An information technology research agenda for public administration. In G. Garson (Ed.), *Modern public information technology systems: Issues and challenges* (pp. 365–392). Hershey, PA: Idea Group Publishing; doi:10.4018/978-1-59904-051-6.ch018

Gasco, M. (2007). Civil servants' resistance towards e-government development. In A. Anttiroiko & M. Malkia (Eds.), *Encyclopedia of digital government* (pp. 190–195). Hershey, PA: Information Science Reference; doi:10.4018/978-1-59140-789-8.ch028

Gasco, M. (2008). Civil servants' resistance towards e-government development. In A. Anttiroiko (Ed.), *Electronic government: Concepts, methodologies, tools, and applications* (pp. 2580–2588). Hershey, PA: Information Science Reference; doi:10.4018/978-1-59904-947-2.ch190

Ghere, R. K. (2010). Accountability and information technology enactment: Implications for social empowerment. In E. Ferro, Y. Dwivedi, J. Gil-Garcia, & M. Williams (Eds.), *Handbook of research on overcoming digital divides: Constructing an equitable and competitive information society* (pp. 515–532). Hershey, PA: Information Science Reference; doi:10.4018/978-1-60566-699-0.ch028

Gibson, I. W. (2012). Simulation modeling of healthcare delivery. In A. Kolker & P. Story (Eds.), *Management engineering for effective healthcare delivery: Principles and applications* (pp. 69–89). Hershey, PA: Medical Information Science Reference; doi:10.4018/978-1-60960-872-9.ch003

Gil-Garcia, J. R. (2007). Exploring e-government benefits and success factors. In A. Anttiroiko & M. Malkia (Eds.), *Encyclopedia of digital government* (pp. 803–811). Hershey, PA: Information Science Reference; doi:10.4018/978-1-59140-789-8.ch122

Gil-Garcia, J. R., & González Miranda, F. (2010). E-government and opportunities for participation: The case of the Mexican state web portals. In C. Reddick (Ed.), *Politics, democracy and e-government: Participation and service delivery* (pp. 56–74). Hershey, PA: Information Science Reference; doi:10.4018/978-1-61520-933-0.ch004

Goldfinch, S. (2012). Public trust in government, trust in e-government, and use of e-government. In Z. Yan (Ed.), *Encyclopedia of cyber behavior* (pp. 987–995). Hershey, PA: Information Science Reference; doi:10.4018/978-1-4666-0315-8.ch081

Goodyear, M. (2012). Organizational change contributions to e-government project transitions. In S. Aikins (Ed.), *Managing e-government projects: Concepts, issues, and best practices* (pp. 1–21). Hershey, PA: Information Science Reference; doi:10.4018/978-1-4666-0086-7.ch001

Gordon, S., & Mulligan, P. (2003). Strategic models for the delivery of personal financial services: The role of infocracy. In S. Gordon (Ed.), *Computing information technology: The human side* (pp. 220–232). Hershey, PA: IRM Press; doi:10.4018/978-1-93177-752-0.ch014

Gordon, T. F. (2007). Legal knowledge systems. In A. Anttiroiko & M. Malkia (Eds.), *Encyclopedia of digital government* (pp. 1161–1166). Hershey, PA: Information Science Reference; doi:10.4018/978-1-59140-789-8.ch175

Graham, J. E., & Semich, G. W. (2008). Integrating technology to transform pedagogy: Revisiting the progress of the three phase TUI model for faculty development. In L. Tomei (Ed.), *Adapting information and communication technologies for effective education* (pp. 1–12). Hershey, PA: Information Science Reference; doi:10.4018/978-1-59904-922-9.ch001

Grandinetti, L., & Pisacane, O. (2012). Web services for healthcare management. In D. Prakash Vidyarthi (Ed.), *Technologies and protocols for the future of internet design: Reinventing the web* (pp. 60–94). Hershey, PA: Information Science Reference; doi:10.4018/978-1-4666-0203-8.ch004

Groenewegen, P., & Wagenaar, F. P. (2008). VO as an alternative to hierarchy in the Dutch police sector. In G. Putnik & M. Cruz-Cunha (Eds.), *Encyclopedia of networked and virtual organizations* (pp. 1851–1857). Hershey, PA: Information Science Reference; doi:10.4018/978-1-59904-885-7.ch245

Gronlund, A. (2001). Building an infrastructure to manage electronic services. In S. Dasgupta (Ed.), *Managing internet and intranet technologies in organizations: Challenges and opportunities* (pp. 71–103). Hershey, PA: Idea Group Publishing; doi:10.4018/978-1-878289-95-7.ch006

Gronlund, A. (2002). Introduction to electronic government: Design, applications and management. In Å. Grönlund (Ed.), *Electronic government: Design, applications and management* (pp. 1–21). Hershey, PA: Idea Group Publishing; doi:10.4018/978-1-930708-19-8.ch001

Gupta, A., Woosley, R., Crk, I., & Sarnikar, S. (2009). An information technology architecture for drug effectiveness reporting and post-marketing surveillance. In J. Tan (Ed.), *Medical informatics: Concepts, methodologies, tools, and applications* (pp. 631–646). Hershey, PA: Medical Information Science Reference; doi:10.4018/978-1-60566-050-9.ch047

Hallin, A., & Lundevall, K. (2007). mCity: User focused development of mobile services within the city of Stockholm. In I. Kushchu (Ed.), Mobile government: An emerging direction in e-government (pp. 12-29). Hershey, PA: Idea Group Publishing. doi:10.4018/978-1-59140-884-0.ch002

Hallin, A., & Lundevall, K. (2009). mCity: User focused development of mobile services within the city of Stockholm. In S. Clarke (Ed.), Evolutionary concepts in end user productivity and performance: Applications for organizational progress (pp. 268-280). Hershey, PA: Information Science Reference. doi:10.4018/978-1-60566-136-0.ch017

Hallin, A., & Lundevall, K. (2009). mCity: User focused development of mobile services within the city of Stockholm. In D. Taniar (Ed.), Mobile computing: Concepts, methodologies, tools, and applications (pp. 3455-3467). Hershey, PA: Information Science Reference. doi:10.4018/978-1-60566-054-7.ch253

Hanson, A. (2005). Overcoming barriers in the planning of a virtual library. In M. Khosrow-Pour (Ed.), *Encyclopedia of information science and technology* (pp. 2255–2259). Hershey, PA: Information Science Reference; doi:10.4018/978-1-59140-553-5.ch397

Haque, A. (2008). Information technology and surveillance: Implications for public administration in a new word order. In T. Loendorf & G. Garson (Eds.), *Patriotic information systems* (pp. 177–185). Hershey, PA: IGI Publishing; doi:10.4018/978-1-59904-594-8.ch008

Hauck, R. V., Thatcher, S. M., & Weisband, S. P. (2012). Temporal aspects of information technology use: Increasing shift work effectiveness. In J. Wang (Ed.), *Advancing the service sector with evolving technologies: Techniques and principles* (pp. 87–104). Hershey, PA: Business Science Reference; doi:10.4018/978-1-4666-0044-7.ch006

Hawk, S., & Witt, T. (2006). Telecommunications courses in information systems programs. *International Journal of Information and Communication Technology Education*, *2*(1), 79–92. doi:10.4018/jicte.2006010107

Helms, M. M., Moore, R., & Ahmadi, M. (2009). Information technology (IT) and the healthcare industry: A SWOT analysis. In J. Tan (Ed.), *Medical informatics: Concepts, methodologies, tools, and applications* (pp. 134–152). Hershey, PA: Medical Information Science Reference; doi:10.4018/978-1-60566-050-9.ch012

Hendrickson, S. M., & Young, M. E. (2014). Electronic records management at a federally funded research and development center. In J. Krueger (Ed.), *Cases on electronic records and resource management implementation in diverse environments* (pp. 334–350). Hershey, PA: Information Science Reference; doi:10.4018/978-1-4666-4466-3.ch020

Henman, P. (2010). Social policy and information communication technologies. In J. Martin & L. Hawkins (Eds.), *Information communication technologies for human services education and delivery: Concepts and cases* (pp. 215–229). Hershey, PA: Information Science Reference; doi:10.4018/978-1-60566-735-5.ch014

Hismanoglu, M. (2011). Important issues in online education: E-pedagogy and marketing. In U. Demiray & S. Sever (Eds.), *Marketing online education programs: Frameworks for promotion and communication* (pp. 184–209). Hershey, PA: Information Science Reference; doi:10.4018/978-1-60960-074-7.ch012

Ho, K. K. (2008). The e-government development, IT strategies, and portals of the Hong Kong SAR government. In A. Anttiroiko (Ed.), *Electronic government: Concepts, methodologies, tools, and applications* (pp. 715–733). Hershey, PA: Information Science Reference; doi:10.4018/978-1-59904-947-2.ch060

Holden, S. H. (2003). The evolution of information technology management at the federal level: Implications for public administration. In G. Garson (Ed.), *Public information technology: Policy and management issues* (pp. 53–73). Hershey, PA: Idea Group Publishing; doi:10.4018/978-1-59140-060-8.ch003

Holden, S. H. (2007). The evolution of federal information technology management literature: Does IT finally matter? In G. Garson (Ed.), *Modern public information technology systems: Issues and challenges* (pp. 17–34). Hershey, PA: Idea Group Publishing; doi:10.4018/978-1-59904-051-6.ch002

Holland, J. W. (2009). Automation of American criminal justice. In M. Khosrow-Pour (Ed.), *Encyclopedia of information science and technology* (2nd ed., pp. 300–302). Hershey, PA: Information Science Reference; doi:10.4018/978-1-60566-026-4.ch051

Holloway, K. (2013). Fair use, copyright, and academic integrity in an online academic environment. In *Digital rights management: Concepts, methodologies, tools, and applications* (pp. 917–928). Hershey, PA: Information Science Reference; doi:10.4018/978-1-4666-2136-7.ch044

Horiuchi, C. (2005). E-government databases. In L. Rivero, J. Doorn, & V. Ferraggine (Eds.), *Encyclopedia of database technologies and applications* (pp. 206–210). Hershey, PA: Information Science Reference; doi:10.4018/978-1-59140-560-3.ch035

Horiuchi, C. (2006). Creating IS quality in government settings. In E. Duggan & J. Reichgelt (Eds.), *Measuring information systems delivery quality* (pp. 311–327). Hershey, PA: Idea Group Publishing; doi:10.4018/978-1-59140-857-4.ch014

Hsiao, N., Chu, P., & Lee, C. (2012). Impact of e-governance on businesses: Model development and case study. In *Digital democracy: Concepts, methodologies, tools, and applications* (pp. 1407–1425). Hershey, PA: Information Science Reference; doi:10.4018/978-1-4666-1740-7.ch070

Huang, T., & Lee, C. (2010). Evaluating the impact of e-government on citizens: Cost-benefit analysis. In C. Reddick (Ed.), *Citizens and e-government: Evaluating policy and management* (pp. 37–52). Hershey, PA: Information Science Reference; doi:10.4018/978-1-61520-931-6.ch003

Hunter, M. G., Diochon, M., Pugsley, D., & Wright, B. (2002). Unique challenges for small business adoption of information technology: The case of the Nova Scotia ten. In S. Burgess (Ed.), *Managing information technology in small business: Challenges and solutions* (pp. 98–117). Hershey, PA: Idea Group Publishing; doi:10.4018/978-1-930708-35-8.ch006

Hurskainen, J. (2003). Integration of business systems and applications in merger and alliance: Case metso automation. In T. Reponen (Ed.), *Information technology enabled global customer service* (pp. 207–225). Hershey, PA: Idea Group Publishing; doi:10.4018/978-1-59140-048-6.ch012

Iazzolino, G., & Pietrantonio, R. (2011). The soveria.it project: A best practice of e-government in southern Italy. In D. Piaggesi, K. Sund, & W. Castelnovo (Eds.), *Global strategy and practice of e-governance: Examples from around the world* (pp. 34–56). Hershey, PA: Information Science Reference; doi:10.4018/978-1-60960-489-9.ch003

Imran, A., & Gregor, S. (2012). A process model for successful e-government adoption in the least developed countries: A case of Bangladesh. In F. Tan (Ed.), *International comparisons of information communication technologies: Advancing applications* (pp. 321–350). Hershey, PA: Information Science Reference; doi:10.4018/978-1-61350-480-2.ch014

Inoue, Y., & Bell, S. T. (2005). Electronic/digital government innovation, and publishing trends with IT. In M. Khosrow-Pour (Ed.), *Encyclopedia of information science and technology* (pp. 1018–1023). Hershey, PA: Information Science Reference; doi:10.4018/978-1-59140-553-5.ch180

Islam, M. M., & Ehsan, M. (2013). Understanding e-governance: A theoretical approach. In M. Islam & M. Ehsan (Eds.), *From government to e-governance: Public administration in the digital age* (pp. 38–49). Hershey, PA: Information Science Reference; doi:10.4018/978-1-4666-1909-8.ch003

Jaeger, B. (2009). E-government and e-democracy in the making. In M. Khosrow-Pour (Ed.), *Encyclopedia of information science and technology* (2nd ed., pp. 1318–1322). Hershey, PA: Information Science Reference; doi:10.4018/978-1-60566-026-4.ch208

Jain, R. B. (2007). Revamping the administrative structure and processes in India for online diplomacy. In A. Anttiroiko & M. Malkia (Eds.), *Encyclopedia of digital government* (pp. 1418–1423). Hershey, PA: Information Science Reference; doi:10.4018/978-1-59140-789-8.ch217

Jain, R. B. (2008). Revamping the administrative structure and processes in India for online diplomacy. In A. Anttiroiko (Ed.), *Electronic government: Concepts, methodologies, tools, and applications* (pp. 3142–3149). Hershey, PA: Information Science Reference; doi:10.4018/978-1-59904-947-2.ch233

Jauhiainen, J. S., & Inkinen, T. (2009). E-governance and the information society in periphery. In C. Reddick (Ed.), *Handbook of research on strategies for local e-government adoption and implementation: Comparative studies* (pp. 497–514). Hershey, PA: Information Science Reference; doi:10.4018/978-1-60566-282-4.ch026

Jensen, M. J. (2009). Electronic democracy and citizen influence in government. In C. Reddick (Ed.), *Handbook of research on strategies for local e-government adoption and implementation: Comparative studies* (pp. 288–305). Hershey, PA: Information Science Reference; doi:10.4018/978-1-60566-282-4.ch015

Jiao, Y., Hurson, A. R., Potok, T. E., & Beckerman, B. G. (2009). Integrating mobile-based systems with healthcare databases. In J. Erickson (Ed.), *Database technologies: Concepts, methodologies, tools, and applications* (pp. 484–504). Hershey, PA: Information Science Reference; doi:10.4018/978-1-60566-058-5.ch031

Joia, L. A. (2002). A systematic model to integrate information technology into metabusinesses: A case study in the engineering realms. In F. Tan (Ed.), *Advanced topics in global information management* (Vol. 1, pp. 250–267). Hershey, PA: Idea Group Publishing; doi:10.4018/978-1-930708-43-3.ch016

Jones, T. H., & Song, I. (2000). Binary equivalents of ternary relationships in entity-relationship modeling: A logical decomposition approach. *Journal of Database Management*, *11*(2), 12–19. doi:10.4018/jdm.2000040102

Juana-Espinosa, S. D. (2007). Empirical study of the municipalitites' motivations for adopting online presence. In L. Al-Hakim (Ed.), *Global e-government: Theory, applications and benchmarking* (pp. 261–279). Hershey, PA: Idea Group Publishing; doi:10.4018/978-1-59904-027-1.ch015

Jun, K., & Weare, C. (2012). Bridging from e-government practice to e-government research: Past trends and future directions. In K. Bwalya & S. Zulu (Eds.), *Handbook of research on e-government in emerging economies: Adoption, e-participation, and legal frameworks* (pp. 263–289). Hershey, PA: Information Science Reference; doi:10.4018/978-1-4666-0324-0.ch013

Junqueira, A., Diniz, E. H., & Fernandez, M. (2010). Electronic government implementation projects with multiple agencies: Analysis of the electronic invoice project under PMBOK framework. In J. Cordoba-Pachon & A. Ochoa-Arias (Eds.), *Systems thinking and e-participation: ICT in the governance of society* (pp. 135–153). Hershey, PA: Information Science Reference; doi:10.4018/978-1-60566-860-4.ch009

Juntunen, A. (2009). Joint service development with the local authorities. In C. Reddick (Ed.), *Handbook of research on strategies for local e-government adoption and implementation: Comparative studies* (pp. 902–920). Hershey, PA: Information Science Reference; doi:10.4018/978-1-60566-282-4.ch047

Kamel, S. (2001). *Using DSS for crisis management.* Hershey, PA: IGI Global; doi:10.4018/978-1-87828-961-2.ch020

Kamel, S. (2006). DSS for strategic decision making. In M. Khosrow-Pour (Ed.), *Cases on information technology and organizational politics & culture* (pp. 230–246). Hershey, PA: Idea Group Publishing; doi:10.4018/978-1-59904-411-8.ch015

Kamel, S. (2009). The software industry in Egypt as a potential contributor to economic growth. In M. Khosrow-Pour (Ed.), *Encyclopedia of information science and technology* (2nd ed., pp. 3531–3537). Hershey, PA: Information Science Reference; doi:10.4018/978-1-60566-026-4.ch562

Kamel, S., & Hussein, M. (2008). Xceed: Pioneering the contact center industry in Egypt. *Journal of Cases on Information Technology*, *10*(1), 67–91. doi:10.4018/jcit.2008010105

Kamel, S., & Wahba, K. (2003). The use of a hybrid model in web-based education: "The Global campus project. In A. Aggarwal (Ed.), *Web-based education: Learning from experience* (pp. 331–346). Hershey, PA: Information Science Publishing; doi:10.4018/978-1-59140-102-5.ch020

Kardaras, D. K., & Papathanassiou, E. A. (2008). An exploratory study of the e-government services in Greece. In G. Garson & M. Khosrow-Pour (Eds.), *Handbook of research on public information technology* (pp. 162–174). Hershey, PA: Information Science Reference; doi:10.4018/978-1-59904-857-4.ch016

Kassahun, A. E., Molla, A., & Sarkar, P. (2012). Government process reengineering: What we know and what we need to know. In *Digital democracy: Concepts, methodologies, tools, and applications* (pp. 1730–1752). Hershey, PA: Information Science Reference; doi:10.4018/978-1-4666-1740-7.ch086

Khan, B. (2005). Technological issues. In B. Khan (Ed.), *Managing e-learning strategies: Design, delivery, implementation and evaluation* (pp. 154–180). Hershey, PA: Information Science Publishing; doi:10.4018/978-1-59140-634-1.ch004

Khasawneh, A., Bsoul, M., Obeidat, I., & Al Azzam, I. (2012). Technology fears: A study of e-commerce loyalty perception by Jordanian customers. In J. Wang (Ed.), *Advancing the service sector with evolving technologies: Techniques and principles* (pp. 158–165). Hershey, PA: Business Science Reference; doi:10.4018/978-1-4666-0044-7.ch010

Khatibi, V., & Montazer, G. A. (2012). E-research methodology. In A. Juan, T. Daradoumis, M. Roca, S. Grasman, & J. Faulin (Eds.), *Collaborative and distributed e-research: Innovations in technologies, strategies and applications* (pp. 62–81). Hershey, PA: Information Science Reference; doi:10.4018/978-1-4666-0125-3.ch003

Kidd, T. (2011). The dragon in the school's backyard: A review of literature on the uses of technology in urban schools. In L. Tomei (Ed.), *Online courses and ICT in education: Emerging practices and applications* (pp. 242–257). Hershey, PA: Information Science Reference; doi:10.4018/978-1-60960-150-8.ch019

Kidd, T. T. (2010). My experience tells the story: Exploring technology adoption from a qualitative perspective - A pilot study. In H. Song & T. Kidd (Eds.), *Handbook of research on human performance and instructional technology* (pp. 247–262). Hershey, PA: Information Science Reference; doi:10.4018/978-1-60566-782-9.ch015

Kieley, B., Lane, G., Paquet, G., & Roy, J. (2002). e-Government in Canada: Services online or public service renewal? In Å. Grönlund (Ed.), Electronic government: Design, applications and management (pp. 340-355). Hershey, PA: Idea Group Publishing. doi:10.4018/978-1-930708-19-8.ch016

Kim, P. (2012). "Stay out of the way! My kid is video blogging through a phone!": A lesson learned from math tutoring social media for children in underserved communities. In *Wireless technologies: Concepts, methodologies, tools and applications* (pp. 1415–1428). Hershey, PA: Information Science Reference; doi:10.4018/978-1-61350-101-6.ch517

Kirlidog, M. (2010). Financial aspects of national ICT strategies. In S. Kamel (Ed.), *E-strategies for technological diffusion and adoption: National ICT approaches for socioeconomic development* (pp. 277–292). Hershey, PA: Information Science Reference; doi:10.4018/978-1-60566-388-3.ch016

Kisielnicki, J. (2006). Transfer of information and knowledge in the project management. In E. Coakes & S. Clarke (Eds.), *Encyclopedia of communities of practice in information and knowledge management* (pp. 544–551). Hershey, PA: Information Science Reference; doi:10.4018/978-1-59140-556-6.ch091

Kittner, M., & Van Slyke, C. (2006). Reorganizing information technology services in an academic environment. In M. Khosrow-Pour (Ed.), *Cases on the human side of information technology* (pp. 49–66). Hershey, PA: Idea Group Publishing; doi:10.4018/978-1-59904-405-7.ch004

Knoell, H. D. (2008). Semi virtual workplaces in German financial service enterprises. In P. Zemliansky & K. St.Amant (Eds.), *Handbook of research on virtual workplaces and the new nature of business practices* (pp. 570–581). Hershey, PA: Information Science Reference; doi:10.4018/978-1-59904-893-2.ch041

Koh, S. L., & Maguire, S. (2009). Competing in the age of information technology in a developing economy: Experiences of an Indian bank. In S. Koh & S. Maguire (Eds.), *Information and communication technologies management in turbulent business environments* (pp. 326–350). Hershey, PA: Information Science Reference; doi:10.4018/978-1-60566-424-8.ch018

Kollmann, T., & Häsel, M. (2009). Competence of information technology professionals in internet-based ventures. In I. Lee (Ed.), *Electronic business: Concepts, methodologies, tools, and applications* (pp. 1905–1919). Hershey, PA: Information Science Reference; doi:10.4018/978-1-60566-056-1.ch118

Kollmann, T., & Häsel, M. (2009). Competence of information technology professionals in internet-based ventures. In A. Cater-Steel (Ed.), *Information technology governance and service management: Frameworks and adaptations* (pp. 239–253). Hershey, PA: Information Science Reference; doi:10.4018/978-1-60566-008-0.ch013

Kollmann, T., & Häsel, M. (2010). Competence of information technology professionals in internet-based ventures. In *Electronic services: Concepts, methodologies, tools and applications* (pp. 1551–1565). Hershey, PA: Information Science Reference; doi:10.4018/978-1-61520-967-5.ch094

Kraemer, K., & King, J. L. (2006). Information technology and administrative reform: Will e-government be different? *International Journal of Electronic Government Research*, *2*(1), 1–20. doi:10.4018/jegr.2006010101

Kraemer, K., & King, J. L. (2008). Information technology and administrative reform: Will e-government be different? In D. Norris (Ed.), *E-government research: Policy and management* (pp. 1–20). Hershey, PA: IGI Publishing; doi:10.4018/978-1-59904-913-7.ch001

Lampathaki, F., Tsiakaliaris, C., Stasis, A., & Charalabidis, Y. (2011). National interoperability frameworks: The way forward. In Y. Charalabidis (Ed.), *Interoperability in digital public services and administration: Bridging e-government and e-business* (pp. 1–24). Hershey, PA: Information Science Reference; doi:10.4018/978-1-61520-887-6.ch001

Lan, Z., & Scott, C. R. (1996). The relative importance of computer-mediated information versus conventional non-computer-mediated information in public managerial decision making. *Information Resources Management Journal, 9*(1), 27–0. doi:10.4018/irmj.1996010103

Law, W. (2004). *Public sector data management in a developing economy*. Hershey, PA: IGI Global; doi:10.4018/978-1-59140-259-6.ch034

Law, W. K. (2005). Information resources development challenges in a cross-cultural environment. In M. Khosrow-Pour (Ed.), *Encyclopedia of information science and technology* (pp. 1476–1481). Hershey, PA: Information Science Reference; doi:10.4018/978-1-59140-553-5.ch259

Law, W. K. (2009). Cross-cultural challenges for information resources management. In M. Khosrow-Pour (Ed.), *Encyclopedia of information science and technology* (2nd ed., pp. 840–846). Hershey, PA: Information Science Reference; doi:10.4018/978-1-60566-026-4.ch136

Law, W. K. (2011). Cross-cultural challenges for information resources management. In *Global business: Concepts, methodologies, tools and applications* (pp. 1924–1932). Hershey, PA: Business Science Reference; doi:10.4018/978-1-60960-587-2.ch704

Malkia, M., & Savolainen, R. (2004). eTransformation in government, politics and society: Conceptual framework and introduction. In M. Malkia, A. Anttiroiko, & R. Savolainen (Eds.), eTransformation in governance: New directions in government and politics (pp. 1-21). Hershey, PA: Idea Group Publishing. doi:10.4018/978-1-59140-130-8.ch001

Management Association. I. (2010). Information resources management: Concepts, methodologies, tools and applications (4 Volumes). Hershey, PA: IGI Global. doi:10.4018/978-1-61520-965-1

Management Association. I. (2010). Electronic services: Concepts, methodologies, tools and applications (3 Volumes). Hershey, PA: IGI Global. doi:10.4018/978-1-61520-967-5

Mandujano, S. (2011). Network manageability security. In D. Kar & M. Syed (Eds.), *Network security, administration and management: Advancing technology and practice* (pp. 158–181). Hershey, PA: Information Science Reference; doi:10.4018/978-1-60960-777-7.ch009

Marich, M. J., Schooley, B. L., & Horan, T. A. (2012). A normative enterprise architecture for guiding end-to-end emergency response decision support. In M. Jennex (Ed.), *Managing crises and disasters with emerging technologies: Advancements* (pp. 71–87). Hershey, PA: Information Science Reference; doi:10.4018/978-1-4666-0167-3.ch006

Markov, R., & Okujava, S. (2008). Costs, benefits, and risks of e-government portals. In G. Putnik & M. Cruz-Cunha (Eds.), *Encyclopedia of networked and virtual organizations* (pp. 354–363). Hershey, PA: Information Science Reference; doi:10.4018/978-1-59904-885-7.ch047

Martin, N., & Rice, J. (2013). Evaluating and designing electronic government for the future: Observations and insights from Australia. In V. Weerakkody (Ed.), *E-government services design, adoption, and evaluation* (pp. 238–258). Hershey, PA: Information Science Reference; doi:10.4018/978-1-4666-2458-0.ch014

i. Martinez, A. C. (2008). Accessing administration's information via internet in Spain. In F. Tan (Ed.), *Global information technologies: Concepts, methodologies, tools, and applications* (pp. 2558–2573). Hershey, PA: Information Science Reference; doi:10.4018/978-1-59904-939-7.ch186

Mbarika, V. W., Meso, P. N., & Musa, P. F. (2006). A disconnect in stakeholders' perceptions from emerging realities of teledensity growth in Africa's least developed countries. In M. Hunter & F. Tan (Eds.), *Advanced topics in global information management* (Vol. 5, pp. 263–282). Hershey, PA: Idea Group Publishing; doi:10.4018/978-1-59140-923-6.ch012

Mbarika, V. W., Meso, P. N., & Musa, P. F. (2008). A disconnect in stakeholders' perceptions from emerging realities of teledensity growth in Africa's least developed countries. In F. Tan (Ed.), *Global information technologies: Concepts, methodologies, tools, and applications* (pp. 2948–2962). Hershey, PA: Information Science Reference; doi:10.4018/978-1-59904-939-7.ch209

Means, T., Olson, E., & Spooner, J. (2013). Discovering ways that don't work on the road to success: Strengths and weaknesses revealed by an active learning studio classroom project. In A. Benson, J. Moore, & S. Williams van Rooij (Eds.), *Cases on educational technology planning, design, and implementation: A project management perspective* (pp. 94–113). Hershey, PA: Information Science Reference; doi:10.4018/978-1-4666-4237-9.ch006

Melitski, J., Holzer, M., Kim, S., Kim, C., & Rho, S. (2008). Digital government worldwide: An e-government assessment of municipal web sites. In G. Garson & M. Khosrow-Pour (Eds.), *Handbook of research on public information technology* (pp. 790–804). Hershey, PA: Information Science Reference; doi:10.4018/978-1-59904-857-4.ch069

Memmola, M., Palumbo, G., & Rossini, M. (2009). Web & RFID technology: New frontiers in costing and process management for rehabilitation medicine. In L. Al-Hakim & M. Memmola (Eds.), *Business web strategy: Design, alignment, and application* (pp. 145–169). Hershey, PA: Information Science Reference; doi:10.4018/978-1-60566-024-0.ch008

Meng, Z., Fahong, Z., & Lei, L. (2008). Information technology and environment. In Y. Kurihara, S. Takaya, H. Harui, & H. Kamae (Eds.), *Information technology and economic development* (pp. 201–212). Hershey, PA: Information Science Reference; doi:10.4018/978-1-59904-579-5.ch014

Mentzingen de Moraes, A. J., Ferneda, E., Costa, I., & Spinola, M. D. (2011). Practical approach for implementation of governance process in IT: Information technology areas. In N. Shi & G. Silvius (Eds.), *Enterprise IT governance, business value and performance measurement* (pp. 19–40). Hershey, PA: Information Science Reference; doi:10.4018/978-1-60566-346-3.ch002

Merwin, G. A. Jr, McDonald, J. S., & Odera, L. C. (2008). Economic development: Government's cutting edge in IT. In M. Raisinghani (Ed.), *Handbook of research on global information technology management in the digital economy* (pp. 1–37). Hershey, PA: Information Science Reference; doi:10.4018/978-1-59904-875-8.ch001

Meso, P., & Duncan, N. (2002). Can national information infrastructures enhance social development in the least developed countries? An empirical investigation. In M. Dadashzadeh (Ed.), *Information technology management in developing countries* (pp. 23–51). Hershey, PA: IRM Press; doi:10.4018/978-1-931777-03-2.ch002

Meso, P. N., & Duncan, N. B. (2002). Can national information infrastructures enhance social development in the least developed countries? In F. Tan (Ed.), *Advanced topics in global information management* (Vol. 1, pp. 207–226). Hershey, PA: Idea Group Publishing; doi:10.4018/978-1-930708-43-3.ch014

Middleton, M. (2008). Evaluation of e-government web sites. In G. Garson & M. Khosrow-Pour (Eds.), *Handbook of research on public information technology* (pp. 699–710). Hershey, PA: Information Science Reference; doi:10.4018/978-1-59904-857-4.ch063

Mingers, J. (2010). Pluralism, realism, and truth: The keys to knowledge in information systems research. In D. Paradice (Ed.), *Emerging systems approaches in information technologies: Concepts, theories, and applications* (pp. 86–98). Hershey, PA: Information Science Reference; doi:10.4018/978-1-60566-976-2.ch006

Mital, K. M. (2012). ICT, unique identity and inclusive growth: An Indian perspective. In A. Manoharan & M. Holzer (Eds.), *E-governance and civic engagement: Factors and determinants of e-democracy* (pp. 584–612). Hershey, PA: Information Science Reference; doi:10.4018/978-1-61350-083-5.ch029

Mizell, A. P. (2008). Helping close the digital divide for financially disadvantaged seniors. In F. Tan (Ed.), *Global information technologies: Concepts, methodologies, tools, and applications* (pp. 2396–2402). Hershey, PA: Information Science Reference; doi:10.4018/978-1-59904-939-7.ch173

Molinari, F., Wills, C., Koumpis, A., & Moumtzi, V. (2011). A citizen-centric platform to support networking in the area of e-democracy. In H. Rahman (Ed.), *Cases on adoption, diffusion and evaluation of global e-governance systems: Impact at the grass roots* (pp. 282–302). Hershey, PA: Information Science Reference; doi:10.4018/978-1-61692-814-8.ch014

Molinari, F., Wills, C., Koumpis, A., & Moumtzi, V. (2013). A citizen-centric platform to support networking in the area of e-democracy. In H. Rahman (Ed.), *Cases on progressions and challenges in ICT utilization for citizen-centric governance* (pp. 265–297). Hershey, PA: Information Science Reference; doi:10.4018/978-1-4666-2071-1.ch013

Monteverde, F. (2010). The process of e-government public policy inclusion in the governmental agenda: A framework for assessment and case study. In J. Cordoba-Pachon & A. Ochoa-Arias (Eds.), *Systems thinking and e-participation: ICT in the governance of society* (pp. 233–245). Hershey, PA: Information Science Reference; doi:10.4018/978-1-60566-860-4.ch015

Moodley, S. (2008). Deconstructing the South African government's ICT for development discourse. In A. Anttiroiko (Ed.), *Electronic government: Concepts, methodologies, tools, and applications* (pp. 622–631). Hershey, PA: Information Science Reference; doi:10.4018/978-1-59904-947-2.ch053

Moodley, S. (2008). Deconstructing the South African government's ICT for development discourse. In C. Van Slyke (Ed.), *Information communication technologies: Concepts, methodologies, tools, and applications* (pp. 816–825). Hershey, PA: Information Science Reference; doi:10.4018/978-1-59904-949-6.ch052

Mora, M., Cervantes-Perez, F., Gelman-Muravchik, O., Forgionne, G. A., & Mejia-Olvera, M. (2003). DMSS implementation research: A conceptual analysis of the contributions and limitations of the factor-based and stage-based streams. In G. Forgionne, J. Gupta, & M. Mora (Eds.), *Decision-making support systems: Achievements and challenges for the new decade* (pp. 331–356). Hershey, PA: Idea Group Publishing; doi:10.4018/978-1-59140-045-5.ch020

Mörtberg, C., & Elovaara, P. (2010). Attaching people and technology: Between e and government. In S. Booth, S. Goodman, & G. Kirkup (Eds.), *Gender issues in learning and working with information technology: Social constructs and cultural contexts* (pp. 83–98). Hershey, PA: Information Science Reference; doi:10.4018/978-1-61520-813-5.ch005

Murphy, J., Harper, E., Devine, E. C., Burke, L. J., & Hook, M. L. (2011). Case study: Lessons learned when embedding evidence-based knowledge in a nurse care planning and documentation system. In A. Cashin & R. Cook (Eds.), *Evidence-based practice in nursing informatics: Concepts and applications* (pp. 174–190). Hershey, PA: Medical Information Science Reference; doi:10.4018/978-1-60960-034-1.ch014

Mutula, S. M. (2013). E-government's role in poverty alleviation: Case study of South Africa. In H. Rahman (Ed.), *Cases on progressions and challenges in ICT utilization for citizen-centric governance* (pp. 44–68). Hershey, PA: Information Science Reference; doi:10.4018/978-1-4666-2071-1.ch003

Nath, R., & Angeles, R. (2005). Relationships between supply characteristics and buyer-supplier coupling in e-procurement: An empirical analysis.[IJEBR]. *International Journal of E-Business Research*, *1*(2), 40–55. doi:10.4018/jebr.2005040103

Nissen, M. E. (2006). Application cases in government. In M. Nissen (Ed.), *Harnessing knowledge dynamics: Principled organizational knowing & learning* (pp. 152–181). Hershey, PA: IRM Press; doi:10.4018/978-1-59140-773-7.ch008

Norris, D. F. (2003). Leading-edge information technologies and American local governments. In G. Garson (Ed.), *Public information technology: Policy and management issues* (pp. 139–169). Hershey, PA: Idea Group Publishing; doi:10.4018/978-1-59140-060-8.ch007

Norris, D. F. (2008). Information technology among U.S. local governments. In G. Garson & M. Khosrow-Pour (Eds.), *Handbook of research on public information technology* (pp. 132–144). Hershey, PA: Information Science Reference; doi:10.4018/978-1-59904-857-4.ch013

Northrop, A. (1999). The challenge of teaching information technology in public administration graduate programs. In G. Garson (Ed.), *Information technology and computer applications in public administration: Issues and trends* (pp. 1–22). Hershey, PA: Information Science Reference; doi:10.4018/978-1-87828-952-0.ch001

Northrop, A. (2003). Information technology and public administration: The view from the profession. In G. Garson (Ed.), *Public information technology: Policy and management issues* (pp. 1–19). Hershey, PA: Idea Group Publishing; doi:10.4018/978-1-59140-060-8.ch001

Northrop, A. (2007). Lip service? How PA journals and textbooks view information technology. In G. Garson (Ed.), *Modern public information technology systems: Issues and challenges* (pp. 1–16). Hershey, PA: Idea Group Publishing; doi:10.4018/978-1-59904-051-6.ch001

Null, E. (2013). Legal and political barriers to municipal networks in the United States. In A. Abdelaal (Ed.), *Social and economic effects of community wireless networks and infrastructures* (pp. 27–56). Hershey, PA: Information Science Reference; doi:10.4018/978-1-4666-2997-4.ch003

Okunoye, A., Frolick, M., & Crable, E. (2006). ERP implementation in higher education: An account of pre-implementation and implementation phases. *Journal of Cases on Information Technology, 8*(2), 110–132. doi:10.4018/jcit.2006040106

Olasina, G. (2012). A review of egovernment services in Nigeria. In A. Tella & A. Issa (Eds.), *Library and information science in developing countries: Contemporary issues* (pp. 205–221). Hershey, PA: Information Science Reference; doi:10.4018/978-1-61350-335-5.ch015

Orgeron, C. P. (2008). A model for reengineering IT job classes in state government. In G. Garson & M. Khosrow-Pour (Eds.), *Handbook of research on public information technology* (pp. 735–746). Hershey, PA: Information Science Reference; doi:10.4018/978-1-59904-857-4.ch066

Owsinski, J. W., & Pielak, A. M. (2011). Local authority websites in rural areas: Measuring quality and functionality, and assessing the role. In Z. Andreopoulou, B. Manos, N. Polman, & D. Viaggi (Eds.), *Agricultural and environmental informatics, governance and management: Emerging research applications* (pp. 39–60). Hershey, PA: Information Science Reference; doi:10.4018/978-1-60960-621-3.ch003

Owsiński, J. W., Pielak, A. M., Sęp, K., & Stańczak, J. (2014). Local web-based networks in rural municipalities: Extension, density, and meaning. In Z. Andreopoulou, V. Samathrakis, S. Louca, & M. Vlachopoulou (Eds.), *E-innovation for sustainable development of rural resources during global economic crisis* (pp. 126–151). Hershey, PA: Business Science Reference; doi:10.4018/978-1-4666-4550-9.ch011

Pagani, M., & Pasinetti, C. (2008). Technical and functional quality in the development of t-government services. In A. Anttiroiko (Ed.), *Electronic government: Concepts, methodologies, tools, and applications* (pp. 2943–2965). Hershey, PA: Information Science Reference; doi:10.4018/978-1-59904-947-2.ch220

Pani, A. K., & Agrahari, A. (2005). On e-markets in emerging economy: An Indian experience. In M. Khosrow-Pour (Ed.), *Advanced topics in electronic commerce* (Vol. 1, pp. 287–299). Hershey, PA: Idea Group Publishing; doi:10.4018/978-1-59140-819-2.ch015

Papadopoulos, T., Angelopoulos, S., & Kitsios, F. (2011). A strategic approach to e-health interoperability using e-government frameworks. In A. Lazakidou, K. Siassiakos, & K. Ioannou (Eds.), *Wireless technologies for ambient assisted living and healthcare: Systems and applications* (pp. 213–229). Hershey, PA: Medical Information Science Reference; doi:10.4018/978-1-61520-805-0.ch012

Papadopoulos, T., Angelopoulos, S., & Kitsios, F. (2013). A strategic approach to e-health interoperability using e-government frameworks. In *User-driven healthcare: Concepts, methodologies, tools, and applications* (pp. 791–807). Hershey, PA: Medical Information Science Reference; doi:10.4018/978-1-4666-2770-3.ch039

Papaleo, G., Chiarella, D., Aiello, M., & Caviglione, L. (2012). Analysis, development and deployment of statistical anomaly detection techniques for real e-mail traffic. In T. Chou (Ed.), *Information assurance and security technologies for risk assessment and threat management: Advances* (pp. 47–71). Hershey, PA: Information Science Reference; doi:10.4018/978-1-61350-507-6.ch003

Papp, R. (2003). Information technology & FDA compliance in the pharmaceutical industry. In M. Khosrow-Pour (Ed.), *Annals of cases on information technology* (Vol. 5, pp. 262–273). Hershey, PA: Information Science Reference; doi:10.4018/978-1-59140-061-5.ch017

Parsons, T. W. (2007). Developing a knowledge management portal. In A. Tatnall (Ed.), *Encyclopedia of portal technologies and applications* (pp. 223–227). Hershey, PA: Information Science Reference; doi:10.4018/978-1-59140-989-2.ch039

Passaris, C. E. (2007). Immigration and digital government. In A. Anttiroiko & M. Malkia (Eds.), *Encyclopedia of digital government* (pp. 988–994). Hershey, PA: Information Science Reference; doi:10.4018/978-1-59140-789-8.ch148

Pavlichev, A. (2004). The e-government challenge for public administration. In A. Pavlichev & G. Garson (Eds.), *Digital government: Principles and best practices* (pp. 276–290). Hershey, PA: Idea Group Publishing; doi:10.4018/978-1-59140-122-3.ch018

Penrod, J. I., & Harbor, A. F. (2000). Designing and implementing a learning organization-oriented information technology planning and management process. In L. Petrides (Ed.), *Case studies on information technology in higher education: Implications for policy and practice* (pp. 7–19). Hershey, PA: Idea Group Publishing; doi:10.4018/978-1-878289-74-2.ch001

Planas-Silva, M. D., & Joseph, R. C. (2011). Perspectives on the adoption of electronic resources for use in clinical trials. In M. Guah (Ed.), *Healthcare delivery reform and new technologies: Organizational initiatives* (pp. 19–28). Hershey, PA: Information Science Reference; doi:10.4018/978-1-60960-183-6.ch002

Pomazalová, N., & Rejman, S. (2013). The rationale behind implementation of new electronic tools for electronic public procurement. In N. Pomazalová (Ed.), *Public sector transformation processes and internet public procurement: Decision support systems* (pp. 85–117). Hershey, PA: Engineering Science Reference; doi:10.4018/978-1-4666-2665-2.ch006

Postorino, M. N. (2012). City competitiveness and airport: Information science perspective. In M. Bulu (Ed.), *City competitiveness and improving urban subsystems: Technologies and applications* (pp. 61–83). Hershey, PA: Information Science Reference; doi:10.4018/978-1-61350-174-0.ch004

Poupa, C. (2002). Electronic government in Switzerland: Priorities for 2001-2005 - Electronic voting and federal portal. In Å. Grönlund (Ed.), *Electronic government: Design, applications and management* (pp. 356–369). Hershey, PA: Idea Group Publishing; doi:10.4018/978-1-930708-19-8.ch017

Powell, S. R. (2010). Interdisciplinarity in telecommunications and networking. In *Networking and telecommunications: Concepts, methodologies, tools and applications* (pp. 33–40). Hershey, PA: Information Science Reference; doi:10.4018/978-1-60566-986-1.ch004

Priya, P. S., & Mathiyalagan, N. (2011). A study of the implementation status of two e-governance projects in land revenue administration in India. In M. Shareef, V. Kumar, U. Kumar, & Y. Dwivedi (Eds.), *Stakeholder adoption of e-government services: Driving and resisting factors* (pp. 214–230). Hershey, PA: Information Science Reference; doi:10.4018/978-1-60960-601-5.ch011

Prysby, C., & Prysby, N. (2000). Electronic mail, employee privacy and the workplace. In L. Janczewski (Ed.), *Internet and intranet security management: Risks and solutions* (pp. 251–270). Hershey, PA: Idea Group Publishing; doi:10.4018/978-1-878289-71-1.ch009

Prysby, C. L., & Prysby, N. D. (2003). Electronic mail in the public workplace: Issues of privacy and public disclosure. In G. Garson (Ed.), *Public information technology: Policy and management issues* (pp. 271–298). Hershey, PA: Idea Group Publishing; doi:10.4018/978-1-59140-060-8.ch012

Prysby, C. L., & Prysby, N. D. (2007). You have mail, but who is reading it? Issues of e-mail in the public workplace. In G. Garson (Ed.), *Modern public information technology systems: Issues and challenges* (pp. 312–336). Hershey, PA: Idea Group Publishing; doi:10.4018/978-1-59904-051-6.ch016

Radl, A., & Chen, Y. (2005). Computer security in electronic government: A state-local education information system. *International Journal of Electronic Government Research*, *1*(1), 79–99. doi:10.4018/jegr.2005010105

Rahman, H. (2008). Information dynamics in developing countries. In C. Van Slyke (Ed.), *Information communication technologies: Concepts, methodologies, tools, and applications* (pp. 104–114). Hershey, PA: Information Science Reference; doi:10.4018/978-1-59904-949-6.ch008

Ramanathan, J. (2009). Adaptive IT architecture as a catalyst for network capability in government. In P. Saha (Ed.), *Advances in government enterprise architecture* (pp. 149–172). Hershey, PA: Information Science Reference; doi:10.4018/978-1-60566-068-4.ch007

Ramos, I., & Berry, D. M. (2006). Social construction of information technology supporting work. In M. Khosrow-Pour (Ed.), *Cases on information technology: Lessons learned* (Vol. 7, pp. 36–52). Hershey, PA: Idea Group Publishing; doi:10.4018/978-1-59140-673-0.ch003

Ray, D., Gulla, U., Gupta, M. P., & Dash, S. S. (2009). Interoperability and constituents of interoperable systems in public sector. In V. Weerakkody, M. Janssen, & Y. Dwivedi (Eds.), *Handbook of research on ICT-enabled transformational government: A global perspective* (pp. 175–195). Hershey, PA: Information Science Reference; doi:10.4018/978-1-60566-390-6.ch010

Reddick, C. G. (2007). E-government and creating a citizen-centric government: A study of federal government CIOs. In G. Garson (Ed.), *Modern public information technology systems: Issues and challenges* (pp. 143–165). Hershey, PA: Idea Group Publishing; doi:10.4018/978-1-59904-051-6.ch008

Reddick, C. G. (2010). Citizen-centric e-government. In C. Reddick (Ed.), *Homeland security preparedness and information systems: Strategies for managing public policy* (pp. 45–75). Hershey, PA: Information Science Reference; doi:10.4018/978-1-60566-834-5.ch002

Reddick, C. G. (2010). E-government and creating a citizen-centric government: A study of federal government CIOs. In C. Reddick (Ed.), *Homeland security preparedness and information systems: Strategies for managing public policy* (pp. 230–250). Hershey, PA: Information Science Reference; doi:10.4018/978-1-60566-834-5.ch012

Reddick, C. G. (2010). Perceived effectiveness of e-government and its usage in city governments: Survey evidence from information technology directors. In C. Reddick (Ed.), *Homeland security preparedness and information systems: Strategies for managing public policy* (pp. 213–229). Hershey, PA: Information Science Reference; doi:10.4018/978-1-60566-834-5.ch011

Reddick, C. G. (2012). Customer relationship management adoption in local governments in the United States. In S. Chhabra & M. Kumar (Eds.), *Strategic enterprise resource planning models for e-government: Applications and methodologies* (pp. 111–124). Hershey, PA: Information Science Reference; doi:10.4018/978-1-60960-863-7.ch008

Reeder, F. S., & Pandy, S. M. (2008). Identifying effective funding models for e-government. In A. Anttiroiko (Ed.), *Electronic government: Concepts, methodologies, tools, and applications* (pp. 1108–1138). Hershey, PA: Information Science Reference; doi:10.4018/978-1-59904-947-2.ch083

Riesco, D., Acosta, E., & Montejano, G. (2003). An extension to a UML activity graph from workflow. In L. Favre (Ed.), *UML and the unified process* (pp. 294–314). Hershey, PA: IRM Press; doi:10.4018/978-1-93177-744-5.ch015

Ritzhaupt, A. D., & Gill, T. G. (2008). A hybrid and novel approach to teaching computer programming in MIS curriculum. In S. Negash, M. Whitman, A. Woszczynski, K. Hoganson, & H. Mattord (Eds.), *Handbook of distance learning for real-time and asynchronous information technology education* (pp. 259–281). Hershey, PA: Information Science Reference; doi:10.4018/978-1-59904-964-9.ch014

Roche, E. M. (1993). International computing and the international regime. *Journal of Global Information Management, 1*(2), 33–44. doi:10.4018/jgim.1993040103

Rocheleau, B. (2007). Politics, accountability, and information management. In G. Garson (Ed.), *Modern public information technology systems: Issues and challenges* (pp. 35–71). Hershey, PA: Idea Group Publishing; doi:10.4018/978-1-59904-051-6.ch003

Rodrigues Filho, J. (2010). E-government in Brazil: Reinforcing dominant institutions or reducing citizenship? In C. Reddick (Ed.), *Politics, democracy and e-government: Participation and service delivery* (pp. 347–362). Hershey, PA: Information Science Reference; doi:10.4018/978-1-61520-933-0.ch021

Rodriguez, S. R., & Thorp, D. A. (2013). eLearning for industry: A case study of the project management process. In A. Benson, J. Moore, & S. Williams van Rooij (Eds.), Cases on educational technology planning, design, and implementation: A project management perspective (pp. 319-342). Hershey, PA: Information Science Reference. doi:10.4018/978-1-4666-4237-9.ch017

Roman, A. V. (2013). Delineating three dimensions of e-government success: Security, functionality, and transformation. In J. Gil-Garcia (Ed.), *E-government success factors and measures: Theories, concepts, and methodologies* (pp. 171–192). Hershey, PA: Information Science Reference; doi:10.4018/978-1-4666-4058-0.ch010

Ross, S. C., Tyran, C. K., & Auer, D. J. (2008). Up in smoke: Rebuilding after an IT disaster. In H. Nemati (Ed.), *Information security and ethics: Concepts, methodologies, tools, and applications* (pp. 3659–3675). Hershey, PA: Information Science Reference; doi:10.4018/978-1-59904-937-3.ch248

Ross, S. C., Tyran, C. K., Auer, D. J., Junell, J. M., & Williams, T. G. (2005). Up in smoke: Rebuilding after an IT disaster. *Journal of Cases on Information Technology*, *7*(2), 31–49. doi:10.4018/jcit.2005040103

Roy, J. (2008). Security, sovereignty, and continental interoperability: Canada's elusive balance. In T. Loendorf & G. Garson (Eds.), *Patriotic information systems* (pp. 153–176). Hershey, PA: IGI Publishing; doi:10.4018/978-1-59904-594-8.ch007

Rubeck, R. F., & Miller, G. A. (2009). vGOV: Remote video access to government services. In A. Scupola (Ed.), Cases on managing e-services (pp. 253-268). Hershey, PA: Information Science Reference. doi:10.4018/978-1-60566-064-6.ch017

Saekow, A., & Boonmee, C. (2011). The challenges of implementing e-government interoperability in Thailand: Case of official electronic correspondence letters exchange across government departments. In Y. Charalabidis (Ed.), *Interoperability in digital public services and administration: Bridging e-government and e-business* (pp. 40–61). Hershey, PA: Information Science Reference; doi:10.4018/978-1-61520-887-6.ch003

Saekow, A., & Boonmee, C. (2012). The challenges of implementing e-government interoperability in Thailand: Case of official electronic correspondence letters exchange across government departments. In *Digital democracy: Concepts, methodologies, tools, and applications* (pp. 1883–1905). Hershey, PA: Information Science Reference; doi:10.4018/978-1-4666-1740-7.ch094

Sagsan, M., & Medeni, T. (2012). Understanding "knowledge management (KM) paradigms" from social media perspective: An empirical study on discussion group for KM at professional networking site. In M. Cruz-Cunha, P. Gonçalves, N. Lopes, E. Miranda, & G. Putnik (Eds.), *Handbook of research on business social networking: Organizational, managerial, and technological dimensions* (pp. 738–755). Hershey, PA: Business Science Reference; doi:10.4018/978-1-61350-168-9.ch039

Sahi, G., & Madan, S. (2013). Information security threats in ERP enabled e-governance: Challenges and solutions. In *Enterprise resource planning: Concepts, methodologies, tools, and applications* (pp. 825–837). Hershey, PA: Business Science Reference; doi:10.4018/978-1-4666-4153-2.ch048

Sanford, C., & Bhattacherjee, A. (2008). IT implementation in a developing country municipality: A sociocognitive analysis. *International Journal of Technology and Human Interaction*, *4*(3), 68–93. doi:10.4018/jthi.2008070104

Schelin, S. H. (2003). E-government: An overview. In G. Garson (Ed.), *Public information technology: Policy and management issues* (pp. 120–138). Hershey, PA: Idea Group Publishing; doi:10.4018/978-1-59140-060-8.ch006

Schelin, S. H. (2004). Training for digital government. In A. Pavlichev & G. Garson (Eds.), *Digital government: Principles and best practices* (pp. 263–275). Hershey, PA: Idea Group Publishing; doi:10.4018/978-1-59140-122-3.ch017

Schelin, S. H. (2007). E-government: An overview. In G. Garson (Ed.), *Modern public information technology systems: Issues and challenges* (pp. 110–126). Hershey, PA: Idea Group Publishing; doi:10.4018/978-1-59904-051-6.ch006

Schelin, S. H., & Garson, G. (2004). Theoretical justification of critical success factors. In G. Garson & S. Schelin (Eds.), *IT solutions series: Humanizing information technology: Advice from experts* (pp. 4–15). Hershey, PA: CyberTech Publishing; doi:10.4018/978-1-59140-245-9.ch002

Scime, A. (2002). Information systems and computer science model curricula: A comparative look. In M. Dadashzadeh, A. Saber, & S. Saber (Eds.), *Information technology education in the new millennium* (pp. 146–158). Hershey, PA: IRM Press; doi:10.4018/978-1-931777-05-6.ch018

Scime, A. (2009). Computing curriculum analysis and development. In M. Khosrow-Pour (Ed.), *Encyclopedia of information science and technology* (2nd ed., pp. 667–671). Hershey, PA: Information Science Reference; doi:10.4018/978-1-60566-026-4.ch108

Scime, A., & Wania, C. (2008). Computing curricula: A comparison of models. In C. Van Slyke (Ed.), *Information communication technologies: Concepts, methodologies, tools, and applications* (pp. 1270–1283). Hershey, PA: Information Science Reference; doi:10.4018/978-1-59904-949-6.ch088

Seidman, S. B. (2009). An international perspective on professional software engineering credentials. In H. Ellis, S. Demurjian, & J. Naveda (Eds.), *Software engineering: Effective teaching and learning approaches and practices* (pp. 351–361). Hershey, PA: Information Science Reference; doi:10.4018/978-1-60566-102-5.ch018

Seifert, J. W. (2007). E-government act of 2002 in the United States. In A. Anttiroiko & M. Malkia (Eds.), *Encyclopedia of digital government* (pp. 476–481). Hershey, PA: Information Science Reference; doi:10.4018/978-1-59140-789-8.ch072

Seifert, J. W., & Relyea, H. C. (2008). E-government act of 2002 in the United States. In A. Anttiroiko (Ed.), *Electronic government: Concepts, methodologies, tools, and applications* (pp. 154–161). Hershey, PA: Information Science Reference; doi:10.4018/978-1-59904-947-2.ch013

Seufert, S. (2002). E-learning business models: Framework and best practice examples. In M. Raisinghani (Ed.), *Cases on worldwide e-commerce: Theory in action* (pp. 70–94). Hershey, PA: Idea Group Publishing; doi:10.4018/978-1-930708-27-3.ch004

Shareef, M. A., & Archer, N. (2012). E-government service development. In M. Shareef, N. Archer, & S. Dutta (Eds.), *E-government service maturity and development: Cultural, organizational and technological perspectives* (pp. 1–14). Hershey, PA: Information Science Reference; doi:10.4018/978-1-60960-848-4.ch001

Shareef, M. A., & Archer, N. (2012). E-government initiatives: Review studies on different countries. In M. Shareef, N. Archer, & S. Dutta (Eds.), *E-government service maturity and development: Cultural, organizational and technological perspectives* (pp. 40–76). Hershey, PA: Information Science Reference; doi:10.4018/978-1-60960-848-4.ch003

Shareef, M. A., Kumar, U., & Kumar, V. (2011). E-government development: Performance evaluation parameters. In M. Shareef, V. Kumar, U. Kumar, & Y. Dwivedi (Eds.), *Stakeholder adoption of e-government services: Driving and resisting factors* (pp. 197–213). Hershey, PA: Information Science Reference; doi:10.4018/978-1-60960-601-5.ch010

Shareef, M. A., Kumar, U., Kumar, V., & Niktash, M. (2012). Electronic-government vision: Case studies for objectives, strategies, and initiatives. In M. Shareef, N. Archer, & S. Dutta (Eds.), *E-government service maturity and development: Cultural, organizational and technological perspectives* (pp. 15–39). Hershey, PA: Information Science Reference; doi:10.4018/978-1-60960-848-4.ch002

Shukla, P., Kumar, A., & Anu Kumar, P. B. (2013). Impact of national culture on business continuity management system implementation. *International Journal of Risk and Contingency Management, 2*(3), 23–36. doi:10.4018/ijrcm.2013070102

Shulman, S. W. (2007). The federal docket management system and the prospect for digital democracy in U S rulemaking. In G. Garson (Ed.), *Modern public information technology systems: Issues and challenges* (pp. 166–184). Hershey, PA: Idea Group Publishing; doi:10.4018/978-1-59904-051-6.ch009

Simonovic, S. (2007). Problems of offline government in e-Serbia. In A. Anttiroiko & M. Malkia (Eds.), *Encyclopedia of digital government* (pp. 1342–1351). Hershey, PA: Information Science Reference; doi:10.4018/978-1-59140-789-8.ch205

Simonovic, S. (2008). Problems of offline government in e-Serbia. In A. Anttiroiko (Ed.), *Electronic government: Concepts, methodologies, tools, and applications* (pp. 2929–2942). Hershey, PA: Information Science Reference; doi:10.4018/978-1-59904-947-2.ch219

Singh, A. M. (2005). Information systems and technology in South Africa. In M. Khosrow-Pour (Ed.), *Encyclopedia of information science and technology* (pp. 1497–1502). Hershey, PA: Information Science Reference; doi:10.4018/978-1-59140-553-5.ch263

Singh, S., & Naidoo, G. (2005). Towards an e-government solution: A South African perspective. In W. Huang, K. Siau, & K. Wei (Eds.), *Electronic government strategies and implementation* (pp. 325–353). Hershey, PA: Idea Group Publishing; doi:10.4018/978-1-59140-348-7.ch014

Snoke, R., & Underwood, A. (2002). Generic attributes of IS graduates: An analysis of Australian views. In F. Tan (Ed.), *Advanced topics in global information management* (Vol. 1, pp. 370–384). Hershey, PA: Idea Group Publishing; doi:10.4018/978-1-930708-43-3.ch023

Sommer, L. (2006). Revealing unseen organizations in higher education: A study framework and application example. In A. Metcalfe (Ed.), *Knowledge management and higher education: A critical analysis* (pp. 115–146). Hershey, PA: Information Science Publishing; doi:10.4018/978-1-59140-509-2.ch007

Song, H., Kidd, T., & Owens, E. (2011). Examining technological disparities and instructional practices in English language arts classroom: Implications for school leadership and teacher training. In L. Tomei (Ed.), *Online courses and ICT in education: Emerging practices and applications* (pp. 258–274). Hershey, PA: Information Science Reference; doi:10.4018/978-1-60960-150-8.ch020

Speaker, P. J., & Kleist, V. F. (2003). Using information technology to meet electronic commerce and MIS education demands. In A. Aggarwal (Ed.), *Web-based education: Learning from experience* (pp. 280–291). Hershey, PA: Information Science Publishing; doi:10.4018/978-1-59140-102-5.ch017

Spitler, V. K. (2007). Learning to use IT in the workplace: Mechanisms and masters. In M. Mahmood (Ed.), *Contemporary issues in end user computing* (pp. 292–323). Hershey, PA: Idea Group Publishing; doi:10.4018/978-1-59140-926-7.ch013

Stellefson, M. (2011). Considerations for marketing distance education courses in health education: Five important questions to examine before development. In U. Demiray & S. Sever (Eds.), *Marketing online education programs: Frameworks for promotion and communication* (pp. 222–234). Hershey, PA: Information Science Reference; doi:10.4018/978-1-60960-074-7.ch014

Straub, D. W., & Loch, K. D. (2006). Creating and developing a program of global research. *Journal of Global Information Management, 14*(2), 1–28. doi:10.4018/jgim.2006040101

Straub, D. W., Loch, K. D., & Hill, C. E. (2002). Transfer of information technology to the Arab world: A test of cultural influence modeling. In M. Dadashzadeh (Ed.), *Information technology management in developing countries* (pp. 92–134). Hershey, PA: IRM Press; doi:10.4018/978-1-931777-03-2.ch005

Straub, D. W., Loch, K. D., & Hill, C. E. (2003). Transfer of information technology to the Arab world: A test of cultural influence modeling. In F. Tan (Ed.), *Advanced topics in global information management* (Vol. 2, pp. 141–172). Hershey, PA: Idea Group Publishing; doi:10.4018/978-1-59140-064-6.ch009

Suki, N. M., Ramayah, T., Ming, M. K., & Suki, N. M. (2013). Factors enhancing employed job seekers intentions to use social networking sites as a job search tool. In A. Mesquita (Ed.), *User perception and influencing factors of technology in everyday life* (pp. 265–281). Hershey, PA: Information Science Reference; doi:10.4018/978-1-4666-1954-8.ch018

Suomi, R. (2006). Introducing electronic patient records to hospitals: Innovation adoption paths. In T. Spil & R. Schuring (Eds.), *E-health systems diffusion and use: The innovation, the user and the use IT model* (pp. 128–146). Hershey, PA: Idea Group Publishing; doi:10.4018/978-1-59140-423-1.ch008

Swim, J., & Barker, L. (2012). Pathways into a gendered occupation: Brazilian women in IT. *International Journal of Social and Organizational Dynamics in IT, 2*(4), 34–51. doi:10.4018/ijsodit.2012100103

Tarafdar, M., & Vaidya, S. D. (2006). Adoption and implementation of IT in developing nations: Experiences from two public sector enterprises in India. In M. Khosrow-Pour (Ed.), *Cases on information technology planning, design and implementation* (pp. 208–233). Hershey, PA: Idea Group Publishing; doi:10.4018/978-1-59904-408-8.ch013

Tarafdar, M., & Vaidya, S. D. (2008). Adoption and implementation of IT in developing nations: Experiences from two public sector enterprises in India. In G. Garson & M. Khosrow-Pour (Eds.), *Handbook of research on public information technology* (pp. 905–924). Hershey, PA: Information Science Reference; doi:10.4018/978-1-59904-857-4.ch076

Thesing, Z. (2007). Zarina thesing, pumpkin patch. In M. Hunter (Ed.), *Contemporary chief information officers: Management experiences* (pp. 83–94). Hershey, PA: IGI Publishing; doi:10.4018/978-1-59904-078-3.ch007

Thomas, J. C. (2004). Public involvement in public administration in the information age: Speculations on the effects of technology. In M. Malkia, A. Anttiroiko, & R. Savolainen (Eds.), *eTransformation in governance: New directions in government and politics* (pp. 67–84). Hershey, PA: Idea Group Publishing; doi:10.4018/978-1-59140-130-8.ch004

Treiblmaier, H., & Chong, S. (2013). Trust and perceived risk of personal information as antecedents of online information disclosure: Results from three countries. In F. Tan (Ed.), *Global diffusion and adoption of technologies for knowledge and information sharing* (pp. 341–361). Hershey, PA: Information Science Reference; doi:10.4018/978-1-4666-2142-8.ch015

van Grembergen, W., & de Haes, S. (2008). IT governance in practice: Six case studies. In W. van Grembergen & S. De Haes (Eds.), *Implementing information technology governance: Models, practices and cases* (pp. 125–237). Hershey, PA: IGI Publishing; doi:10.4018/978-1-59904-924-3.ch004

van Os, G., Homburg, V., & Bekkers, V. (2013). Contingencies and convergence in European social security: ICT coordination in the back office of the welfare state. In M. Cruz-Cunha, I. Miranda, & P. Gonçalves (Eds.), *Handbook of research on ICTs and management systems for improving efficiency in healthcare and social care* (pp. 268–287). Hershey, PA: Medical Information Science Reference; doi:10.4018/978-1-4666-3990-4.ch013

Velloso, A. B., Gassenferth, W., & Machado, M. A. (2012). Evaluating IBMEC-RJ's intranet usability using fuzzy logic. In M. Cruz-Cunha, P. Gonçalves, N. Lopes, E. Miranda, & G. Putnik (Eds.), *Handbook of research on business social networking: Organizational, managerial, and technological dimensions* (pp. 185–205). Hershey, PA: Business Science Reference; doi:10.4018/978-1-61350-168-9.ch010

Villablanca, A. C., Baxi, H., & Anderson, K. (2009). Novel data interface for evaluating cardiovascular outcomes in women. In A. Dwivedi (Ed.), *Handbook of research on information technology management and clinical data administration in healthcare* (pp. 34–53). Hershey, PA: Medical Information Science Reference; doi:10.4018/978-1-60566-356-2.ch003

Villablanca, A. C., Baxi, H., & Anderson, K. (2011). Novel data interface for evaluating cardiovascular outcomes in women. In *Clinical technologies: Concepts, methodologies, tools and applications* (pp. 2094–2113). Hershey, PA: Medical Information Science Reference; doi:10.4018/978-1-60960-561-2.ch806

Virkar, S. (2011). Information and communication technologies in administrative reform for development: Exploring the case of property tax systems in Karnataka, India. In J. Steyn, J. Van Belle, & E. Mansilla (Eds.), *ICTs for global development and sustainability: Practice and applications* (pp. 127–149). Hershey, PA: Information Science Reference; doi:10.4018/978-1-61520-997-2.ch006

Virkar, S. (2013). Designing and implementing e-government projects: Actors, influences, and fields of play. In S. Saeed & C. Reddick (Eds.), *Human-centered system design for electronic governance* (pp. 88–110). Hershey, PA: Information Science Reference; doi:10.4018/978-1-4666-3640-8.ch007

Wallace, A. (2009). E-justice: An Australian perspective. In A. Martínez & P. Abat (Eds.), *E-justice: Using information communication technologies in the court system* (pp. 204–228). Hershey, PA: Information Science Reference; doi:10.4018/978-1-59904-998-4.ch014

Wang, G. (2012). E-democratic administration and bureaucratic responsiveness: A primary study of bureaucrats' perceptions of the civil service e-mail box in Taiwan. In K. Kloby & M. D'Agostino (Eds.), *Citizen 2.0: Public and governmental interaction through web 2.0 technologies* (pp. 146–173). Hershey, PA: Information Science Reference; doi:10.4018/978-1-4666-0318-9.ch009

Wangpipatwong, S., Chutimaskul, W., & Papasratorn, B. (2011). Quality enhancing the continued use of e-government web sites: Evidence from e-citizens of Thailand. In V. Weerakkody (Ed.), *Applied technology integration in governmental organizations: New e-government research* (pp. 20–36). Hershey, PA: Information Science Reference; doi:10.4018/978-1-60960-162-1.ch002

Wedemeijer, L. (2006). Long-term evolution of a conceptual schema at a life insurance company. In M. Khosrow-Pour (Ed.), *Cases on database technologies and applications* (pp. 202–226). Hershey, PA: Idea Group Publishing; doi:10.4018/978-1-59904-399-9.ch012

Whybrow, E. (2008). Digital access, ICT fluency, and the economically disadvantages: Approaches to minimize the digital divide. In F. Tan (Ed.), *Global information technologies: Concepts, methodologies, tools, and applications* (pp. 1409–1422). Hershey, PA: Information Science Reference; doi:10.4018/978-1-59904-939-7.ch102

Whybrow, E. (2008). Digital access, ICT fluency, and the economically disadvantages: Approaches to minimize the digital divide. In C. Van Slyke (Ed.), *Information communication technologies: Concepts, methodologies, tools, and applications* (pp. 764–777). Hershey, PA: Information Science Reference; doi:10.4018/978-1-59904-949-6.ch049

Wickramasinghe, N., & Geisler, E. (2010). Key considerations for the adoption and implementation of knowledge management in healthcare operations. In M. Saito, N. Wickramasinghe, M. Fuji, & E. Geisler (Eds.), *Redesigning innovative healthcare operation and the role of knowledge management* (pp. 125–142). Hershey, PA: Medical Information Science Reference; doi:10.4018/978-1-60566-284-8.ch009

Wickramasinghe, N., & Geisler, E. (2012). Key considerations for the adoption and implementation of knowledge management in healthcare operations. In *Organizational learning and knowledge: Concepts, methodologies, tools and applications* (pp. 1316–1328). Hershey, PA: Business Science Reference; doi:10.4018/978-1-60960-783-8.ch405

Wickramasinghe, N., & Goldberg, S. (2007). A framework for delivering m-health excellence. In L. Al-Hakim (Ed.), *Web mobile-based applications for healthcare management* (pp. 36–61). Hershey, PA: IRM Press; doi:10.4018/978-1-59140-658-7.ch002

Wickramasinghe, N., & Goldberg, S. (2008). Critical success factors for delivering m-health excellence. In N. Wickramasinghe & E. Geisler (Eds.), *Encyclopedia of healthcare information systems* (pp. 339–351). Hershey, PA: Medical Information Science Reference; doi:10.4018/978-1-59904-889-5.ch045

Wyld, D. (2009). Radio frequency identification (RFID) technology. In J. Symonds, J. Ayoade, & D. Parry (Eds.), *Auto-identification and ubiquitous computing applications* (pp. 279–293). Hershey, PA: Information Science Reference; doi:10.4018/978-1-60566-298-5.ch017

Yaghmaei, F. (2010). Understanding computerised information systems usage in community health. In J. Rodrigues (Ed.), *Health information systems: Concepts, methodologies, tools, and applications* (pp. 1388–1399). Hershey, PA: Medical Information Science Reference; doi:10.4018/978-1-60566-988-5.ch088

Yee, G., El-Khatib, K., Korba, L., Patrick, A. S., Song, R., & Xu, Y. (2005). Privacy and trust in e-government. In W. Huang, K. Siau, & K. Wei (Eds.), *Electronic government strategies and implementation* (pp. 145–190). Hershey, PA: Idea Group Publishing; doi:10.4018/978-1-59140-348-7.ch007

Yeh, S., & Chu, P. (2010). Evaluation of e-government services: A citizen-centric approach to citizen e-complaint services. In C. Reddick (Ed.), *Citizens and e-government: Evaluating policy and management* (pp. 400–417). Hershey, PA: Information Science Reference; doi:10.4018/978-1-61520-931-6.ch022

Young-Jin, S., & Seang-tae, K. (2008). E-government concepts, measures, and best practices. In A. Anttiroiko (Ed.), *Electronic government: Concepts, methodologies, tools, and applications* (pp. 32–57). Hershey, PA: Information Science Reference; doi:10.4018/978-1-59904-947-2.ch004

Yun, H. J., & Opheim, C. (2012). New technology communication in American state governments: The impact on citizen participation. In K. Bwalya & S. Zulu (Eds.), *Handbook of research on e-government in emerging economies: Adoption, e-participation, and legal frameworks* (pp. 573–590). Hershey, PA: Information Science Reference; doi:10.4018/978-1-4666-0324-0.ch029

Zhang, N., Guo, X., Chen, G., & Chau, P. Y. (2011). User evaluation of e-government systems: A Chinese cultural perspective. In F. Tan (Ed.), *International enterprises and global information technologies: Advancing management practices* (pp. 63–84). Hershey, PA: Information Science Reference; doi:10.4018/978-1-60960-605-3.ch004

Zuo, Y., & Hu, W. (2011). Trust-based information risk management in a supply chain network. In J. Wang (Ed.), *Supply chain optimization, management and integration: Emerging applications* (pp. 181–196). Hershey, PA: Business Science Reference; doi:10.4018/978-1-60960-135-5.ch013

Compilation of References

Aach, T., Mayntz, C., Rongen, P. M., Schmitz, G., & Stegehuis, H. (2002, May). Spatiotemporal multiscale vessel enhancement for coronary angiograms. In *Medical Imaging 2002* (pp. 1010–1021). International Society for Optics and Photonics. doi:10.1117/12.467056

Acharjee, S., Dey, N., Biswas, D., Das, P., & Chaudhuri, S. S. (2012, November). A novel Block Matching Algorithmic Approach with smaller block size for motion vector estimation in video compression. In *Intelligent Systems Design and Applications (ISDA), 2012. 12th International Conference on* (pp. 668-672). IEEE. doi:10.1109/ISDA.2012.6416617

Aguiar, E. D., Stoll, C., Theobalt, C., Ahmed, N., Seidel, H. P., & Thrun, S. (2008). Performance capture from sparse multi-view video. Association for Computing Machinery Transactions on Graphics, 27(3), 98:1-98:10.

Ahmed, N. (2012). A system for 360-degree acquisition and 3D animation reconstruction using multiple RGB-D cameras. In *Proceedings of the 25th International Conference on Computer Animation and Social Agents* (vol. 1, pp. 9-12). Wiley.

Ahmed, N., Theobalt, C., Rossl, C., Thrun, S., & Seidel, H. P. (2008). Dense correspondence finding for parametrization-free animation reconstruction from video. *Proceedings of Computer Vision and Pattern Recognition, 1*, 1–8.

Akramullah, S. (2014). Digital video concepts, methods, and metrics: Quality, compression, performance, and power trade-off analysis. *Apress.*

Alain, T., & Shoji, T. (2008). Color in image and video processing: Most recent trends and future research directions. *EURASIP Journal on Image and Video Processing.*

Aldoma, A., Blodow, N., Gossow, D., Gedikli, S., Rusu, R. B., Vincze, M., & Bradski, G. (2011). CAD-Model Recognition and 6 DOF Pose Estimation. In *Proceedings of International Conference on Computer Vision 3D Representation and Recognition workshop* (vol. 1, pp. 585-592). IEEE.

Alhamzi, K., Elmogy, M., & Barakat, S. (2014). 3D Object Recognition Based on Image Features: A Survey. *International Journal of Computer and Information Technology, 3*(3).

Alhwarin, F., Wang, C., Ristic-Durrant, D., & Gräser, A. (2008, September). Improved SIFT-Features Matching for Object Recognition. In *British Computer Society International Academic Conference* (pp. 178-190).

All India Council for Technical Education. (2014). *Approval process handbook (2013 – 2014).* Retrieved from http://www.informindia.co.in/education/Approval_Process_Handbook_091012.pdf

Allan, I., Loeb, M. R., & Moxon, E. R. (1987). Limited genetic diversity of Haemophilus influenzae (type b). *Microbial Pathogenesis, 2*(2), 139–145. doi:10.1016/0882-4010(87)90105-7 PMID:3509858

Alnihoud, J. (2008). An efficient region-based approach for object recognition and retrieval based on mathematical morphology and correlation coefficient. *International Arab Journal Information Technology, 5*(2).

Aloimonos, Y. (Ed.). (1993). *Active Perception*. Hillsdale, NJ: Erlbaum.

Altuwaijri, M., & Bayoumi, M. (1995). A new thinning algorithm for Arabic characters using self-organizing neural network. In *IEEE International Symposium on Circuits and Systems* (pp. 1824–1827). doi:10.1109/ISCAS.1995.523769

Andrei, S., Mark, A. L., William, F. G., Gentaro, H., Mary, C. W., Etta, D. P., & Henry, F. (1996). Technologies for augmented reality systems: Realizing ultrasound guided needle biopsies. In *Proceeding from, 23rd Annual Conference On Computer Graphics and Interactive Techniques*.

Angulo, J.M., & Avilés, R. (1989). *Curso de robótica*. Madrid, España: Paraninfo.

ANSI Z535.1 (2011). *Safety Colors*. New York: American National Standards Institute.

Arcelli, C. (1981). Pattern thinning by contour tracing. *Computer Graphics and Image Processing*, *17*(2), 130–144. doi:10.1016/0146-664X(81)90021-6

Arcelli, C., & Baja, G. S. D. (1989). A one-pass two-operation process to detect the skeletal pixels on the 4-distance transform. *IEEE Transactions on Pattern Analysis and Machine Intelligence*, *11*(4), 411–414. doi:10.1109/34.19037

Arcelli, C., Cordella, L., & Levialdi, S. (1975). Parallel thinning of binary pictures. *Electronics Letters*, *11*(7), 148–149. doi:10.1049/el:19750113

Aslani, S., & Mahdavi-Nasab, H. (2013). Optical flow based moving object detection and tracking for traffic surveillance. *International Journal of Electrical, Electronics, Communication Energy Science and Engineering*, *7*(9), 789–793.

Aspiring Mind's National Employability Report- Engineering Graduate. (2011). Retrieved from http://www.aspiringminds.in/docs/national_employability_report_engineers_2011.pdf

Atasoy, S., Mateus, D., Lallemand, J., Meining, A., Yang, G. Z., & Navab, N. (2010). Endoscopic video manifolds. In *Medical Image Computing and Computer-Assisted Intervention–MICCAI 2010* (pp. 437–445). Springer Berlin Heidelberg. doi:10.1007/978-3-642-15745-5_54

Azarbayejani, A., Starner, T., Horowitz, B., & Pentland, A. (1993). Visually controlled graphics. *IEEE Transactions on Pattern Analysis and Machine Intelligence*, *15*(6), 602–605. doi:10.1109/34.216730

Baak, A., Muller, M., Bharaj, G., Seidel, H. P., & Theobalt, C. (2011). A data-driven approach for real-time full body pose reconstruction from a depth camera. In *Proceedings of International Conference on Computer Vision* (vol. 1, pp. 1092-1099). IEEE. doi:10.1109/ICCV.2011.6126356

Baccichet, P., & Chimienti, A. (2004). A low complexity concealment algorithm for the whole-frame loss in H.264/ AVC. In *Proceedings of the 6th IEEE Workshop on Multimedia Signal Processing* (pp. 279– 282). IEEE. doi:10.1109/ MMSP.2004.1436547

Bachir, B. M., Tarek, B., SenLin, L., & Hocine, L. (2014). Weighted Samples Based Background Modeling for the Task of Motion Detection in Video Sequences. *TELKOMNIKA Indonesian Journal of Electrical Engineering*, *12*(11), 7778–7784. doi:10.11591/telkomnika.v12i11.6545

Badrinath, G. S., Nigam, A., & Gupta, P. (2011, November). An efficient finger-knuckle-print based recognition system fusing sift and surf matching scores. In *International Conference on Information and Communications Security* (pp. 374-387). Springer Berlin Heidelberg. doi:10.1007/978-3-642-25243-3_30

Bae, S., Lee, J., Lee, J., Kim, E., Lee, S., Yu, J., & Kang, Y. (2010). Antimicrobial resistance in Haemophilus influenzae respiratory tract isolates in Korea: Results of a nationwide acute respiratory infections surveillance. *Antimicrobial Agents and Chemotherapy*, *54*(1), 65–71. doi:10.1128/AAC.00966-09 PMID:19884366

Bag, S., & Harit, G. (2011). An improved contour-based thinning method for character images Skeletonizing. *Pattern Recognition Letters*, *32*(14), 1836–1842. doi:10.1016/j.patrec.2011.07.001

Bag, S., & Harit, G. (2011). Skeletonizing character images using a modified medial axis-based strategy. *International Journal of Pattern Recognition and Artificial Intelligence*, *25*(07), 1035–1054. doi:10.1142/S0218001411009020

Bai, M. R., Krishna, V. V., & Sree Devi, J. (2010). A new morphological approach for noise removal cum edge detection. *International Journal of Computer Science Issues*, *7*(6), 187–190.

Bai, X., Latecki, L. J., & Liu, W. Y. (2007). Skeleton pruning by contour partitioning with discrete curve evolution. *IEEE Transactions on Pattern Analysis and Machine Intelligence*, *29*(3), 449–462. doi:10.1109/TPAMI.2007.59 PMID:17224615

Bajanca, P., & Caniça, M. (2004). Emergence of nonencapsulated and encapsulated non-b-type invasive Haemophilus influenzae isolates in Portugal (19892001). *Journal of Clinical Microbiology*, *42*(2), 807–810. doi:10.1128/JCM.42.2.807-810.2004 PMID:14766857

Baker, R., & Yacef, K. (2009). The state of educational data mining in 2009: A review and future visions. *Journal of Educational Data Mining, 1*(1), 3-17.

Baker, R., & Yacef, K. (2009). The state of educational data mining in 2009: A review and future visions. J. Educ. *DataMining*, *1*(1), 3–17.

Banerjee, M., & Kundu, M. K. (2008). Handling of impreciseness in gray level corner detection using fuzzy set theoretic approach. *Applied Soft Computing*, *8*(4), 1680–1691. doi:10.1016/j.asoc.2007.09.001

Banks, W.P., & Krajicek, D. (1991). Perception. *Annual Review of Psychology*. DOI: 10.1146/annurev.ps.42.020191.001513

Barenkamp, S. J., Munson, R. S., & Granoff, D. M. (1981). Subtyping isolates of Haemophilus influenzae type b by outer-membrane protein profiles. *The Journal of Infectious Diseases*, *143*(5), 668–676. doi:10.1093/infdis/143.5.668 PMID:6972422

Basu, M. (2002). Gaussian based edge-detection methods: A survey. *IEEE Transactions on System, Man, and Cybernetics Part C: Application and Reviews, 32*(3).

Bay, H., Ess, A., Tuytelaars, T., & Van Gool, L. (2008). Speeded-up robust features (SURF). *Computer Vision and Image Understanding*, *110*(3), 346–359. doi:10.1016/j.cviu.2007.09.014

Beaudet, P. R. (1978, November). Rotationally invariant image operators. In *International Joint Conference on Pattern Recognition* (Vol. 579, p. 583).

Belfiore, S., Grangetto, M., Magli, E., & Olmo, G. (2005). Concealment of whole-frame losses for wireless low bit-rate video based on multiframe optical flow estimation. *IEEE Transactions on Multimedia*, *7*(2), 316–329. doi:10.1109/TMM.2005.843347

Belfiore, S., Grangetto, M., Magli, E., & Olmo, G., G. (2003). An error concealment algorithm for streaming video. In *Proceedings of the 2003 International Conference on Image Processing (ICIP 2003)*. IEEE. doi:10.1109/ICIP.2003.1247328

Ben-Gharbia, K., Maciejewski, A., & Roberts, R. (2014). A kinematic analysis and evaluation of planar robots designed from optimally fault-tolerant Jacobians. *IEEE Transactions on Robotics*, *30*(2), 516–524. doi:10.1109/TRO.2013.2291615

Berger, K., Ruhl, K., Schroeder, Y., Bruemmer, C., Scholz, A., & Magnor, M. A. (2011). Markerless motion capture using multiple color-depth sensors. *Proceedings of Vision Modelling and Visualization*, *1*, 317–324.

Bergholm, F. (1987). Edge focusing. *IEEE Transactions on Pattern Analysis and Machine Intelligence*, *9*(6), 726–741. doi:10.1109/TPAMI.1987.4767980 PMID:21869435

Bertrand, G., & Couprie, M. (2006). New 2D parallel thinning algorithms based on critical kernels. In *International Workshop on Combinatorial Image Analysis* (pp. 45–59). doi:10.1007/11774938_5

Beun, M. (1973). A flexible method for automatic reading of handwritten numerals. *Philips Technical Review*, *31*, 89–101.

Bezdek, J. C., Chandrasekhar, R., & Attikiouzel, Y. (1998). A geometric approach to edge detection. *IEEE Transactions on Fuzzy Systems*, *6*(1), 52–75. doi:10.1109/91.660808

Bhandarkar, S. M., Zhang, Y., & Potter, W. D. (1994). An edge detection technique using genetic algorithm-based optimization. *Pattern Recognition*, *27*(9), 1159–1180. doi:10.1016/0031-3203(94)90003-5

Bharath, R., Kumar, P., Dusa, C., Akkala, V., Puli, S., Ponduri, H., & Desai, U. B. et al. (2015). FPGA-Based Portable Ultrasound Scanning System with Automatic Kidney Detection. *Journal of Imaging*, *1*(1), 193–219. doi:10.3390/jimaging1010193

Bhowmik, M. K., Shil, S., & Saha, P. (2013). Feature Points Extraction of Thermal Face Using Harris Interest Point Detection. *Procedia Technology*, *10*, 724–730. doi:10.1016/j.protcy.2013.12.415

Biederman, I. (1987). Recognition-by components: A theory of human image understanding. *Psychological Review*, *94*(2), 115–147. doi:10.1037/0033-295X.94.2.115 PMID:3575582

Biederman, I. (1990). Higher-level vision. In *Visual Cognition and Action*. The MIT Press.

Biederman, I., & Bar, M. (1999). One-shot viewpoint invariance in matching novel objects. *Vision Research*, *39*(17), 2885–2899. doi:10.1016/S0042-6989(98)00309-5 PMID:10492817

Biederman, I., & Ju, G. (1988). Surface vs. edge-based determinants of visual recognition. *Cognitive Psychology*, *20*(1), 38–64. doi:10.1016/0010-0285(88)90024-2 PMID:3338267

Birgmajer, B., Kovacic, Z., & Postruzin, Z. (2006). Integrated visión system for supervisión and guidance of a steam generator tuve inspection manipulator. In *Proceedings of the 2006 IEEE International Conference on Control Applications*. Munich, Germany: IEEE.

Birinci, M., Diaz-de-Maria, F., & Abdollahian, G. (2011). Neighborhood matching for object recognition algorithms based on local image features. In *IEEE Digital Signal Processing Workshop and IEEE Signal Processing Education Workshop* (pp. 157–162). DSP/SPE. doi:10.1109/DSP-SPE.2011.5739204

Blain, A., MacNeil, J., Wang, X., Bennett, N., Farley, M. M., Harrison, L. H., . . . Reingold, A. (2014, September). Invasive Haemophilus influenzae Disease in Adults≥ 65 Years, United States, 2011. In Open forum infectious diseases (Vol. 1, No. 2). Oxford University Press.

Blom, A. & Hiroshi, S. (2011). *Employability and skill set of newly graduated engineers in India*. Academic Press.

Bloomenthal, J. (1983, July). Edge inference with applications to antialiasing. *Computer Graphics*, *17*(3), 157–162. doi:10.1145/964967.801145

Booy, R., Heath, P. T., Slack, M. P., Begg, N., & Moxon, E. R. (1997). Vaccine failures after primary immunisation with Haemophilus influenzae type-b conjugate vaccine without booster. *Lancet*, *349*(9060), 1197–1202. doi:10.1016/S0140-6736(96)06392-1 PMID:9130940

Bose, S., Chowdhury, S. R., Sen, C., Chakraborty, S., Redha, T., & Dey, N. (2014, November). Multi-thread video watermarking: A biomedical application. In *Circuits, Communication, Control and Computing (I4C), 2014 International Conference on* (pp. 242-246). IEEE. doi:10.1109/CIMCA.2014.7057798

Briere, E. C., Jackson, M., Shah, S. G., Cohn, A. C., Anderson, R. D., MacNeil, J. R., & Messonnier, N. E. et al. (2012). Haemophilus influenzae type b disease and vaccine booster dose deferral, United States, 1998–2009. *Pediatrics*, *130*(3), 414–420. doi:10.1542/peds.2012-0266 PMID:22869828

Briere, E. C., Rubin, L., Moro, P. L., Cohn, A., Clark, T., & Messonnier, N. (2014). Prevention and control of haemophilus influenzae type b disease: Recommendations of the Advisory Committee on Immunization Practices (ACIP). *MMWR. Recommendations and Reports*, *63*(RR-01), 1–14. PMID:24572654

Bui, H. H., Sidney, J., Li, W., Fusseder, N., & Sette, A. (2007). Development of an epitope conservancy analysis tool to facilitate the design of epitope-based diagnostics and vaccines. *BMC Bioinformatics*, *8*(1), 1. doi:10.1186/1471-2105-8-361 PMID:17897458

Byeong, H. K. (2007). A review on image and video processing. *International Journal of Multimedia and Ubiquitous Engineering*, *2*(2).

Cagnoni, S., Mordonini, M., & Sartori, J. (2007, April). Particle swarm optimization for object detection and segmentation. In *Workshops on Applications of Evolutionary Computation* (pp. 241–250). Springer Berlin Heidelberg. doi:10.1007/978-3-540-71805-5_27

Canny, J. (1986). A computational approach to edge detection. *IEEE Transactions on Pattern Analysis and Machine Intelligence*, *8*(6), 679–698. doi:10.1109/TPAMI.1986.4767851 PMID:21869365

Caputo, A. C., Pelagagge, P. M., & Salini, P. (2013). AHP-based methodology for selecting safety devices of industrial machinery. *Safety Science*, *53*, 202–218. doi:10.1016/j.ssci.2012.10.006

Carranza, J., Theobalt, C., Magnor, M. A., & Seidel, H. P. (2003). Free-viewpoint video of human actors. *ACM Transactions on Graphics*, *22*(3), 569–577. doi:10.1145/882262.882309

Castaneda, V., Mateus, D., & Navab, N. (2011). Stereo time-of-flight. In *Proceedings of International Conference on Computer Vision* (vol. 1, pp. 650-657). IEEE.

Chaminda, T. E., Maria, G. M., & Nabeel, K. (2011). 3D medical video transmission over 4g networks. In *Proceedings of the 4th International Symposium on Applied Sciences in Biomedical and Communication Technologies*.

Chen, C. M., & Chen, L. H. (2015). A novel method for slow motion replay detection in broadcast basketball video. *Multimedia Tools and Applications*, *74*(21), 9573–9593. doi:10.1007/s11042-014-2137-5

Chen, Y., Keman, Y., & Jiang, L. (2004). An error concealment algorithm for entire frame loss in video transmission. In *Proceedings of the 2004 Picture Coding Symposium (2004 PCS)*.

Choi, W. P., Lam, K. M., & Siu, W. C. (2003). Extraction of the Euclidean skeleton based on a connectivity criterion. *Pattern Recognition*, *36*(3), 721–729. doi:10.1016/S0031-3203(02)00098-5

Chong, J., Satish, N., Catanzaro, B. C., Ravindran, K., & Keutzer, K. (2007). Efficient parallelization of H.264 decoding with macro block level scheduling. In *Proceedings of the 2007 IEEE International Conference on Multimedia and Expo (ICME 2007)* (pp. 1874-1877). doi:10.1109/ICME.2007.4285040

Choudhury, Z. H., & Mehata, K. M. (2012). Robust facial Marks detection method Using AAM and SURF. *International Journal Engineering Research and Applications*, *2*(6), 708–715.

Chowdhury, R. A. K. (2014). Video processing—An overview. In Image, Video Processing and Analysis, Hardware, Audio, Acoustic and Speech Processing. Chennai: Academic Press.

Cieszynski, J. (2007). *Closed circuit television* (3rd ed.). Oxford, UK: Newnes.

Coelho, A. M., Estrela, V. V., Fernandes, S. R., & do Carmo, F. P. (2012b). Error concealment by means of motion refinement and regularized Bregman divergence. In H. Yin, J. A. F. Costa & G. Barreto (Eds.), *Proceedings of the Intelligent Data Engineering and Automated Learning - IDEAL 2012* (Vol. 7435, pp. 650-657). doi:10.1007/978-3-642-32639-4_78

Coelho, A. M., & Estrela, V. V. (2012a). EM-based mixture models applied to video event detection. In P. Sanguansat (Ed.), *Principal component analysis – Engineering applications* (pp. 101–124). InTech. doi:10.5772/38129

Coelho, A. M., & Estrela, V. V. (2013). State-of-the art motion estimation in the context of 3D TV. In R. A. Farrugia & C. J. Debono (Eds.), *Multimedia networking and coding* (pp. 148–173). doi:10.4018/978-1-4666-2660-7.ch006

Couprie, M. (2005). *Note on fifteen 2D parallel thinning algorithms.* Internal Report, Universit´e de Marne-la-Vall´ee, IGM2006-01.

Craig, J. J. (2006). *Robótica, tercera edición.* Naucalpan de Juárez, México: Pearson Educación.

Cucchiara, R., Grana, C., Neri, G., Piccardi, M., & Prati, A. (2002). The Sakbot system for moving object detection and tracking. In Video-Based Surveillance Systems (pp. 145-157). Springer US. doi:10.1007/978-1-4615-0913-4_12

Cucchiara, R., Grana, C., Piccardi, M., Prati, A., & Sirotti, S. (2001). Improving shadow suppression in moving object detection with HSV color information. In Intelligent Transportation Systems, 2001. Proceedings. 2001 IEEE (pp. 334-339). IEEE. doi:10.1109/ITSC.2001.948679

Cucchiara, R., Grana, C., Piccardi, M., & Prati, A. (2003). Detecting moving objects, ghosts, and shadows in video streams. *Pattern Analysis and Machine Intelligence. IEEE Transactions on*, *25*(10), 1337–1342.

Cui, Y., Deng, Z., & Ren, W. (2009). Novel temporal error concealment algorithm based on residue restoration, In *Proceedings of the 2009 EEE International Conference on Wireless Communications, Networking and Mobile Computing* (pp. 1-4). doi:10.1109/WICOM.2009.5302239

Damasio, A. R., Tranel, D., & Damasio, H. (1990). Face agnosia and the neural substrate of memory. *Annual Review of Neuroscience*, *13*(1), 89–109. doi:10.1146/annurev.ne.13.030190.000513 PMID:2183687

Daniela, G. T., Luciana, P., & Benoit, M. (2008). *2008 ACM Symposium on Applied Computing*. ACM.

Datta, A., & Parui, S. K. (1994). A robust parallel thinning algorithm for binary images. *Pattern Recognition*, *27*(9), 1181–1192. doi:10.1016/0031-3203(94)90004-3

Davis, M. (1992). The role of amygdala in fear and anxiety. *Annual Review of Neuroscience*, *15*(1), 352–375. doi:10.1146/annurev.ne.15.030192.002033 PMID:1575447

Debevec, P. E., Hawkins, T., Tchou, C., Duiker, H. P., Sarokin, W., & Sagar, M. (2000). Acquiring the reflectance field of a human face. *Proceedings of SIGGRAPH*, *1*, 145–156.

Department of Higher Education. (2010). *Annual Report 2009-2010*. Ministry of HRD, 151.

Dey, N., Das, P., Roy, A. B., Das, A., & Chaudhuri, S. S. (2012, October). DWT-DCT-SVD based intravascular ultrasound video watermarking. In *Information and Communication Technologies (WICT), 2012 World Congress on* (pp. 224-229). IEEE. doi:10.1109/WICT.2012.6409079

Dey, N., Nandi, P., Barman, N., Das, D., & Chakraborty, S. (2012). A comparative study between Moravec and Harris corner detection of noisy images using adaptive wavelet thresholding technique. *arXiv preprint arXiv:1209.1558.*

Dey, N., Bose, S., Das, A., Chaudhuri, S. S., Saba, L., Shafique, S., & Suri, J. S. et al. (2016). Effect of watermarking on diagnostic preservation of atherosclerotic ultrasound video in stroke telemedicine. *Journal of Medical Systems*, *40*(4), 1–14. doi:10.1007/s10916-016-0451-3 PMID:26860914

Dey, N., Das, S., & Rakshit, P. (2011). A novel approach of obtaining features using wavelet based image fusion and Harris corner detection. *Int J Mod Eng Res*, *1*(2), 396–399.

Dipanda, A., & Woo, S. (2005). Towards a real-time 3D shape reconstruction using a structured light system. *Pattern Recognition*, *38*(10), 1632–1650. doi:10.1016/j.patcog.2005.01.006

Djara, T., Assogba, M. K., Naït-Ali, A., & Vianou, A. (2013). Comparison of Harris Detector and ridge bifurcation points in the process of fingerprint registration using supervised contactless biometric system. *International Journal of Innovative Technology and Exploring Engineering*, *2*(6).

Doychev, Z. (1985). *Edge detection and feature extraction*. Springer Verlag.

Edward, R., & Drummond, T. (2006). Machine learning for high-speed corner detection. In *Proceedings European Conference on Computer Vision*. Cambridge, MAMIT Press.

Egorov, V., & Sarvazyan, A. P. (2008). Mechanical imaging of the breast. *IEEE Transactions on Medical Imaging*, *27*(9), 1275–1287. doi:10.1109/TMI.2008.922192 PMID:18753043

Egorov, V., Van, R. H., & Sarvazyan, A. P. (2010). Vaginal tactile imaging. *IEEE Transactions on Bio-Medical Engineering*, *57*(7), 1736–1744. doi:10.1109/TBME.2010.2045757 PMID:20483695

Einarsson, P., Chabert, C. F., Jones, A., Ma, W. C., Lamond, B., Hawkins, T., & Debevec, P. E. et al. (2006). Relighting human locomotion with flowed reflectance fields. In *Proceedings of Eurographics Symposium on Rendering* (vol. 1, pp. 183-194). Eurographics Association.

Elgammal, A., Duraiswami, R., Harwood, D., & Davis, L. S. (2002). Background and foreground modeling using non-parametric kernel density estimation for visual surveillance. *Proceedings of the IEEE*, *90*(7), 1151–1163. doi:10.1109/JPROC.2002.801448

Elhabian, S. Y., El-Shayed, K. M., & Ahmed, S. H. (2008). Moving object detection in spatial domain using background removal techniques - State of art. *Recent Patents on Computer Science*, *1*(1), 32–54. doi:10.2174/1874479610801010032

Erickson, B. J., & Jack, C. R. Jr. (n.d.). Correlation of single photon emission CT with MR image data using fiduciary markers. *AJNR. American Journal of Neuroradiology*, *14*(3), 713–720. PMID:8517364

European Commission. (2006). Directive 2006/42/EC of the European Parliament and of the Council of 17 May 2006 on machinery. *Official Journal of the European Union, L, 157*, 24–86.

Farah, M. J. (1990). *Visual agnosia*. Cambridge, MA: MIT Press.

Farah, M. J., McMullen, P. A., & Meyer, M. M. (1991). Can recognition of living things be selectively impaired? *Neuropsychologia*, *29*(2), 185–193. doi:10.1016/0028-3932(91)90020-9 PMID:2027434

Farn, E. J., Chen, L. H., & Liou, J. H. (2003). A new slow-motion replay extractor for soccer game videos. *International Journal of Pattern Recognition and Artificial Intelligence*, *17*(08), 1467–1481. doi:10.1142/S0218001403002964

Fleury, M., Altaf, M., Moiron, S., Qadri, N., & Ghanbari, M. (2013). Source coding methods for robust wireless video streaming. In R. A. Farrugia & C. J. Debono (Eds.), *Multimedia networking and coding*. Retreived from http://www.igi-global.com/chapter/source-coding-methods-robust-wireless/73139

Florea, L., Halld, B., Kohlbacher, O., Schwartz, R., Hoffman, S., & Istrail, S. (2003, August). Epitope prediction algorithms for peptide-based vaccine design. In *Bioinformatics Conference, 2003. CSB 2003. Proceedings of the 2003 IEEE* (pp. 17-26). IEEE. doi:10.1109/CSB.2003.1227293

Förstner, W., & Gülch, E. (1987, June). A fast operator for detection and precise location of distinct points, corners and centres of circular features. In *Proceedings of International Society for Photogrammetry and Remote Sensing intercommission conference on fast processing of photogrammetric data* (pp. 281-305).

Foxwell, A. R., Kyd, J. M., & Cripps, A. W. (1998). Nontypeable Haemophilus influenzae: Pathogenesis and prevention. *Microbiology and Molecular Biology Reviews*, *62*(2), 294–308. PMID:9618443

Frangi, A. F., Niessen, W. J., Vincken, K. L., & Viergever, M. A. (1998, October). Multiscale vessel enhancement filtering. In *International Conference on Medical Image Computing and Computer-Assisted Intervention* (pp. 130-137). Springer Berlin Heidelberg.

Galil, K., Singleton, R., Levine, O. S., Fitzgerald, M. A., Bulkow, L., Getty, M., & Parkinson, A. et al. (1999). Reemergence of invasive Haemophilus influenzae type b disease in a well-vaccinated population in remote Alaska. *The Journal of Infectious Diseases*, *179*(1), 101–106. doi:10.1086/314569 PMID:9841828

Gauglitz, S., Höllerer, T., & Turk, M. (2011). Evaluation of interest point detectors and feature descriptors for visual tracking. *International Journal of Computer Vision*, *94*(3), 335–360. doi:10.1007/s11263-011-0431-5

Gerig, G., Koller, T., Székely, G., Brechbühler, C., & Kübler, O. (1993, June). Symbolic description of 3-D structures applied to cerebral vessel tree obtained from MR angiography volume data. In *Biennial International Conference on Information Processing in Medical Imaging* (pp. 94-111). Springer Berlin Heidelberg. doi:10.1007/BFb0013783

Ghosh, S., & Bag, S. (2013). A modified thinning strategy to handle junction point distortion for Bangla. In *IEEE Students'Technology Symposium* (pp. 52–56).

Ghosh, S., & Bag, S. (2013). An improvement on thinning to handle characters with noisy contour. In *National Conference on Computer Vision, Pattern Recognition, Image Processing and Graphics* (pp. 1–4). doi:10.1109/NCVPRIPG.2013.6776178

Giraudon, G. (1985). Edge detection from local negative maximum of second derivative. In *Proceedings of IEEE, International Conference on Computer Vision and Pattern Recognition*, (pp. 643-645).

Girshick, R., Shotton, J., Kohli, P., Criminisi, A., & Fitzgibbon, A. (2011). Efficient regression of general-activity human poses from depth images. In *Proceedings of International Conference on Computer Vision* (vol. 1, pp. 856-863). IEEE. doi:10.1109/ICCV.2011.6126270

Goernemann, O., & Stubenrauch, H. J. (2013). *Electro-sensitive protective devices (ESPE) for safe machines*. Retrieved March 15, 2016, from http://www.sick.com/group/DE/home/service/ safemachinery/standards_and_regulations/Documents/8016058_WP_ESPE_en_20130328_WEB.pdf

González, R. C., & Woods, R. E. (2008). *Digital image processing* (3rd ed.). Upper Saddle River, NJ: Prentice Hall.

Greenberg-Kushnir, N., Haskin, O., Yarden-Bilavsky, H., Amir, J., & Bilavsky, E. (2012). Haemophilus influenzae type b meningitis in the short period after vaccination: a reminder of the phenomenon of apparent vaccine failure. *Case reports in infectious diseases, 2012.*

Griisser, O. J., & Landis, T. (1991). *Visual agnosias and other disturbances of visual perception and cognition*. London: Macmillan.

Gross, C. G. (1973). Visual functions of infero temporal cortex. In R. Jung (Ed.), Handbook of Sensory Physiology. Berlin: Springer-Verlag.

Gross, C. G., Rodman, H. R., & Cochin, P. M. (1993). Inferior temporal cortex as a pattern recognition device. In *Proceedings of 3rd Necrotizing Enterocolitis Research Symposium*. Slam: NEC Res.

Grossberg, S., & Mingolla, E. (1985). Neural dynamics of form perception: Boundary completion, illusory figures and neon color spreading. *Psychological Review*, *2*(2), 173–211. doi:10.1037/0033-295X.92.2.173 PMID:3887450

Guo, D., Ju, H., & Yao, Y. (2009). *Research of manipulator motion planning algorithm based on vision*. Paper presented at 2009 Sixth International Conference on Fuzzy Systems and Knowledge Discovery, Chengdu, China. doi:10.1109/FSKD.2009.664

Gupta, A., Vaishnav, H., & Garg, H. (2015). Image processing using Xilinx System Generator (XSG) in FPGA. *IJRSI, 2*(9).

Gu, X., Yu, D., & Zhang, L. (2004). Image thinning using pulse coupled neural network. *Pattern Recognition Letters*, *25*(9), 1075–1084. doi:10.1016/j.patrec.2004.03.005

Hameed, M., Sharif, M., Raza, M., Haider, S. W., & Iqbal, M. (2012). Framework for the comparison of classifiers for medical image segmentation with transform and moment based features. *Research Journal of Recent Sciences*, 2277-2502.

Hamelryck, T. (2005). An amino acid has two sides: A new 2D measure provides a different view of solvent exposure. *Proteins: Structure, Function, and Bioinformatics*, *59*(1), 38–48. doi:10.1002/prot.20379 PMID:15688434

Hamelryck, T., & Manderick, B. (2003). PDB file parser and structure class implemented in Python. *Bioinformatics (Oxford, England)*, *19*(17), 2308–2310. doi:10.1093/bioinformatics/btg299 PMID:14630660

Hangzai, L., & Jianping, F. (2006). Building concept ontology for medical video annotation. In *Proceedings from14th ACM International Conference on Multimedia*.

Han, N. H., La, C. W., & Rhee, P. K. (1997). An efficient fully parallel thinning algorithm. In *International Conference on document Analysis and Recognition* (pp. 137–141).

Haralick, R. M. (1983). Ridge and valley on digital images. *Computer Vision Graphics and Image Processing*, *22*(1), 28–38. doi:10.1016/0734-189X(83)90094-4

Haralick, R. M. (1984). Digital step edges from zero-crossing of second directional derivatives. *IEEE Transactions on Pattern Analysis and Machine Intelligence*, *6*(1), 58–68. doi:10.1109/TPAMI.1984.4767475 PMID:21869165

Hardie, R. C., & Boncelet, C. G. (1994). Gradient-based edge detection using nonlinear edge enhancing prefilters. *IEEE Transactions on Image Processing, 4*(11), 1572-1577.

Hariharakrishnan, K., Schonfeld, D., Raffy, P., & Yassa, F. (2003, September). Video tracking using block matching. In *Image Processing, 2003. ICIP 2003. Proceedings. 2003 International Conference on* (Vol. 3). IEEE. doi:10.1109/ICIP.2003.1247402

Harinarayan, R., Pannerselvam, R., Ali, M. M., & Tripathi, D. K. (2011, March). Feature extraction of Digital Aerial Images by FPGA based implementation of edge detection algorithms. In *Emerging Trends in Electrical and Computer Technology (ICETECT), 2011 International Conference on* (pp. 631-635). IEEE. doi:10.1109/ICETECT.2011.5760194

Harris, C., & Stephens, M. (1988). A combined corner and edge detector. In *Proceedings of the 4th Alvey Vision Conference*, (pp. 147–151).

Hasan, S., Yakovlev, A., & Boussakta, S. (2010, July). Performance efficient FPGA implementation of parallel 2-D MRI image filtering algorithms using Xilinx system generator. In *Communication Systems Networks and Digital Signal Processing (CSNDSP), 2010 7th International Symposium on* (pp. 765-769). IEEE.

Hasler, N., Rosenhahn, B., Thormahlen, T., Wand, M., Gall, J., & Seidel, H. P. (2009). Markerless motion capture with unsynchronized moving cameras. *Proceedings of Computer Vision and Pattern Recognition, 1*, 65–73.

Haste Andersen, P., Nielsen, M., & Lund, O. (2006). Prediction of residues in discontinuous B-cell epitopes using protein 3D structures. *Protein Science, 15*(11), 2558–2567. doi:10.1110/ps.062405906 PMID:17001032

Hawkins, T., Einarsson, P., & Debevec, P. E. (2005). A dual light stage. In *Proceedings of Eurographics Symposium on Rendering* (vol. 1, pp. 91-98). Eurographics Association.

Helmer, S., & Lowe, D. G. (2004, June). Object class recognition with many local features. In *Computer Vision and Pattern Recognition Workshop, 2004. CVPRW'04. Conference on* (pp. 187-187). IEEE. doi:10.1109/CVPR.2004.409

Heric, D., & Zazulam, D. (2007). Combined edge detection using wavelet transform and signal registration. *Elsevier Journal of Image and Vision Computing, 25*(5), 652–662. doi:10.1016/j.imavis.2006.05.008

Hernandez, J., Morita, H., Nakano-Miytake, M., & Perez-Meana, H. (2009). *Movement detection and tracking using video frames*. Progress in Pattern Recognition, Image Analysis, Computer Vision, and Applications. doi:10.1007/978-3-642-10268-4_123

Hess, M., & Martinez, G. (2004, December). Facial feature extraction based on the smallest univalue segment assimilating nucleus (susan) algorithm. In *Proceedings of Picture Coding Symposium* (Vol. 1, pp. 261-266).

Hilaire, X., & Tombre, K. (2002). Improving the accuracy of skeleton-based vectorization. In *International Workshop on Graphics Recognition* (pp. 273–288).

Hsieh, J. W., Chen, L. C., & Chen, D. Y. (2014). Symmetrical surf and its applications to vehicle detection and vehicle make and model recognition. *IEEE Transactions on Intelligent Transportation Systems, 15*(1), 6–20. doi:10.1109/TITS.2013.2294646

Huang, L., Wan, G., & Liu, C. (2003). An improved parallel thinning algorithm.*In International Conference on document Analysis and Recognition* (pp. 780–783).

Humphreys, G. W. & Riddoch, M. J. (Eds.). (1987). Visual object processing: A cognitive neuropsychological approach. Hillsdale, NJ: Lawrence Erlbaum Associates.

Humphreys, G. W., & Riddoch, M. J. (1987). *To See Bur Not To See: A Case Study of Visual Agnosia*. Hillsdale, NJ: Lawrence Erlbaum Associates.

Hunt, V. D. (1983). *Industrial robotics handbook* (1st ed.). South Norwalk, CT: Industrial Press Inc.

IEC 61496-1. (2012). *Safety of machinery – Electro-sensitive protective equipment – Part 1: General requirements and tests*. Geneva, Switzerland: International Electrotechnical Commission.

IEC 61496-2. (2013). *Safety of machinery – Electro-sensitive protective equipment – Part 2: Particular requirements for equipment using active opto-electronic protective devices (AOPDs)*. Geneva, Switzerland: International Electrotechnical Commission.

IEC 61496-3. (2008). *Safety of machinery – Electro-sensitive protective equipment – Part 3: Particular requirements for Active Opto-electronic Protective Devices responsive to Diffuse Reflection (AOPDDR)*. Geneva, Switzerland: International Electrotechnical Commission.

IEC 61496-4. (2007). *Safety of machinery – Electro-sensitive protective equipment – Part 4: Particular requirements for equipment using vision based protective devices (VBPD)*. Geneva, Switzerland: International Electrotechnical Commission.

IEC 61496-4-2. (2014). *Safety of machinery – Electro-sensitive protective equipment – Part 4-2: Particular requirements for equipment using vision based protective devices (VBPD) – Additional requirements when using reference pattern techniques (VBPDPP)*. Geneva, Switzerland: International Electrotechnical Commission.

IEC 61496-4-3. (2015). *Safety of machinery – Electro-sensitive protective equipment – Part 4-3: Particular requirements for equipment using vision based protective devices (VBPD) – Additional requirements when using stereo vision techniques (VBPDST)*. Geneva, Switzerland: International Electrotechnical Commission.

Ikeda, N., Araki, T., Dey, N., Bose, S., Shafique, S., El-Baz, A., & Suri, J. S. (2014). Automated and accurate carotid bulb detection, its verification and validation in low quality frozen frames and motion video. *International Angiology: A Journal of the International Union of Angiology, 33*(6), 573-589.

Ikeda, N., Araki, T., Dey, N., Bose, S., Shafique, S., El-Baz, A., Cuadrado, G.E., Anzidei, M., Saba, L., & Suri, J. S. (2014). Automated and accurate carotid bulb detection, its verification and validation in low quality frozen frames and motion video. *International Angiology: A Journal of the International Union of Angiology, 33*(6), 573-589.

Ikeda, N., Gupta, A., Dey, N., Bose, S., Shafique, S., Arak, T., & Suri, J. S. et al. (2015). Improved correlation between carotid and coronary atherosclerosis SYNTAX score using automated ultrasound carotid bulb plaque IMT measurement. *Ultrasound in Medicine & Biology*, *41*(5), 1247–1262. doi:10.1016/j.ultrasmedbio.2014.12.024 PMID:25638311

Iñigo, R., & Vidal, E. (2004). *Robots industriales manipuladores*. Ciudad de México, México: Alfaomega.

ISO 12100. (2010). *Safety of machinery – General principles for design – Risk assessment and risk reduction*. Geneva, Switzerland: International Standards Organization.

ISO 13855. (2010). *Safety of machinery – Positioning of safeguards with respect to the approach speeds of parts of the human body*. Geneva, Switzerland: International Standards Organization.

ISO 13856-1. (2013). *Pressure-sensitive protective devices – Part 1: General principles for design and testing of pressure-sensitive mats and pressure-sensitive floors*. Geneva, Switzerland: International Standards Organization.

ISO 13857. (2008). *Safety of machinery – Safety distances to prevent hazard zones being reached by upper and lower limbs*. Geneva, Switzerland: International Standards Organization.

ITU-T. (2013). *Information technology - Generic coding of moving pictures and associated audio information: Video*. Retrieved from http://www.itu.int/rec/T-REC-H.264

James, A. P., & Dasarathy, B. V. (2014, September). Medical image fusion: A survey of state of the art. *Information Fusion*, *19*, 4–19. doi:10.1016/j.inffus.2013.12.002

Jego, B., Robart, M., Saha, K., & Pau, D. P. (2012, September). FAST detector on many-core computers. In *Consumer Electronics-Berlin (ICCE-Berlin), 2012 IEEE International Conference on* (pp. 263-266). IEEE. doi:10.1109/ICCE-Berlin.2012.6336453

Joshi, K. A., & Thakore, D. G. (2012). A survey on moving object detection and tracking in video surveillance system. *International Journal of Soft Computing and Engineering*, *2*(3).

Juan, L., & Gwun, O. (2009). A comparison of SIFT, PCA-SIFT and SURF. *International Journal of Image Processing*, *3*(4), 143–152.

Kabsch, W., & Sander, C. (1983). Dictionary of protein secondary structure: Pattern recognition of hydrogen-bonded and geometrical features. *Biopolymers*, *22*(12), 2577–2637. doi:10.1002/bip.360221211 PMID:6667333

Kaufman, M. (2012). *Mars landing 2012: Inside the NASA curiosity mission.* Washington, DC: National Geographic Society.

Kaul, S. (2006). *Higher education in India: seizing the opportunity*. The Indian Council for Research on International Economic Relations.

Kehtarnavaz, N., Monaco, J., Nimtschek, J., & Weeks, A. (1998). Color image segmentation using multi-scale clustering. In *Image Analysis and Interpretation,IEEE Southwest Symposium*, (pp. 142 – 147). Doi:10.1109/IAI.1998.666875

Kim, D., Rho, S., & Hwang, E. (2012). Local feature based multi-object recognition scheme for surveillance. *Engineering Applications of Artificial Intelligence*, *25*(7), 1373–1380. doi:10.1016/j.engappai.2012.03.005

Kim, S. (2012). Robust corner detection by image-based direct curvature field estimation for mobile robot navigation. *International Journal of Advanced Robotic Systems*, 9.

Kim, Y. M., Chan, D., Theobalt, C., & Thrun, S. (2008). Design and calibration of a multi-view tof sensor fusion system. In *Proceedings of IEEE CVPR Workshop on Time-of-flight Computer Vision* (vol. 1, pp. 1-7). IEEE.

Kim, Y. M., Theobalt, C., Diebel, J., Kosecka, J., Micusik, B., & Thrun, S. (2009). Multi-view image and ToF sensor fusion for dense 3d reconstruction. In *Proceedings of IEEE Workshop on 3-D Digital Imaging and Modeling* (vol. 1, pp. 1542-1549). IEEE. doi:10.1109/ICCVW.2009.5457430

Kirsch, R. (1971). Computer determination of the constituent structure of biological images. *Computers and Biomedical Research, an International Journal*, *4*(3), 315–328. doi:10.1016/0010-4809(71)90034-6 PMID:5562571

Koelstra, S., & Patras, I. (2009, May). The FAST-3D spatio-temporal interest region detector. In *2009 10th Workshop on Image Analysis for Multimedia Interactive Services* (pp. 242-245). IEEE. doi:10.1109/WIAMIS.2009.5031478

Koenderink, J. J. (1990). *Solid shape*. Cambridge, MA: MIT Press.

Kolkeri, V. S., Lee, J. H., & Rao, K. R. (2009). Error concealment techniques in H.264/AVC for wireless video transmission in mobile networks. *International Journal of Advances in Engineering Science, 2*(2).

Konishi, S., Yuille, A. L., Coughlan, J. M., & Zhu, S. C. (2003). Statistical edge detection: Learning and evaluating edge cues. *IEEE Transactions on Pattern Analysis and Machine Intelligence*, *25*(1), 57–74. doi:10.1109/TPAMI.2003.1159946

Koschan, & Abidi, M. (2005). Detection and classification of edges in color images. *IEEE Signal Processing Magazine*, 64-73.

Kothe, U. (1995). Primary Image Segmentation. Springer. Doi:10.1007/978-3-642-79980-8_65

Koundinya, K., & Chanda, B. (1994). Detecting lines in gray level images using search techniques. *Signal Processing*, *37*(2), 287–299. doi:10.1016/0165-1684(94)90110-4

Kouwenberg, J. J., Ulrich, L., Jäkel, O., & Greilich, S. (2016). A 3D feature point tracking method for ion radiation. *Physics in Medicine and Biology*, *61*(11), 4088–4104. doi:10.1088/0031-9155/61/11/4088 PMID:27163162

Kovacic, Z. (2010). Early prediction of student success: Mining students' enrolment data. *Proceedings of Informing Science & IT Education Conference* (InSITE).

Krishnapuram, R., & Chen, L. F. (1993). Implementation of parallel thinning algorithms using recurrent neural networks. *IEEE Transactions on Neural Networks*, *4*(1), 142–147. doi:10.1109/72.182705 PMID:18267712

Kumar, M., & Saxena, R. (2013). Algorithm and technique on various edge detection: A survey. *Signal & Image Processing, 4*(3), 65.

Kumar, P., Ranganath, S., Weimin, H., & Sengupta, K. (2005). Framework for real-time behavior interpretation from traffic video. *Intelligent Transportation Systems. IEEE Transactions on, 6*(1), 43–53.

Kumar, V., & Nagappan, A. (2012). *Study and comparison of various point based feature extraction methods in palmprint authentication system*. Editorial Committees.

Kung, W., Kim, C., & Kuo, C. (2006). Spatial and temporal error concealment techniques for video transmission over noisy channels. *IEEE Transactions on Circuits and Systems for Video Technology, 16*(7), 789–802. doi:10.1109/TCSVT.2006.877391

Lacassagne, L., Manzanera, A., Denoulet, J., & Mérigot, A. (2009). High performance motion detection: Some trends toward new embedded architectures for vision systems. *Journal of Real-Time Image Processing, 4*(2), 127–146. doi:10.1007/s11554-008-0096-7

Lam, L., Lee, S. W., & Suen, C. Y. (1992). Thinning methodologies—A comprehensive survey. *IEEE Transactions on Pattern Analysis and Machine Intelligence, 14*(9), 869–885. doi:10.1109/34.161346

Lam, L., & Suen, C. Y. (1991). A dynamic shape preserving thinning algorithm. *Signal Processing, 22*(2), 199–208. doi:10.1016/0165-1684(91)90050-S

Lee. (2012). Edge Detection Analysis. *International Journal of Computer Science, 1*(5-6).

Lee, C. Y., Wang, H. J., Chen, C. M., Chuang, C. C., Chang, Y. C., & Chou, N. S. (2014). A modified harris corner detection for breast ir Image. *Mathematical Problems in Engineering*.

Leung, W., Ng, C. M., & Yu, P. C. (2000). Contour following parallel thinning for simple binary images. In *International Conference on Systems, Man, and Cybernetics* (pp. 1650–1655).

Li, J., Ding, Y., Shi, Y., & Li, W. (2010). A divide-and-rule scheme for shot boundary detection based on SIFT. *JDCTA, 4*(3), 202–214. doi:10.4156/jdcta.vol4.issue3.20

Li, L., Huang, W., Gu, I. Y., & Tian, Q. (2003, November). Foreground object detection from videos containing complex background. In *Proceedings of the eleventh ACM international conference on Multimedia* (pp. 2-10). ACM. doi:10.1145/957013.957017

Lin, W., Sun, M. T., Li, H., & Ho, H. M. (2010). A new shot change detection method using information from motion estimation. In *PCM'10 Proceedings of the Advances in Multimedia Information Processing, and 11th Pacific Rim conference on Multimedia: Part II* (pp. 264-275). Springer-Verlag Berlin. doi:10.1007/978-3-642-15696-0_25

Liu, C., Torralba, A., Freeman, W. T., Durand, F., & Adelson, E. H. (2005). Motion magnification. *ACM Transactions on Graphics, 24*(3), 519–526. doi:10.1145/1073204.1073223

Li, X. B., & Basu, A. (1991). Variable-resolution character thinning. *Pattern Recognition Letters, 12*(4), 241–248. doi:10.1016/0167-8655(91)90038-N

Loeb, M. R., & Smith, D. H. (1980). Outer membrane protein composition in disease isolates of Haemophilus influenzae: Pathogenic and epidemiological implications. *Infection and Immunity, 30*(3), 709–717. PMID:6971807

Lorenz, C., Carlsen, I. C., Buzug, T. M., Fassnacht, C., & Weese, J. (1997). Multi-scale line segmentation with automatic estimation of width, contrast and tangential direction in 2D and 3D medical images. In CVRMed-MRCAS'97 (pp. 233-242). Springer Berlin Heidelberg.

Lottenbach, K. R., Granoff, D. M., Barenkamp, S. J., Powers, D. C., Kennedy, D., Irby-Moore, S., & Mink, C. M. et al. (2004). Safety and immunogenicity of Haemophilus influenzae type B polysaccharide or conjugate vaccines in an elderly adult population. *Journal of the American Geriatrics Society*, *52*(11), 1883–1887. doi:10.1111/j.1532-5415.2004.52511.x PMID:15507066

Lowe, D. G. (1999). Object recognition from local scale-invariant features. In *Computer vision, 1999. The proceedings of the seventh IEEE international conference on* (Vol. 2, pp. 1150-1157). IEEE. doi:10.1109/ICCV.1999.790410

Lowe, D. G. (1999). Object recognition from local scale-invariant features. In *Proceedings of International Conference on Computer Vision* (vol. 1, pp. 1150-1157). IEEE.

Lowe, D. G. (2004). Distinctive image features from scale-invariant keypoints. *International Journal of Computer Vision*, *60*(2), 91–110. doi:10.1023/B:VISI.0000029664.99615.94

Luan, J. (2002). Data mining and its applications in higher education. *New Directions for Institutional Research*, *113*(113), 17–36. doi:10.1002/ir.35

Lu, S., Su, B., & Tan, C. L. (2010). Document image binarization using background estimation and stroke edges. *International Journal on Document Analysis and Recognition*, *13*(4), 303–314. doi:10.1007/s10032-010-0130-8

Lu, S., Wang, Z., & Shen, J. (2003). Neuro-fuzzy synergism to the intelligent system for edge detection and enhancement. *Elsevier Journal of Pattern Recognition*, *36*(10), 2395–2409. doi:10.1016/S0031-3203(03)00083-9

Mahesh, S. M. N. (2012). Image mosaic using fast corner detection. *International Journal of Advanced Research in Electronics and Communication Engineering*, *1*(6).

Maini, R., & Aggarwal, H. (2009). Study and comparison of various image edge detection techniques. *International Journal of Image Processing*,*3*(1), 1-11.

Mair, E., Hager, G. D., Burschka, D., Suppa, M., & Hirzinger, G. (2010, September). Adaptive and generic corner detection based on the accelerated segment test. In *European conference on Computer vision* (pp. 183-196). Springer Berlin Heidelberg. doi:10.1007/978-3-642-15552-9_14

Manfred, J. P. (2014). Segmentation and indexing of endoscopic videos. In *Proceedings from22nd ACM International Conference on Multimedia.*

Manzanera, A., & Bernard, T. M. (2003). Metrical properties of a collection of 2D parallel thinning algorithms. In *International Workshop on Combinatorial Image Analysis* (pp. 255–266). doi:10.1016/S1571-0653(04)00491-3

Marr, D., & Nishihara, H. K. (1978). Representation and recognition of the spatial organization of three-dimensional shapes. In *Proceedings of the Royal Society of London: Series B,* (pp. 200269-94). doi:10.1098/rspb.1978.0020

Martínez, A., & Fernández, E. (2013). *Learning ROS for robotics programming*. Birmingham, UK: Packt Pub.

Martinez-Perez, M. P., Jimenez, J., & Navalon, J. L. (1987). A thinning algorithm based on contours. *Computer Vision Graphics and Image Processing*, *39*(2), 186–201. doi:10.1016/S0734-189X(87)80165-2

Masud, M. D. (2012). Knowledge-based Image segmentation using swarm intelligence techniques. *International Journal of Innovative Computing and Applications*, *4*(2), 75–99. doi:10.1504/IJICA.2012.046779

Matas, J., Chum, O., Urban, M., & Pajdla, T. (2002). Robust wide baseline stereo from maximally stable extremal regions. In *Proceedings of British Machine Vision Conference*, (pp. 384-396). doi:10.5244/C.16.36

Mayank, V., & Richa, S. (2005). *Video biometrics. In Video Data Management and Information Retrieval* (pp. 149–176). Hershey, PA: IGI Global.

Melhi, M., Ipson, S. S., & Booth, W. (2001). A novel triangulation procedure for thinning hand-written text. *Pattern Recognition Letters*, *22*(10), 1059–1071. doi:10.1016/S0167-8655(01)00038-1

Mian, A., Bennamoun, M., & Owens, R. (2010). On the repeatability and quality of key points for local featurebased 3D object retrieval from cluttered scenes. *International Journal of Computer Vision*, *89*(2-3), 348–361. doi:10.1007/s11263-009-0296-z

Michela, A., Beatrice, L., & Francesco, M. (2005). *2005 ACM Symposium on Applied Computing*. ACM.

Microsoft. (2010). *Kinect for Microsoft windows and Xbox 360.* Retrieved January 14, 2016, from http://www.kinectforwindows.org/

Mikolajczyk, K., & Schmid, C. (2001). Indexing based on scale invariant interest points. In *Computer Vision, 2001. ICCV 2001. Proceedings. Eighth IEEE International Conference on* (Vol. 1, pp. 525-531). IEEE. doi:10.1109/ICCV.2001.937561

Mikolajczyk, K., & Schmid, C. (2004). Scale & affine invariant interest point detectors. *International Journal of Computer Vision*, *60*(1), 63–86. doi:10.1023/B:VISI.0000027790.02288.f2

Mikolajczyk, K., & Schmid, C. (2005). A performance evaluation of local descriptors. *IEEE Transactions on Pattern Analysis and Machine Intelligence*, *27*(10), 1615–1630. doi:10.1109/TPAMI.2005.188 PMID:16237996

Miksik, O., & Mikolajczyk, K. (2012, November). Evaluation of local detectors and descriptors for fast feature matching. In *International Conference on Pattern Recognition (ICPR), 2012 21st International Conference on* (pp. 2681-2684). IEEE.

Mikulka, J., Gescheidtova, E., & Bartusek, K. (2012). Soft-tissues image processing: Comparison of traditional segmentation methods with 2D active contour methods. *Measurement Science Review*, *12*(4).

Ministry of Human Resource Development. (2012). *Report to the people on Education- 2010-2011*. Retrieved from http://mhrd.gov.in/sites/upload_files/mhrd/files/RPE-2010-11.pdf

Mishra, S. (2007). *Quality assurance in higher education: An introduction. Commonwealth of Learning*. National Assessment and Accreditation Council.

Miyashita, Y. (1993). Inferior temporal cortex: Where visual perception meets memory. *Annual Review of Neuroscience*, *16*(1), 245–263. doi:10.1146/annurev.ne.16.030193.001333 PMID:8460893

Mokhtarian, F., & Suomela, R. (1998). Robust image corner detection through curvature scale space. *IEEE Transactions on Pattern Analysis and Machine Intelligence*, *20*(12), 1376–1381. doi:10.1109/34.735812

Monteiro, F. C., & Campilho, A. (2008). Watershed framework to region-based image segmentation. In *Proceedings of International Conference on Pattern Recognition, ICPR 19th*, (pp. 1-4).

Montruccoli, G. C., Montruccoli, S. D., & Casali, F. (2004). A new type of breast contact thermography plate: A preliminary and qualitative investigation of its potentiality on phantoms. *Physica Medica*, *20*(1), 27–31.

Moravec, H. (1980). *Obstacle Avoidance and Navigation in the Real World by a Seeing Robot Rover.* Tech Report CMU-RI-TR-3, Carnegie-Mellon University, Robotics Institute.

Morgan, B. (2010). Interest point detection for reconstruction in high granularity tracking detectors. *Journal of Instrumentation*, *5*(07), P07006. doi:10.1088/1748-0221/5/07/P07006

Naeimeh, D., Beikzadeh, M. D., & Amnuaisuk, S. P. (2005). Application of enhanced analysis model for data mining processes in higher educational system. *Information Technology Based Higher Education and Training, 2005. ITHET 2005. 6th International Conference on*. IEEE.

Nagendraprasad, M. V., Wang, P. S. P., & Gupta, A. (1993). Algorithms for thinning and rethickening binary digital patterns. *Digital Signal Processing*, *3*(2), 97–102. doi:10.1006/dspr.1993.1014

Naik, N., & Purohit, S. (2012). Prediction of Final Result and Placement of Students using Classification Algorithm. *International Journal of Computers and Applications*, *56*(12).

Naor-Raz, G., Tarr, M. J., & Kersten, D. (2003). Is color an intrinsic property of object representation? *Perception*, *32*(6), 667–680. doi:10.1068/p5050 PMID:12892428

Nascimento, J. C., & Marques, J. S. (2006). Performance evaluation of object detection algorithms for video surveillance. *Multimedia. IEEE Transactions on*, *8*(4), 761–774.

Nektarios, K., Vasileios, S., Vassilis, M., Ioannis, P., Nikolaos, A., & Elias, P. (2014). A smart card based software system for surgery specialties. *IJUDH International Journal of User-Driven Health Care*, *4*(1), 48–63. doi:10.4018/ijudh.2014010104

Nghe, T. N., Janecek, P., & Haddawy, P. (2007). A comparative analysis of techniques for predicting academic performance. Frontiers In Education Conference-Global Engineering: Knowledge Without Borders, Opportunities Without Passports, 2007. FIE'07. 37th Annual. IEEE.

Nichani, S., Silver, W., & Schatz, D. A. (2004). *Auto-setup of a video safety curtain system*. US Patent 6,829.371.

Nichani, S., Wolff, R., Silver, W., & Schatz, D. A. (2007). *Video safety detector with projected pattern*. US Patent 7,167,575.

Nielsen, M., Lundegaard, C., Worning, P., Lauemøller, S. L., Lamberth, K., Buus, S., & Lund, O. et al. (2003). Reliable prediction of T-cell epitopes using neural networks with novel sequence representations. *Protein Science*, *12*(5), 1007–1017. doi:10.1110/ps.0239403 PMID:12717023

Ogor, E. N. (2007, September). Student academic performance monitoring and evaluation using data mining techniques. In *Electronics,Robotics and Automotive Mechanics Conference (CERMA 2007)* (pp. 354-359). IEEE. doi:10.1109/CERMA.2007.4367712

Ohm, J.-R., Sullivan, G. J., Schwarz, H., Tan, T. K., & Wiegand, T. (2012). Comparison of the coding efficiency of video coding standards – Including High Efficiency Video Coding (HEVC). *IEEE Transactions on Circuits and Systems for Video Technology*, *22*(12), 1669–1684. doi:10.1109/TCSVT.2012.2221192

Oliva, A., & Schyns, P. (2000). Diagnostic colors mediate scene recognition. *Cognitive Psychology*, *41*(2), 176–210. doi:10.1006/cogp.1999.0728 PMID:10968925

Ong, E. P., & Weise, L. (2009). *Video object segmentation. In Encyclopedia of information communication technology* (pp. 809–816). Hershey, PA: IGI Global. doi:10.4018/978-1-59904-845-1.ch106

Ophir, J., Céspides, I., Ponnekanti, H., & Li, X. (1991). Elastography: A quantitative method for imaging the elasticity of biological tissues. *Ultrasonic Imaging*, *13*(2), 111–134. doi:10.1177/016173469101300201 PMID:1858217

Othman, D. M. F. B., Abdullah, N., & Rusli, N. A. B. A. (2010, October). An overview of MRI brain classification using FPGA implementation. In *Industrial Electronics & Applications (ISIEA), 2010 IEEE Symposium on* (pp. 623-628). IEEE. doi:10.1109/ISIEA.2010.5679389

Ouyang, W., Tombari, F., Mattoccia, S., & Di Stefano, L. (2012). Performance evaluation of full search equivalent pattern matching algorithms. *IEEE Transactions on Pattern Analysis and Machine Intelligence*, *34*(1), 127–143. doi:10.1109/TPAMI.2011.106 PMID:21576734

Ownby, M., & Mahmoud, W. H. (2003, March). A design methodology for implementing DSP with Xilinx System Generator for Matlab. In Southeastern Symposium on System Theory (Vol. 35, pp. 404-408).

Oyelade, O. J., Oladipupo, O. O., & Obagbuwa, I. C. (2010). Application of k-Means Clustering algorithm for prediction of Students' Academic Performance. *International Journal of Computer Science and Information Security*, *7*(1).

Pal, G., Acharjee, S., Rudrapaul, D., Ashour, A. S., & Dey, N. (2015). Video segmentation using minimum ratio similarity measurement. *International Journal of Image Mining*, *1*(1), 87–110. doi:10.1504/IJIM.2015.070027

Panayides, A., Pattichis, M. S., Pattichis, C. S., & Pitsillides, A. (2011). A tutorial for emerging wireless medical video transmission systems. *IEEE Antennas & Propagation Magazine*, *53*(2), 202–213. doi:10.1109/MAP.2011.5949369

Panchal, P. M., Panchal, S. R., & Shah, S. K. (2013). A comparison of SIFT and SURF. *International Journal of Innovative Research in Computer and Communication Engineering*, *1*(2), 323–327.

Paragios, N., & Deriche, R. (2000). Geodesic active contours and level sets for the detection and tracking of moving objects. *Pattern Analysis and Machine Intelligence. IEEE Transactions on*, *22*(3), 266–280.

Parekh, H. S., Thakore, D. G., & Jaliya, U. K. (2014). A survey on object detection and tracking methods. *International Journal of Innovative Research in Computer and Communication Engineering*, *2*(2).

Parker, K. J., Doyley, M. M., & Rubens, D. J. (2011). Imaging the elastic properties of tissue: The 20 year20-year perspective. *Physics in Medicine and Biology*, *56*(2), 513. doi:10.1088/0031-9155/56/2/513 PMID:21119234

Patel, A., Kasat, D. R., Jain, S., & Thakare, V. M. (2014). Performance analysis of various feature detector and descriptor for real-time video based face tracking. *International Journal of Computers and Applications*, *93*(1).

Patil, R., & Jondhale, K. (2010). Edge based technique to estimate number of clusters in k-means color image segmentation. In *Proceedings.3rd IEEE International Conference on Computer Science and Information Technology (ICCSIT)*, (pp. 117-121). doi:10.1109/ICCSIT.2010.5563647

Pavlidis, T. (1982). An asynchronous thinning algorithm. *Computer Graphics and Image Processing*, *20*(2), 133–157. doi:10.1016/0146-664X(82)90041-7

Peng, Q., Yang, T., & Zhu, C. (2002). Block-based temporal error concealment for video packet using motion vector extrapolation. In *Proceedings of the 2002 IEEE International Conference on Communications, Circuits and Systems and West Sino Expositions* (vol. 1, pp. 10–14). IEEE doi:10.1109/ICCCAS.2002.1180560

Perona, P., & Malik, J. (1990). Scale-space and edge detection using anisotropic diffusion. *IEEE Transactions on Pattern Analysis and Machine Intelligence*, *12*(7), 629–639. doi:10.1109/34.56205

Peters, B., Bulik, S., Tampe, R., Van Endert, P. M., & Holzhütter, H. G. (2003). Identifying MHC class I epitopes by predicting the TAP transport efficiency of epitope precursors. *Journal of Immunology (Baltimore, MD.: 1950)*, *171*(4), 1741–1749. doi:10.4049/jimmunol.171.4.1741 PMID:12902473

Philbin, J., Chum, O., Isard, M., Sivic, J., & Zisserman, A. (2007, June). Object retrieval with large vocabularies and fast spatial matching. In *2007 IEEE Conference on Computer Vision and Pattern Recognition* (pp. 1-8). IEEE. doi:10.1109/CVPR.2007.383172

Pilz GmbH & Co. KG (2015) *SafetyEYE Operating Manual*. Retrieved March 15, 2016, from https://www.pilz.com/download/open/PSENse_Operat_Man_21743-EN-18.pdf

Pinker, S. (1985). Visual cognition: An introduction. In S. Pinker (Ed.), *Visual Cognition* (pp. 1–63). Cambridge, MA: MIT Press.

Pless, R. (2003, October). *Image Spaces and video trajectories: Using Isomap to explore video sequences* (Vol. 3). ICCV.

Prati, A., Mikic, I., Trivedi, M. M., & Cucchiara, R. (2003). Detecting moving shadows: Algorithms and evaluation. *Pattern Analysis and Machine Intelligence. IEEE Transactions on*, *25*(7), 918–923.

Qiang, Z., & Baoxin, L. (2011). Video –based motion expertise analysis in simulation based surgical training using hierarchical Dirichlet Process Hidden Markov Model. In *Proceedings from2011 International ACM Workshop on Medical Multimedia Analysis and Retrieval.*

Quadri, M., & Kalyankar, N. V. (2010). Drop out feature of student data for academic performance using decision tree techniques. *Global Journal of Computer Science and Technology, 10*(2).

Quah, C. K., Michael, K., Alex, O., Hook, S. S., & Andre, G. (2009). Video based motion capture for measuring human movement. In Digital sport for performance enhancement and competitive evolution: Intelligent gaming technologies. Hershey, PA: IGI Global.

Rahangdale, K., & Kokate, M. (2016). *Event detection using background subtraction for surveillance systems*. Academic Press.

Rahman, M., Hossain, S., Baqui, A. H., Shoma, S., Rashid, H., Nahar, N., & Khatun, F. et al. (2008). Haemophilus influenzae type-b and non-b-type invasive diseases in urban children (< 5years) of Bangladesh: Implications for therapy and vaccination. *The Journal of Infection*, *56*(3), 191–196. doi:10.1016/j.jinf.2007.12.008 PMID:18280571

Rakesh, K., Sankhayan, C., & Nabendu, C. (2011). Optimizing mobile terminal equipment for video medic services. In *Proceedings of the 1st International Conference on Wireless Technologies for Humanitarian Relief*, (pp. 373-378).

Rakibe, R. S., & Patil, B. D. (2013). Background subtraction algorithm based human motion detection. *International Journal of Scientific and Research Publications, 3*(5).

Ramanath, R., Snyder, W. E., Yoo, Y., & Drew, M. S. (2005). Color image processing pipeline. *IEEE Signal Processing Magazine*, *22*(1), 34–43. doi:10.1109/MSP.2005.1407713

Ramaswami, M., & Bhaskaran, R. (2010). A CHAID Based Performance Prediction Model in Educational Data Mining. *International Journal of Computer Science Issues*, *7*(4).

Raut, N. P., & Gokhale, A. V. (2013). FPGA implementation for image processing algorithms using xilinx system generator. *IOSR Journal of VLSI and Signal Processing, 2*(4).

Ravela, S. (2003). On multi-scale differential features and their representations for image retrieval and recognition. University of Massachusetts Amherst.

Richard, D., Sara, G., Kholood, S., Gary, U., Graham, M., & Janet, E. (2014). Early response markers from video games for rehabilitation strategies. ACM SIGAPP Applied Computing Review, 36-43.

Richardson, I. E. G. (2010). *The H.264 advanced video compression standard*. John Wiley & Sons, Ltd. doi:10.1002/9780470989418

Ridley, J., & Pearce, D. (2006). *Safety with machinery* (2nd ed.). New York: Routledge.

Rockwell Automation, Inc. (Ed.). (2012). *GuardMaster SC300 hand detection safety sensor*. Retrieved March 15, 2016, from http://literature.rockwellautomation.com/idc/groups/literature/documents/pp/442l-pp001_-en-p.pdf

Rodriguez, P., Pattichis, M. S., Pattichis, C. S., Abdallah, R., & Goens, M. B. (2006). Object-based ultrasound video processing for wireless transmission in cardiology. In M-Health (pp. 491-507). Springer US. doi:10.1007/0-387-26559-7_37

Rolls, E. T. (1994). Brain mechanisms for invariant visual recognition and learning. *Behavioural Processes*, *22*(1-2), 113–138. doi:10.1016/0376-6357(94)90062-0 PMID:24925242

Romero, C., & Sebastián, V. (2010). Educational data mining: a review of the state of the art. *Systems, Man, and Cybernetics, Part C: Applications and Reviews. IEEE Transactions on*, *40*(6), 601–618.

Romero, C., & Ventura, S. (2007). Educational data mining: A survey from 1995 to 2005. *Expert Systems with Applications*, *33*(1), 135–146. doi:10.1016/j.eswa.2006.04.005

Roobottom, C. A., Mitchell, G., Morgan-Hughes, G., & Mitchell, M.-H. (2010). Radiation-reduction strategies in cardiac computed tomographic angiography. *Clinical Radiology*, *65*(11), 859–867. doi:10.1016/j.crad.2010.04.021 PMID:20933639

Rosenfeld, A. (1970). Connectivity in digital pictures. *Journal of Alternative and Complementary Medicine (New York, N.Y.)*, *17*, 146–160.

Rosenfeld, A., & Thurston, M. (1971). Edge and curve detection for visual scene analysis. *IEEE Transactions on Computers*, *100*(5), 562–569. doi:10.1109/T-C.1971.223290

Rossion, B., & Pourtois, G. (2004). Revisiting Snodgrass and Vander warts object pictorial set: The role of surface detail in basic-level object recognition. *Perception*, *33*(2), 217–236. doi:10.1068/p5117 PMID:15109163

Rosten, E., & Drummond, T. (2006, May). Machine learning for high-speed corner detection. In *European Conference on Computer Vision* (pp. 430-443). Springer Berlin Heidelberg.

Russo, F. (1999). FIRE operators for image processing. *Fuzzy Sets and Systems*, *103*(2), 265–275. doi:10.1016/S0165-0114(98)00226-7

Rusu, R. B., & Cousins, S. (2011). 3D is here: Point Cloud Library. In *Proceedings of International Conference on Robotics and Automation* (vol. 1., pp. 1-8). IEEE.

Rutovitz, D. (1966). Pattern recognition. *Journal of the Royal Statistical Society. Series A (General)*, *129*(4), 504–530. doi:10.2307/2982255

Sanjeev, A. (1995). *Discovering Enrollment Knowledge in University Databases*. KDD.

Sansoni, G., Trebeschi, M., & Docchio, F. (2009). State-of-the-art and applications of 3D imaging sensors in industry, cultural heritage, medicine, and criminal investigation. *Sensors (Basel, Switzerland)*, *9*(1), 568–601. doi:10.3390/s90100568 PMID:22389618

Santis, A. D., & Sinisgalli, C. (1999). A Bayesian approach to edge detection in noisy images. *IEEE Transactions on Circuits and Systems. I, Fundamental Theory and Applications*, *46*(6), 686–699. doi:10.1109/81.768825

Sarvazyan, A., Hall, T. J., Urban, M. W., Fatemi, M., Aglyamov, S. R., & Garra, B. S. (2011). Overview of elastography–An emerging branch of medical imaging. *Current Medical Imaging Reviews*, *7*(4), 255–282. doi:10.2174/157340511798038684 PMID:22308105

Saxena, A., Chung, S. H., & Ng, A. Y. (2008). 3-D depth reconstruction from a single still image. *International Journal of Computer Vision*, *76*(1), 53–69. doi:10.1007/s11263-007-0071-y

Schatz, D. A., Nichani, S., & Shillman, R. J. (2001). *Video safety curtain*. US Patent 6,297.844.

Schmidt, B., & Wang, L. (2014). Depth camera based collision avoidance via active robot control. *Journal of Manufacturing Systems*, *33*(4), 711–718. doi:10.1016/j.jmsy.2014.04.004

Schunck, B. G. (1987). Edge detection with Gaussian filters at multiple scales. In *Proceedings IEEE Comp. Soc. Work. Comp. Vis.*

Schuon, S., Theobalt, C., Davis, J., & Thrun, S. (2008). High-quality scanning using time of-flight depth superresolution. In *Proceedings of IEEE Computer Society Conference on Computer Vision and Pattern Recognition Workshops, 2008 (CVPRW '08)* (vol. 1, pp. 1-7). Anchorage, AK: IEEE.

Scovanner, P., Ali, S., & Shah, M. (2007). A 3-dimensional SIFT descriptor and its application to action recognition. In *Proceedings of the 15th international conference on Multimedia* (*vol. 1*, pp 357–360). ACM Press New York. doi:10.1145/1291233.1291311

Selig, J. M. (1992). *Introductory robotics*. London: Prentice Hall.

Serby, D., Meier, E. K., & Van Gool, L. (2004, August). Probabilistic object tracking using multiple features. In *Pattern Recognition, 2004. ICPR 2004.Proceedings of the 17th International Conference on* (*Vol. 2*, pp. 184-187). IEEE.

Shang, L., Yi, Z., & Ji, L. (2007). Binary image thinning using autowaves generated by PCNN. *Neural Processing Letters*, *25*(1), 49–62. doi:10.1007/s11063-006-9030-9

Sharma, N., & Kansal, V. (2011). Identifying the capabilities of data mining in providing the Quality in Technical Education. *Proceedings of the 5th National Conference; INDIACom-2011,Computing For Nation Development*. BVICAM.

Sheu, H. T., & Hu, W. C. (1996). A rotationally invariant two-phase scheme for corner detection. *Pattern Recognition*, *29*(5), 819–828. doi:10.1016/0031-3203(95)00121-2

Shi, J., & Tomasi, C. (1994). Good features to track. In *Proceedings of 9th IEEE Conference on Computer vision and Pattern Recognition*. Springer.

Shinfeng, D. L., Wang, C. C., Chuang, C. Y., & Fu, K. R. (2011). A hybrid error concealment technique for H.264/AVC based on boundary distortion estimation. In D. S. L. Javier (Ed.), Advances on video coding. InTech. Retrieved from http://www.intechopen.com/books/recent-advances-on-video-coding/a-hybrid-error-concealment-techniquefor-h-264-avc-based-on-boundary-distortion-estimation

Shin, J., & Kim, D. (2014). Hybrid approach for facial feature detection and tracking under occlusion. *IEEE Signal Processing Letters*, *21*(12), 1486–1490. doi:10.1109/LSP.2014.2338911

Siciliano, B., & Khatib, O. (2008). *Springer handbook of robotics*. Berlin, Germany: Springer-Verlag. doi:10.1007/978-3-540-30301-5

Sick, Inc. (Ed.). (2009). *Operating instructions. V200 Workstation Extended. V300 Workstation Extended*. Retrieved March 15, 2016, from https://www.sick.com/media/dox/3/63/763/Operating_ instructions_V200_Work_Station_Extended_V300_Work_Station_Extended_en_IM0026763.PDF

Sinha, R. M. K. (1987). A width-independent algorithm for character skeleton estimation. *Computer Vision Graphics and Image Processing*, *40*(3), 388–397. doi:10.1016/S0734-189X(87)80148-2

Siqueira, A., Terra, M., & Bergerman, M. (2011). *Robust control of robots: Fault tolerant approaches*. London: Springer London. doi:10.1007/978-0-85729-898-0

Siraj, F., & Abdoulha, M. A. (2007). Mining enrolment data using predictive and descriptive approaches. *Knowledge-Oriented Applications in Data Mining*, 53-72.

Smith, S. M., & Brady, J. M. (1997). SUSAN—A new approach to low level image processing. *International Journal of Computer Vision*, *23*(1), 45–78. doi:10.1023/A:1007963824710

Sobel, I. (2014). *History and definition of the sobel operator, pattern classification and scene analysis* (Vol. 73). John Wiley and Sons.

Song, K., Chung, T., Kim, Y., Oh, Y., & Kim, C. (2007). Error concealment of H.264/AVC video frames for mobile video broadcasting. *IEEE Transactions on Consumer Electronics*, *53*(2), 704–711. doi:10.1109/TCE.2007.381749

Srimani, P. K., & Balaji, K. (2014). A comparative study of different classifiers on search engine based educational data. *International Journal of Conceptions on computing and Information Technology*, *2*, 6-11.

Starck, J., & Hilton, A. (2007). Surface capture for performance-based animation. *IEEE Computer Graphics and Applications*, *27*(3), 21–31. doi:10.1109/MCG.2007.68 PMID:17523359

Stranzl, T., Larsen, M. V., Lundegaard, C., & Nielsen, M. (2010). NetCTLpan: Pan-specific MHC class I pathway epitope predictions. *Immunogenetics*, *62*(6), 357–368. doi:10.1007/s00251-010-0441-4 PMID:20379710

Sujaritha, M., & Annadurai, S. (2009). Color image segmentation using binary level-set partitioning approach. *International Journal of Soft Computing*, *4*, 76–84.

Sullivan, G. J., Ohm, J.-R., Han, W.-J., & Wiegand, T. (2012). Overview of the high efficiency video coding (HEVC) standard. *IEEE Transactions on Circuits and Systems for Video Technology*, *22*(12), 1649–1668. doi:10.1109/TCSVT.2012.2221191

Suzuki, N., & Hattori, A. (2008). The road to surgical simulation and surgical navigation. *Virtual Reality (Waltham Cross)*, *12*(4), 281–291. doi:10.1007/s10055-008-0103-0

Sweredoski, M. J., & Baldi, P. (2008). PEPITO: Improved discontinuous B-cell epitope prediction using multiple distance thresholds and half sphere exposure. *Bioinformatics (Oxford, England)*, *24*(12), 1459–1460. doi:10.1093/bioinformatics/btn199 PMID:18443018

Tai, S., Hong, C. S., & Fu, C. (2010). An object-based full frame concealment strategy for H.264/AVC using true motion estimation. In *Proceedings of the 2010 Fourth Pacific-Rim Symposium on Image and Video Technology (PSIVT 2010)* (pp. 214–219). IEEE. doi:10.1109/PSIVT.2010.43

Tanaka, J. W., & Presnell, L. M. (1999). Color diagnosticity in object recognition. *Perception & Psychophysics*, *61*(6), 1140–1153. doi:10.3758/BF03207619 PMID:10497433

Tanaka, J. W., Weiskopf, D., & Williams, P. (2001). Of color and objects: The role of color in high-level vision. *Trends in Cognitive Sciences*, *5*(5), 211–215. doi:10.1016/S1364-6613(00)01626-0 PMID:11323266

Tang, Y. Y., & You, X. G. (2003). Skeletonization of ribbon-like shapes based on a new wavelet function. *IEEE Transactions on Pattern Analysis and Machine Intelligence*, *25*(9), 1118–1133. doi:10.1109/TPAMI.2003.1227987

Tan, J. T. C., & Arai, T. (2011). Triple stereo vision system for safety monitoring of human-robot collaboration in cellular manufacturing. In *Proceedings of 2011 IEEE International Symposium on Assembly and Manufacturing (ISAM)* (vol. 1, pp. 1-6). Tampere, Finland: IEEE. doi:10.1109/ISAM.2011.5942335

Telea, A., Sminchisescu, C., & Dickinson, S. (2004). Optimal inference for hierarchical skeleton abstraction. In *International Conference on Pattern Recognition* (pp. 19–22).

Tevs, A., Berner, A., Wand, M., Ihrke, I., & Seidel, H. P. (2011). Intrinsic Shape Matching by Planned Landmark Sampling. *Computer Graphics Forum*, *30*(2), 543–552. doi:10.1111/j.1467-8659.2011.01879.x

Theobalt, C., Ahmed, N., Ziegler, G., & Seidel, H. P. (2007). High-quality reconstruction of virtual actors from multiview video streams. *IEEE Signal Processing Magazine*, *24*(6), 45–57. doi:10.1109/MSP.2007.905701

Tien, M. C., Wang, Y. T., Chou, C. W., Hsieh, K. Y., Chu, W. T., & Wu, J. L. (2008, June). Event detection in tennis matches based on video data mining. In *2008 IEEE International Conference on Multimedia and Expo* (pp. 1477-1480). doi:10.1109/ICME.2008.4607725

Timor, M., & Erdogan, S. Z. (2005). A data mining application in a student database. *Journal of Aeronautics and Space Technologies, 2*(2), 53-57.

Tomasi, C., & Kanade, T. (2004). Detection and tracking of point features. *Pattern Recognition, 37*, 165–168.

Torres, F., Pomares, J., Gil, P., Puente, S., & Aracil, R. (2002). *Robots y sistemas sensoriales, segunda edición*. Madrid, España: Pearson Educación.

Turk, D. C. (1984). The pathogenicity of Haemophilus influenzae. *Journal of Medical Microbiology, 18*(1), 1–16. doi:10.1099/00222615-18-1-1 PMID:6146721

Turo, D., Otto, P., Egorov, V., Sarvazyan, A., Gerber, L. H., & Sikdar, S. (2012). Elastography and tactile imaging for mechanical characterization of superficial muscles. *Journal of the Acoustical Society of America, 132*(3).

Tuytelaars, T., & Mikolajczyk, K. (2008). Local invariant feature detectors: A survey. *Foundations and Trends in Computer Graphics and Vision, 3*(3), 177–280. doi:10.1561/0600000017

Tuytelaars, T., & Van Gool, L. (2004). Matching widely separated views based on affine invariant regions. *International Journal of Computer Vision, 59*(1), 61–85. doi:10.1023/B:VISI.0000020671.28016.e8

UGC Report. (2008). *Higher education in India - Issues related to Expansion, Inclusiveness, Quality and Finance*. Retrieved from http://www.ugc.ac.in/oldpdf/pub/report/12.pdf

UGC Report. (2011a). *Higher Education in India - Strategies and Schemes during Eleventh Plan Period (2007-2012) for Universities and Colleges*. Retrieved from http://www.ugc.ac.in/oldpdf/pub/he/HEIstategies.pdf

UGC Report. (2011b). *Inclusive and qualitative expansion of higher education 12 Five-Year Plan, 2012-17*. Retrieved from http://www.ugc.ac.in/ugcpdf/740315_12FYP.pdf

UGC Report. (2012). *Higher education in India at a glance*. Retrieved from http://www.ugc.ac.in/ugcpdf/208844_HEglance2012.pdf

UGC Report. (2013). *Higher education in India at a glance*. Retrieved from http://www.ugc.ac.in/pdfnews/6805988_HEglance2013.pdf

Ullman, S. (1989). Aligning pictorial descriptions: An approach to object recognition. *Cognition, 32*(3), 193–254. doi:10.1016/0010-0277(89)90036-X PMID:2752709

Urrea, C., & Coltters, J. P. (2015). Design and implementation of a graphic 3D simulator for the study of control techniques applied to cooperative robots. *International Journal of Control Automation and Systems, 13*(6), 1476–1485. doi:10.1007/s12555-014-0278-y

Urrea, C., & Kern, J. (2016). Trajectory tracking control of a real redundant manipulator of the SCARA type. *Journal of Electrical Engineering & Technology, 11*(1), 215–226. doi:10.5370/JEET.2016.11.1.215

Usharani, C., & Chandrasekaran, R. (2010). Course planning of higher education to meet market demand by using data mining techniques-a case of a Technical University in India. *International Journal of Computer Theory and Engineering, 2*(5), 809–814. doi:10.7763/IJCTE.2010.V2.245

Verez-Bencomo, V., Fernandez-Santana, V., Hardy, E., Toledo, M. E., Rodríguez, M. C., Heynngnezz, L., & Villar, A. et al. (2004). A synthetic conjugate polysaccharide vaccine against Haemophilus influenzae type b. *Science*, *305*(5683), 522–525. doi:10.1126/science.1095209 PMID:15273395

Vesna, Z. (2014). *Illumination independent moving object detection algorithm. In Video surveillance techniques and technologies.* IGI Global.

Villringer, A., & Chance, B. (1997). Non-invasive optical spectroscopy and imaging of human brain function. *Trends in Neurosciences*, *20*(10), 435–442. doi:10.1016/S0166-2236(97)01132-6 PMID:9347608

Vincze, M., & K˝ov'ari, B. (2009). Comparative survey of thinning algorithms. In *International Symposium of Hungarian Researchers on Computational Intelligence and Informatics* (pp. 173–184).

Vlasic, D., Baran, I., Matusik, W., & Popovic, J. (2007). Articulated mesh animation from multi-view silhouettes. Association for Computing Machinery Transactions on Graphics, 27(3), 97:1-97:9.

Waiyamai, K. (2003). *Improving Quality of Graduate Students by Data Mining*. Bangkok: Department of Computer Engineering, Faculty of Engineering, Kasetsart University.

Wang, H., Mohamad, D., & Ismail. (2014). An efficient parameters selection for object recognition based colour features in traffic. *International Arab Journal of Information Technology, 11*(3).

Wang, H., Zou, X., Liu, C., Lu, J., & Liu, T. (2008). *Study on behavior simulation for picking manipulator in virtual environment based on binocular stereo vision.* Paper presented at System Simulation and Scientific Computing, 2008. ICSC 2008. Asia Simulation Conference, Beijing, China.

Wang, L., Liu, X., Lin, S., Xu, G., & Shum, H. Y. (2004, October). Generic slow-motion replay detection in sports video. In *Image Processing, 2004. ICIP'04. 2004 International Conference on* (Vol. 3, pp. 1585-1588). IEEE.

Wang, C.-C., Chuang, C.-Y., & Lin, S. D. (2010). An integrated spatial error concealment technique for H.264/AVC based-on boundary distortion estimation, In *Proceedings of the 2010 Fifth International Conference on Innovative Computing, Information and Control* (pp. 1-4).

Wang, P. S. P., & Zhang, Y. Y. (1989). A fast and flexible thinning algorithm. *IEEE Transactions on Computers*, *38*(5), 741–745. doi:10.1109/12.24276

Ward, A. D., & Hamarneh, G. (2010). The groupwise medial axis transform for fuzzy skeletonization and pruning. *IEEE Transactions on Pattern Analysis and Machine Intelligence*, *32*(6), 1084–1096. doi:10.1109/TPAMI.2009.81 PMID:20431133

Warhade, K. K., Merchant, S. N., & Desai, U. B. (2013). Shot boundary detection in the presence of illumination and motion. *Signal. Image and Video Processing*, *7*(3), 581–592. doi:10.1007/s11760-011-0262-4

Weiss, A., Hirshberg, D., & Black, M. J. (2011). Home 3D body scans from noisy image and range data. In *Proceedings of International Conference on Computer Vision* (vol. 1, pp. 1951-1958). IEEE. doi:10.1109/ICCV.2011.6126465

Weiss, R. E., Egorov, V., Ayrapetyan, S., Sarvazyan, N., & Sarvazyan, A. (2008). Prostate mechanical imaging: A new method for prostate assessment. *Urology*, *71*(3), 425–429. doi:10.1016/j.urology.2007.11.021 PMID:18342178

Wells, P. N. T., & Liang, H.-D. (2011). Medical ultrasound: Imaging of soft tissue strain and elasticity. *Journal of the Royal Society, Interface*, *8*(64), 1521–1549. doi:10.1098/rsif.2011.0054 PMID:21680780

Wiegand, T., Sullivan, G. J., Bjontegaard, G., & Luthra, A. (2003). A.: Overview of theH.264/AVC video coding standard. *IEEE Transactions on Circuits and Systems for Video Technology*, *13*(7), 560–576. doi:10.1109/TCSVT.2003.815165

Witkin, A. P. (1983). Scale-space filtering. In *Proceedings.International Joint Conference on Artificial Intelligence.*

Wöhler, C., Progscha, W., Krüger, L., Döttling, D., & Wendler, M. (2010). *Method and device for safeguarding a hazardous area*. US Patent 7,729,511.

Wolf, J., & Daley, A. J. (2007). Microbiological aspects of bacterial lower respiratory tract illness in children: Typical pathogens. *Paediatric Respiratory Reviews*, *8*(3), 204–211. doi:10.1016/j.prrv.2007.08.002 PMID:17868918

Wuestefeld, M. (2007). *Optoelectronic Protection Device*. US Patent Application Publication US 2007/0280670.

Wu, H., Inada, J., Shioyama, T., Chen, Q., & Simada, T. (2001, June). Automatic facial feature points detection with susan operator. In *Proceedings of the Scandinavian Conference on Image Analysis* (pp. 257-263).

Wu, J., Yin, Z., & Xiong, Y. (2007). The fast multilevel fuzzy edge detection of blurry images. *IEEE Signal Processing Letters*, *14*(5), 344–347. doi:10.1109/LSP.2006.888087

Wu, Y., Shivakumara, P., Wei, W., Lu, T., & Pal, U. (2015). A new ring radius transform-based thinning method for multi-oriented video characters. *International Journal on Document Recognition and Analysis*, *18*(2), 137–151. doi:10.1007/s10032-015-0238-y

Yadav, S. K., & Pal, S. (2012). Data mining application in enrollment management: A case study. *International Journal of Computers and Applications*, *41*(5).

Yan, B., & Gharavi, H. (2010). A hybrid frame concealment algorithm for H.264/AVC. *IEEE Transactions on Image Processing*, *19*(1), 98–107. doi:10.1109/TIP.2009.2032311 PMID:19758866

Yathongchai, W., Yathongchai, C., Kerdprasop, K., & Kerdprasop, N. (2003). *Factor Analysis with Data Mining Technique in Higher Educational Student Drop Out*. Latest Advances in Educational Technologies.

Ye, M., & Yang, R. (2014). Real-time simultaneous pose and shape estimation for articulated objects using a single depth camera. In *Proceedings of IEEE Conference on Computer Vision and Pattern Recognition* (vol. 1, pp. 2345-2352). IEEE. doi:10.1109/CVPR.2014.301

Ye, Q., Gao, W., & Zeng, A. W. (2003). Color image segmentation using density based clustering.*International Conference on Acoustics, Speech and Signal Processing* (Vol. 3, pp. 345-348). Doi:10.1109/ICASSP.2003.1199480

Yoneyama, A., Nakajima, Y., Yanagihara, H., & Sugano, M. (1999). Moving object detection from MPEG video stream. *Systems and Computers in Japan*, *30*(13), 1–12. doi:10.1002/(SICI)1520-684X(19991130)30:13<1::AID-SCJ1>3.0.CO;2-G

Younes, L., Romaniuk, B., & Bittar, E. (2012). A comprehensive and comparative survey of the SIFT algorithm - Feature detection, description, and characterization. In *Proceedings of the International Conference on Computer Vision Theory and Applications (VISAPP)*. SciTePress.

Yu, Y., & Chang, C. (2006). A new edge detection approach based on image context analysis. *Elsevier Journal of Image and Vision Computing*, *24*(10), 1090–1102. doi:10.1016/j.imavis.2006.03.006

Yu, Z., & Chen, Y. (2009). A real-time motion detection algorithm for traffic monitoring systems based on consecutive temporal difference.*Proceedings of the 7th Asian Control Conference.*

Zeno, A., Michael, R., Pal, H., Jiang, Z., Carsten, G., Ilangko, B., & Cathal, G. (2015). Expert Driven Semi-Supervised Elucidation Tool for Medical Endoscopic Videos. In *Proceedings from the 6th ACM Multimedia Systems Conference.*

Zhang, T. Y., & Suen, C. Y. (1984). A fast parallel algorithm for thinning digital patterns. *Communications of the ACM*, *27*(3), 236–239. doi:10.1145/357994.358023

Zhang, Y. Y., & Wang, P. S. P. (1994). A new parallel thinning methodology. *International Journal of Pattern Recognition and Artificial Intelligence*, *8*(05), 999–1011. doi:10.1142/S0218001494000504

Zhang, Y. Y., & Wang, P. S. P. (1995). Analysis and design of parallel thinning algorithms—A generic approach. *International Journal of Pattern Recognition and Artificial Intelligence*, *9*(05), 735–752. doi:10.1142/S0218001495000298

Zhan, X., & Zhu, X. (2009). Refined spatial error concealment with directional entropy, In *Proceedings of the 2009 IEEE International Conference on Wireless Communications, Networking and Mobile Computing* (pp. 1-4). doi:10.1109/WICOM.2009.5302608

Zheng, J. H., & Chau, L. P. (2003). Motion vector recovery algorithm for digital video using Lagrange Interpolation. *IEEE Transactions on Broadcasting*, *49*(4), 383–389. doi:10.1109/TBC.2003.819050

Zheng, S., Liu, J., & Tian, J. W. (2004). A new efficient SVM-based edge detection method. *Elsevier Journal of Pattern Recognition Letters*, *25*(10), 1143–1154. doi:10.1016/j.patrec.2004.03.009

Zhu, X., & Zhang, S. (2008). A shape-adaptive thinning method for binary images. In *International Conference on Cyberworlds* (pp. 721–724). doi:10.1109/CW.2008.133

Ziou, D., & Tabbone, S. (1998). *Edge detection techniques – An overview, pattern recognition and image analysis* (Vol. 8). Nauka: Interperiodica Publishing.

Zollhöfer, M., Nießner, M., Izadi, S., Rehmann, C., Zach, C., Fisher, M., Wu, C., Fitzgibbon, A., Loop, C., Theobalt, C., & Stamminger, M. (2014). Real-time non-rigid reconstruction using an RGB-D camera. *Association for Computing Machinery Transactions on Graphics, 33*(4), 156:1-156:12.

About the Contributors

Nilanjan Dey, Ph.D., is an Asst. Professor in the Department of Information Technology in Techno India College of Technology, Rajarhat, Kolkata, India. He holds an honorary position of Visiting Scientist at Global Biomedical Technologies Inc., CA, USA and Research Scientist of Laboratory of Applied Mathematical Modeling in Human Physiology, Territorial Organization of- Sgientifig and Engineering Unions, BULGARIA, Associate Researcher of Laboratoire RIADI, University of Manouba, TUNISIA. He is the Editor-in-Chief of International Journal of Ambient Computing and Intelligence (IGI Global), US, International Journal of Rough Sets and Data Analysis (IGI Global), US, and the International Journal of Synthetic Emotions (IJSE), IGI Global, US. He is Series Editor of Advances in Geospatial Technologies (AGT) Book Series, (IGI Global), US, Executive Editor of International Journal of Image Mining (IJIM), Inderscience, Regional Editor-Asia of International Journal of Intelligent Engineering Informatics (IJIEI), Inderscience and Associated Editor of International Journal of Service Science, Management, Engineering, and Technology, IGI Global. His research interests include: Medical Imaging, Soft computing, Data mining, Machine learning, Rough set, Mathematical Modeling and Computer Simulation, Modeling of Biomedical Systems, Robotics and Systems, Information Hiding, Security, Computer Aided Diagnosis, Atherosclerosis. He has 8 books and 170 international conferences and journal papers. He is a life member of IE, UACEE, ISOC etc. https://sites.google.com/site/nilanjandeyprofile/.

Amira S. Ashouris an Assistant Professor in the Department of Electronics and Electrical Communication Engineering, Faculty of Engineering, Tanta University Egypt. She was the Vice Chair of Computer Engineering Department, Computers and Information Technology College, Taif University, KSA for one year. She has been the vice chair of CS department, CIT college, Taif University, KSA for 5 years. She is in the Electronics and Electrical Communications Engineering, Faculty of Engineering, Tanta University, Egypt. She received her PhD in the Smart Antenna (2005) from the Electronics and Electrical Communications Engineering, Tanta University, Egypt. Her research interests include: image processing, Medical imaging, Machine learning, Biomedical Systems, Pattern recognition, Signal/image/video processing, Image analysis, Computer vision, and Optimization. She has 4 books and about 70 published journal papers. She is an Editor-in-Chief of the International Journal of Synthetic Emotions (IJSE), IGI Global, US. She is an Associate Editor for the IJRSDA, IGI Global, US as well as the IJACI, IGI Global, US. She is an Editorial Board Member of the International Journal of Image Mining (IJIM), Inderscience.

Prasenjit Kumar Patra is an Asst. Professor in the Department of computer science in Bengal College of Engineering and Technology, Durgapur, India. He acquired the B. Tech degree on 2009 and M. Tech on 2013 in Computer Science Engineering. His work has spanned a seeming widely diverse set of topics, Data mining, scheduling techniques in distributed environments, Fault tolerance in distributed computing environments, availability for services and high-performance with the help of proper resource allocation, sensor networks.

***

Ram Kumar obtained his B. Tech in Electronics and Communication Engineering from Rajasthan Institute of Engineering and Technology, Jaipur in 2010. He obtained his M. Tech in microelectronics and VLSI from National Institute of Technology, Silchar (India) in 2013. Currently he is a Research Scholar in National Institute of Technology, Silchar. His area of interest is biomedical engineering and VLSI circuits.

Jyotsna Rani obtained her High School degree from Cotton College, Guwahati. Currently she is pursuing her B.Tech degree in Electronics and Communication Engineering from National Institute of Technology, Silchar (India) (expecting to graduate in 2016). Her area of interest is Biomedical Engineer and Digital Image Processing.

Abahan Sarkar obtained his B.Tech degree in Electronics and Communication Engineering from NERIST, Nirjuli, Arunachal Pradesh, India in 2003. He has completed his M. Tech in Computer Science and Technology from Tripura University, Suryamaninagar, Tripura, India. Mr. Sarkar worked as Faculty member in Faculty of Science and Technology, The ICFAI University Tripura from 2007 to 2012. Currently he is a Research Scholar in the department of Electrical Engineering in National Institute of Technology Silchar, Assam, India. His research interests include, Industrial automation, and Image processing and Machine vision.

Fazal A. Talukdar obtained his B. E (Hons) from REC Silchar (Now National Institute of Technology, Silchar (India)) in 1987. He obtained his M. Tech in 1993 and PhD in 2002-2003 from Indian Institute of Technology Delhi and Jadavpur University respectively. Currently he is working as Professor in ECE department, National Institute of Technology, Silchar (India). His area of interest is image processing and VLSI circuits.

Index

G

H

I

K

L

M

O

P

Q

R

S

T

V

W

X